Core Collections of Plant Genetic Resources

Core Collections of Plant Genetic Resources

Edited by

T. Hodgkin, A.H.D. Brown, Th.J.L. van Hintum and E.A.V. Morales

JOHN WILEY & SONS

Chichester - New York - Brisbane - Toronto - Singapore

A Co-Publication with the International Plant Genetic Resources Institute (IPGRI) and Sayce Publishing (United Kingdom)

Published by John Wiley & Sons
Baffins Lane, Chichester
West Sussex PO19 1UD, United Kingdom

Other Wiley Editorial Offices

John Wiley & Sons, Inc., 605 Third Avenue,
New York, NY 10158-0012, USA

Jacaranda Wiley Ltd., G.P.O. Box 859, Brisbane,
Queensland 4001, Australia

John Wiley & Sons (Canada) Ltd., 22 Worcester Road,
Rexdale, Ontario M9W 1L1, Canada

John Wiley & Sons (SEA) Pte Ltd., 37 Jalam Pemumpin 05-04,
Block B, Union Industrial Building, Singapore 2057

Co-Publishers

International Plant Genetic Resources Institute (IPGRI)
Via delle Sette Chiese 142, 00145 Rome, Italy

Sayce Publishing, Indio House, Bovey Tracey,
Devon TQ13 9BG, United Kingdom

A catalogue record for this book is available from the British Library

ISBN 471 95545 0

Printed by Colorcraft Ltd., North Point, Hong Kong

Contents

Preface

Work on conserving crop plant genetic resources has always been concerned as much with the effective use of the resources as with conserving variation *per se*. Core collections can make an important contribution to both objectives and, over the past decade, have become an increasingly important part of discussions on conserving and using crop germplasm more effectively. The IBPGR/CGN/CENARGEN workshop on 'Core Collections: Improving the Management and Use of Plant Germplasm Collections', held in Brasília in August 1992, was convened in response to this upsurge of interest in core collections among plant germplasm experts in order to bring together in a single volume a wide range of relevant information on the subject. CGN had been at the centre of the action, described in this volume, to develop a barley core collection and IBPGR (renamed IPGRI in 1994) had received many requests for advice on the use of a core collection and for information on how to construct one. CENARGEN, one of the world's largest and most active national genetic resources institutions, was among the organisations giving serious consideration to doing so.

In the late 1980s, IBPGR had worked on the development of a position paper on core collections, based on literature then available. That effort became bogged down in contentious academic views on sampling theory and was not pursued. By 1992 there was a much greater sense of urgency about the matter because many genetic resources institutions were looking to the core collection as a means of addressing some urgent management problems. The aim of the workshop and of this publication, therefore, was not merely to refine the concept but to examine how it has been applied in order to develop some practical solutions to the problems facing gene bank managers and genetic resources users in the 1990s.

The problems for which core collections might offer solutions are felt to be essentially of two kinds, both having their origin in the way in which genetic resources work has developed over the past 20 years. In the early 1970s the loss of traditional cultivars and landraces in the wake of new agricultural developments seemed to be the most urgent problem and a massive collecting effort was made to address it. In the first place, more genebanks were established — many more than could have been foreseen at that time. The volume of the collections that have been assembled worldwide has far outgrown the management resources and regimes of these facilities, collectively. Second, the use that has been made of these collections to bring about economically significant improvements in yield and profit has been patchy and often not up to the expectations of governments. Generally, this has been because gene bank managers have been unable to cope with the major task of evaluating their material to enable crop breeders to select from it the samples most likely to meet their needs.

As originally described by O.H. Frankel in his article on 'Genetic perspectives of germplasm conservation (in Arber, W. et al. *Genetic Manipulation: Impact on Man and Society*, published in 1984 by Cambridge University Press, UK), a core collection is a limited set of accessions of a crop species and its wild relatives which would represent, with a minimum of repetitiveness, the genetic diversity of a crop species and its wild relatives. This subset of the whole collection would provide potential users with a large amount of the available genetic variation of the crop genepool in a workable number of accessions. It would therefore be useful to plant breeders seeking new characters which require screening techniques not possible with a large collection. Because each of the accessions in a core collection is, to some extent, representative of a number of accessions (from a particular area of the world or with some shared characters), the core can also be used as a point of entry to the active or base collections of a crop. Detailed research can be carried out on a core to obtain an effective picture of the characteristics of the gene pool as a whole.

The core collection concept has aroused considerable worldwide interest and debate within the plant genetic resources community. It has been welcomed as a way of making an existing collection more accessible through designating a small group of accessions that would be the focus of evaluation and use and would provide an entry point to the larger collection which it aims to represent. However, concerns remain that the available knowledge of genetic diversity is still not sufficient, in any crop, to enable a meaningful core to be selected and that many of the useful characters occur at such a low frequency that they would almost always be omitted from a small core collection, no matter how it was selected. A more general concern that has been expressed is that the development of a core collection will lead to a neglect of the rest of the collection and a reduction in the resources available to work on non-core accessions.

In 1989 IBPGR carried out a worldwide survey of institutions and researchers known to be involved in developing core collections. Over 20 projects were identified involving grain legume, vegetable and fruit crops, and descriptions of established core collections of okra, wild *Glycine* species and winter wheat already existed. Since that time there has been increasing interest in core collections and further projects have been initiated. The analysis of genetic diversity in collections, to assist in their management and use, has also increased.

The IBPGR survey clearly showed how important the core collection approach was becoming to the management and use of plant genetic resources. However, it also revealed a number of problems in effectively developing this approach. After considering the nature and variety of the work in progress, the potential importance of core collections in developing countries, and the problems of taking the concept further, the co-sponsors identified the pressing need for a publication to provide a theoretical and practical basis for the further development of cores to improve genebank management and promote the use of germplasm collections.

The workshop in Brasília provided a timely opportunity to engage in healthy scientific debate focused on the issues involved in developing core collections and the contributions such collections might make to current germplasm management problems. A long period of preparation went into the event. CENARGEN kindly offered to make available its excellent facilities in Brasília and a programme committee representing the three sponsoring organisations selected speakers and topics to approach the subject from various perspectives so as to throw light on such questions as: Can a core collection be a useful tool for streamlining the management of a large genetic resources collection? If so, how should such a collection be constructed and how can it be made most effective as a management tool? Can a core collection help to facilitate effective access to a large collection by users and thus help to channel germplasm more efficiently from conservation into breeding programmes, biotechnology programmes, etc? If the potential is there for a core collection to be used in this way, how can it be realised in practice?

It was no surprise that final, definitive answers to these questions did not emerge from the week of deliberations in Brasília. However, the presence of experts working on different facets of these questions, and in a variety of crops, had the very positive outcome of bringing together the numerous strands of relevant knowledge, stimulating debate and establishing collaboration. The publication of the contributions, revised in the light of workshop discussions, should help to refine the theory and practice of core collections in particular, and to develop more effective ways of conserving and using plant genetic resources in general.

T. Hodgkin, A.H.D. Brown, Th.J.L. van Hintum, E.A.V. Morales and A. McCusker

Contributors

F. Anthony, CATIE, Aparto 59, 7170 Turrialba, Costa Rica

S. Beebe, Bean Program, CIAT, AA 6713, Cali, Colombia

A.H.D. Brown, Division of Plant Industry, CSIRO, GPO Box 1600, Canberra, ACT 2601, Australia

C.M.T. Cordeiro, CENARGEN, Brasília, Distrito Federal, Brazil

I.R.S. Costa, CENARGEN, Brasília, Distrito Federal, Brazil

J. Crossa, Biometrics and Statistics Unit, CIMMYT, Apdo. Postal 6-641, 06600 México DF, Mexico

I.H. DeLacy, Department of Agriculture, University of Queensland, QLD 4072, Australia

P. Ferreira, CATIE, 7170 Turrialba, Costa Rica

P.L. Forsline, Plant Genetic Resources Unit, USDA-ARS, Cornell University, Geneva, New York 14456-0462, USA

N.W. Galwey, Department of Genetics, University of Cambridge, Downing Street, Cambridge CB2 3EH, UK

P. Gepts, Department of Agronomy and Range Science, University of California, Davis, CA 95616-8515, USA

S. Hamon, Laboratoire de Ressources Génétiques et d'Amélioration des Plantes Tropicales, ORSTOM, BP 5045, 34032 Montpellier Cedex, France

Th.J.L. van Hintum, Centre for Plant Breeding and Reproduction Research, Centre for Genetic Resources, P.O. Box 16, 6700 AA Wageningen, Netherlands

T. Hodgkin, IBPGR, Via delle Sette Chiese 142, 00145 Rome, Italy

M. Iwanaga, Genetic Resources Unit, CIAT, AA 6713, Cali, Colombia

M.T. Jackson, IRRI, Los Baños, Philippines

A.A. Jaradat, Department of Plant Sciences, JUST, P.O. Box 3030, Irbid, Jordan (Present address: IPGRI-WANA, c/o ICARDA, P.O. Box 5466, Aleppo, Syria)

P. Jones, Land Use Program, CIAT, AA 6713, Cali, Colombia

H. Knüpffer, Genebank, Institute of Plant Genetics and Crop Plant Research, Corrensstraße 3, D-06466 Gatersleben, Germany

S. Kresovich, Plant Genetic Resources Unit, USDA-ARS, Cornell University, Geneva, New York 14456-0462, USA (Present address: USDA-ARS, Genetic Resources Conservation Unit, University of Georgia, Griffith, GA 30223-1797, USA)

W.F. Lamboy, Plant Genetic Resources Unit, USDA-ARS, Cornell University, Geneva, New York 14456-0462, USA

M.C. Mackay, Australian Winter Cereals Collection, RMB 944, Tamworth, NSW 2340, Australia

A. McCusker, IBPGR, Via delle Sette Chiese 142, 00145 Rome, Italy (Present address: P.O. Box 793, Woden 2606, Australia)

J.R. McFerson, Plant Genetic Resources Unit, USDA-ARS, Cornell University, Geneva, New York 14456-0462, USA

E.A.V. Morales, CENARGEN, Brasília, Distrito Federal, Brazil

H. Morishima, National Institute of Genetics, Mishima 411, Japan

M. Noirot, Laboratoire de Ressources Génétiques et d'Amélioration des Plantes Tropicales, ORSTOM, BP 5045, 34032 Montpellier Cedex, France

T. Nomura, Kyoto Sangyo University, Kyoto 603, Japan

K.E. Prasada Rao, Genetic Resources Unit, ICRISAT, Patancheru, 502 324 Andhra Pradesh, India (Present address: 31-32-29 Dabagardens, Visakhapatnam 530020, India)

C.O. Qualset, Department of Agronomy and Range Science, University of California, Davis, CA 95616-8515, USA

V. Ramanatha Rao, IBPGR, Via delle Sette Chiese 142, 00145 Rome, Italy

D.M.S. Rocha, CENARGEN, Brasília, Distrito Federal, Brazil

D.J. Schoen, Department of Biology, McGill University, 1205 Avenue Dr, Penfield, Montreal, Quebec H3A 1B1, Canada

S. Silva, CNPMF, Cruz das Almas, Bahia, Brazil

P.L. Spagnoletti Zeuli, Dip. di Biologia, Difesa e Biotecnologie Agro-Forestali, Universitá della Basilicata, Potenza, Italy

S. Taba, Maize Germplasm Bank, CIMMYT, Apdo. Postal 6-641, 06600 México DF, Mexico

J. Tohme, Biotechnology Research Unit, CIAT, AA 6713, Cali, Colombia

A.C.C. Valois, CENARGEN, Brasília, Distrito Federal, Brazil

D.A. Vaughan, IRRI, Los Baños, Philippines (Present address: National Institute of Agrobiological Resources, Tsukuba, Ibaraki 305, Japan)

K. Yonezawa, Kyoto Sangyo University, Kyoto 603, Japan

Part 1

INTRODUCTION

Core Collections of Plant Genetic Resources
Edited by T. Hodgkin, A.H.D. Brown, Th.J.L. van Hintum and E.A.V. Morales

A Wiley-Sayce Publication

1.1

The core collection at the crossroads

A.H.D. Brown

Abstract

National gene banks are now entering an era of increased activity and responsibility, particularly for their own indigenous crop germplasm. Their programmes need to cover all phases of germplasm activities, and yet have to operate with limited resources. The core collection offers a way to meet these challenges. A core collection consists of a limited set of accessions derived from an existing collection, chosen to represent the genetic spectrum in that collection. The concept has met criticism in four areas: that the rest of the collection is vulnerable to decay or disposal; that the bias towards representing diversity ignores usefulness; that the system is too inflexible; and that variation within accessions is ignored. Most of these concerns appear to arise from misuse or misunderstanding of the core approach.

Gene banks handling clonal crops, or species with recalcitrant seeds, can gain much from using core collections. Field gene banks are expensive to run and prone to loss. The core approach can guide the choice of accessions for growing in the field and those for developing and using *in vitro* methods. The principles of stratified, representative sampling in the core concept also apply to the choice of populations for conservation *in situ*. The need for choice arises from the cost of such strategies in crop species and the limit on areas for the preservation of wild species. A better use of limited resources will enable specific scientific goals (such as the discovery of new resistances) to be achieved. Large numbers of new deposits could swamp national gene banks. Some selection is unavoidable to prevent the loss of significant components of their collections, and core collections could assist in dealing with excess numbers. Further, a core will render a collection more workable for the user, and thus ensure support for its conservation in the longer term.

Those of us charged with the responsibility of conserving plant genetic resources have challenging work ahead. Recent events that have shaped our current situation include the almost universal adoption of the Food and Agriculture Organisation (FAO) Undertaking on Plant Genetic Resources and the launching of the International Plant Genetic Resources Institute (IPGRI). These developments have brought new enthusiasm for, and emphasis upon, national programmes of plant genetic resources conservation. We are challenged to do a better job assembling, managing, conserving and using collections, particularly of our indigenous genetic resources.

Yet at the same time these tasks must be achieved with limited resources. In addition, it is important to address all phases of gene bank activities — from collection through to actual use of the genetic variation in plant improvement programmes. A hindrance to fulfilling all these functions is the unrestrained growth in the number of accessions in collections and the amorphous nature that collections can take on.

Recognising that the sheer size of a germplasm collection could deter its use, Frankel (1984) proposed that it could be pruned to a 'core collection'. The core collection would represent 'with minimum repetitiveness, the genetic diversity of a crop species and its relatives'. The remaining accessions would not necessarily be discarded but would be managed as a 'reserve collection'.

Since this proposal, we have examined the rationale, purposes and general principles of core collections (Frankel and Brown, 1984; Brown, 1989a, b). As to the size of a core collection, sampling theory of selective neutral alleles in finite populations indicated that about 10% drawn randomly from the whole collection was relatively efficient in retaining its allelic variation (about 70% retained). However, this could be improved by stratifying the collection into groups of related accessions and selecting 10% from each group. A number of core collections have now been formed and others are described in this volume. In addition, there has been much debate as to the reasons for core collections and ways to form them. In this chapter, I will begin by reviewing many of the issues, and then outline the steps and questions involved in selecting cores, with a special focus on clonal crops.

WHAT IS A CORE COLLECTION?

Like many timely and appropriately named ideas, that of the core collection has already begun to diverge. Let us start with the original concept put forward by Frankel and Brown (1984) and Brown (1989a, b):

> A *core collection* consists of a limited set of accessions[1] derived from an existing germplasm collection, chosen to represent the genetic spectrum in the whole collection. The core should include as much as possible of its genetic diversity. The remaining accessions in the collection are called the *reserve collection*[2].

The entries in the core are chosen primarily to be representative. They are ecologically or genetically distinct from one another. Within the primary constraint of covering the genetic spectrum and ecological range, the aim should then be to maximise the genetic diversity. This implies that the core should not contain duplicates and should minimise similarity between its entries. For example, the sampling would include single-spaced points along an ecological gradient (a spectrum) rather than be restricted to a few repeats from both extremes (maximum diversity).

---

1 The term 'accession' is used in this chapter to refer to a sample maintained in the whole collection. The term 'entry' is used to refer to any accession or subline from it that is selected for inclusion in the core.

2 The workshop on which this volume is based resolved that the term 'reserve collection' should not be formally recognised by the International Board for Plant Genetic Resources (IBPGR). However, the concept itself must be discussed in this chapter, and the term is used solely in that context.

While it might be convenient to speak in the singular of a core for a whole group of related species, such as the core for perennial *Glycine* species (Brown et al., 1987), in practice core selection for each species is made separately. Indeed, the sampling procedure and intensity are likely to differ between the crop species, its progenitor or close wild relatives and its distant relatives.

At least two modifications to the core concept have been proposed. One is to alter the term itself to 'core subsets' in order to 'underscore the fact that these are not separate collections, but a set of designated accessions within an existing collection' (National Research Council, 1991). The term 'core set' would be simpler, as the prefix 'sub-' is redundant. This approach may be appropriate for collections that are under no pressure to be reduced in size and for which no physical separation of the core from the rest of the collection is planned. However, separation of the core and some reduction of the total size may be needed in collections with burgeoning numbers, and for these collections the original term is appropriate. More importantly, it has precedence.

A second modification proposes the use of the term 'core' for sets of accessions that a curator has chosen as examples in the collection for a specific purpose, such as a core set for tolerance of acid soils, and another for rust resistance (Mackay, *Chapter 4.4, this volume*). However, this use of the term 'core' is confusing and lacks legitimacy because it has been used previously to denote a genetically representative set (Frankel and Brown, 1984; Brown, 1989a). Why not just incorporate the specific purpose for which a set has been chosen into its name (for example, an 'acid-soil tolerant set' or a 'rust-resistant set')?

A new development of the core concept has arisen from cutting the linkage of a core to a specific, existing germplasm collection and applying the concept to a crop species as a whole. Thus the core collection for that crop would consist of a limited number of entries chosen to represent the genetic diversity of the species and its wild relatives for breeding and research. It would be a synthetic core, assembled from the various cooperating germplasm collections or from fresh sampling of wild or crop populations. The most ambitious project of this type so far is the Barley Core Collection project initiated by the European Cooperative Programme for the Conservation and Exchange of Crop Genetic Resources (Knüpffer, *Chapter 4.1*).

I wish to stress the differences between the synthetic core collection and a core collection attached to a specific gene bank. The first and most important difference is that the synthetic core is of more assistance to germplasm use than to gene bank management. It forms a set of entries in addition to those held in a national collection that will be managed as a new and distinct unit. In contrast, a core collection is taken from an existing collection — it does not have to be gathered from several sources. Second, the synthetic core relies on an international committee of experts on that crop to agree on its composition. In contrast, a curator can implement the original concept directly by nominating its entries from the current holdings.

These and other uses of the term 'core collection' are likely to continue. However, the rest of this chapter focuses on the original conception. From the definition of 'core collection' we now turn to the 'why' and the 'how' in assembling and using core collections.

THE FUNCTIONS OF A CORE COLLECTION

Core collections have many roles to play in the management and use of genetic resources. This is because most activities in gene bank management require the curator to make choices or to set priorities among accessions because of limited resources. It is usually not logical or efficient to start at accession number 1 and work sequentially through all accessions in the collection. The germplasm operations for which a core collection offers distinct advantages are:

- *Addition of new accessions:* The core collection provides a reference set for deciding whether new samples arriving at a gene bank are worth adding to the collection, or even to the core itself. Does the new sample resemble any current core entry? If so, are there enough of this type in the whole collection already? If not, should the sample be a new core entry? The core may help in identifying gaps in the collection.

- *Conservation:* The core contains material of highest priority for conservation. It should have first call on the monitoring of viability by routine seed testing. Curators faced with the task of regenerating and multiplying a large and neglected collection could attend to the core first. Duplicates held at other gene banks for safety should include the core entries. The representative nature of the core makes it suitable for developing new methods of conservation (such as ultra-dry seeds, *in vitro* or cryogenic storage).

- *Characterisation:* The core is the suitable material for developing an adequate list of descriptors. A sufficient number of characters and states should be used to distinguish between its entries.

- *Evaluation:* It is for this task that the core collection has most to offer (Frankel and Brown, 1984). The core enables a logical and efficient two-step procedure to be carried out in sampling the whole collection. Its entries can be the first to be evaluated for expensive or complex traits. It provides a set of materials covering the range of variation in the whole collection for developing new methods of evaluation that would be sound for the whole collection. Further, by focusing evaluation on a restricted set of accessions, the core assists the development of a multivariate database to study the interrelationships between characters and between kinds of data (passport, characterisation and evaluation characters).

- *Germplasm enhancement:* The breeding of desirable characters from alien genetic backgrounds into locally adapted stocks is a lengthy and expensive process. The core forms a reduced set of representative accessions for testing general combining ability with local germplasm in the search for yield enhancement (Frankel and Brown, 1984; Abel and Pollak, 1991; Spagnoletti Zeuli and Qualset, *Chapter 5.1*).

- *Germplasm distribution:* Designation of a core can help to accelerate the response to requests because core samples can be multiplied and packaged in advance, ready for dispatch. More importantly, it provides an opportunity to distribute representative germplasm on a reduced scale.

The general feature of most of these benefits is that the judicious reduction of the number of accessions to be handled in one operation saves resources. These resources are available for a more complete range of activities to be conducted in greater depth and thoroughness because the entries are representative of the collection (Morales, *Chapter 5.3*).

CRITICISMS OF THE CORE COLLECTION CONCEPT

If the core can serve curators in so many ways, why has it not been implemented more widely? Numerous objections to the core concept and stratification of germplasm collections have already appeared or can be foreseen. These objections can be grouped under four headings: vulnerability of the reserve collection; bias towards diversity rather than usefulness; inflexibility of core entries; and lack of validity in sampling variation.

Vulnerability of the reserve collection

Several criticisms revolve around concern for the reserve collection (Marshall, 1989; National Research Council, 1991). This concern is that a core collection might threaten the size of the total collection as administrators seek economies and dispose of the reserve as excess to needs. The approach may lead to combining materials or simply neglecting germplasm that is not part of the core. In general, the division of a collection into core and reserve may threaten the 'integrity' of a carefully assembled collection.

These criticisms assume that the core collection is an entity on its own. We have always stressed that a fundamental role of the core is that it is a guide to the whole collection. Another role is that it fosters better conservation of germplasm (Brown, 1989a). No reduction of collection size is involved. Indeed, the appraisal of all the accessions in hand needed for setting up a core collection can produce evidence for increasing the total collection size through targeted collecting in particular areas. Most collections in gene banks do not have unassailable 'integrity' because their contents reflect uneven or historic sampling effects (Frankel and Soulé, 1981).

The need to reduce the size of some collections because of a shortage of resources will arise whether or not a core has been set up. Indeed, as national programmes become increasingly responsible for local genetic resources, they are in danger of being swamped by large numbers of new deposits. The practical bottlenecks that limit international acquisition (such as costly periodic exploration, shipment and quarantine) are not in place to check such growth. To use Holden's (1984) phrase, national programmes will have to 'deal with the deluge' of local material. Some selection seems inescapable, particularly for clonal crops (*see below*).

Bias towards representing diversity rather than usefulness

The fact that entries in the core are chosen to represent the genetic spectrum of the collection has led some authors to claim that the core forms a suboptimal sample. They feel it ignores the relative ease of making the crosses needed to use a character in breeding and that it is directed more at the needs of the molecular biologist and geneticist than at those of the breeder.

These claims can be disputed. Suppose a new resistance is sought from a gene bank and, from a search of the core, this resistance is found in a wild relative. This does not necessarily mean the breeder would stop screening and start crossing. The decision may well be to examine more material from the rest of the collection of the cultivated species to seek other resistances. The breeder may prefer a weaker resistance than that first found, if it is in a better genetic background. The benefit that the first search of the core has provided in this case is a resistance phenotype as a yardstick, and possibly a useful resource if no other options appear. Consider the case of the resistances to soybean leaf rust found in wild perennial *Glycine* species which have so far been extremely difficult to transfer to soybean. Such resistances have proved very useful because they uncovered cryptic pathotype variation in the pathogen to which the generally susceptible crop species is blind (Burdon and Speer, 1984).

It is true that in many instances the core is unlikely to contain the single most 'useful' source of a character for the breeder. However, as the first sampling of a two-step process, it provides a logical general strategy in identifying the best source. Further, a diverse core is more likely to contain adequate sources of many characters than that selected by other strategies. For example, Vaughan (1991) found that a diverse core of the rice collection at the International Rice Research Institute (IRRI) had appreciably frequent sources of all six resistances he was seeking, whereas random or sequential sets of accessions were less reliable in their contents.

The representation of diversity at the expense of utility implies that the core will contain some entries of little use to a particular user. If these are obvious at the outset, they can be excluded from that project. The attraction of the core to many biologists is its diversity and its frugal use of resources. However, not all biologists will be content with a core; some have specialist needs (for example, for rare morphological, developmental or physiological mutants) which can escape inclusion in a core. One advantage of a diverse core for the breeder is the chance to become acquainted with the diversity of phenotypes in a crop or its related wild species. Such cores can sow the seed in the breeder's imagination of new characters to use in plant improvement.

It should be obvious that the core procedure was never intended to pre-empt all other methods of structuring or sampling from germplasm collections. Nor was it intended to replace all other batches of samples that a curator might dispatch, or that a user might work with. Yet some of the specific requests that a gene bank receives might be better answered with a core than with the varieties requested (Vaughan and Jackson, *Chapter 5.2*).

Inflexibility of the entries in the core

One area of potential difficulty is the rate at which changes should be made in the composition of the core. Clearly, some evolution of the core contents should be anticipated (Brown, 1989a) because the core primarily represents the diversity of a collection that is itself changing. Yet a disadvantage of change is the reduced scope for studying interactions among attributes of all kinds that accrue as the evaluation of a limited set of entries proceeds. Some balance can be struck between the contending needs for change and for stability. Disagreements about contents can be resolved at the working level. Of course, the curator or the user is not bound to work with the complete core set and either may choose to employ a subset for a specific purpose. In all these issues what is important is not to take too inflexible a view of core contents.

Lack of general validity in sampling variation

Several concerns have been voiced about whether stratified random sampling is the optimal strategy for finding needed variation. The supposed problems for the core as a sampling strategy are: the available knowledge of genetic diversity in any crop may be insufficient to enable a reliable core to be developed; the characters for which breeders turn to gene banks often are very rare variants or combinations, partly because the characters that are common in a crop species are likely to be in a breeder's working collection already (Duvick, 1984); and the core procedure deals with whole accessions and appears to take no explicit account of polymorphism within accessions.

Information requirement

The National Research Council (1991) claimed that the lack of basic information on accessions was a major constraint in designating core entries. This is arguable. Complete documentation and evaluation of a collection would be the ideal basis for a fully deterministic choice of a core set. However, there is no need to delay establishing a core for lack of complete data. Second, some comfort can be drawn from the robust nature of 'naive random sampling' (Brown, 1989b; Schoen and Brown, *Chapter 2.3*), provided that a highly uneven distribution of polymorphism or of identity by descent (redundancy) is not present in the collection. Even the most limited and uneven amount of data can be used to improve upon the basic efficiency of random sampling. Yet such a use of the data in hand does not validate the criticism voiced by Hawkes (1992) that 'the core idea is basically flawed' because the choice of the core requires evaluating the whole collection. In contrast, the choice of the core should precede, rather than follow, extensive evaluation.

The rare variant

Any particular extremely rare variant or specific combination of characters is unlikely to be in the core. The core will, of course, contain a sample of many of the other rare variants from the collection, but the problem of the specific rare variant has long bedevilled discussions of sampling strategy (Marshall and Brown, 1975, 1981; Chang, 1989). An oft-cited example is the finding of resistance to grassy stunt virus as a rare variant in one population of the wild species *Oryza nivara*. Whether one has a specific rare variant in a collection and finds it depends upon luck and total sample size alone, and is largely unaffected by strategy. Such variants are thus irrelevant to and not a paradigm for strategic considerations. The core concept is not aimed at them. However, evaluating a core does have the following benefits in addressing the problem of the rare variant:

- The user might not know *a priori* whether the desired variant is a rare one in the species. Checking the core first is the most efficient way of establishing this.
- If the core has been studied for other similar traits (for example, resistance to an array of specific pathotypes of a disease), 'hot spots' of genetic diversity may be detected, even if the desired variant, a specific resistance, was absent. The accessions from these hot spots in the reserve collection would be the next logical ones to screen. Such hot spots of diversity are a particular feature of inbreeding species (Schoen and Brown, 1991).
- Absence of the variant in the core might encourage one to test a core of wild related species. If the variant was found there, the cost of extracting it from exotic germplasm could be weighed against the cost of screening more of the reserve collection of the cultivated species. This sequence would be relevant to the case of resistance to the white-backed planthopper and the brown planthopper in rice. These resistances are rare in the cultivated species (about 1%) but common in a wild relative (about 50%) (Chang, 1989).

Polymorphism

The procedures for sampling of polymorphism within an accession differ markedly between species, depending primarily upon their breeding system.

For outbreeding species, the sample size for each accession must be kept up to avoid inbreeding depression and genetic drift. In some outbreeding species (for example, maize) the collection may contain many accessions that are relatively homogeneous inbred lines, while other accessions are open-pollinated populations. In these cases the core should include a set of entries representing the inbred lines, as distinct from another set of entries representing the populations. The entries in such a core, with different mating systems, are likely to differ markedly from one another in their level of polymorphism. In screening the core for specific genes such as rust resistance, the number (S) of plants from each entry with inbreeding coefficient F required to find genes of a given frequency (p) in the accession with given certainty (say, 95%) is readily formulated (Marshall and Brown, 1975):

$$S = 3 / \{(F - 2).\log_e[1 - p]\}$$

The core should thus be structured (inbred entries, population entries and other such groupings) to alert the user to genetic heterogeneities in the collection. The user can then choose a sample size for each accession which is appropriate for overcoming the problem of intra-accession polymorphism.

For self-pollinated species, one option to consider is that of reducing the core entry to a pure line. Homogeneity has some appeal in fixing the entries of species cores such as the Barley Core Collection referred to earlier. However, it biases the representative nature of such a core and reduces its total variation and information content (Brown, 1992). Generally, the amount of intra-accession polymorphism is likely to differ greatly among the accessions of autogamous species in a gene bank. This implies that both the mean of an accession (for example, for tolerance or resistance scores of individual plants) and the variance within it need to be measured if new phenotypes are to be found. Such studies require more work on fewer accessions, a situation for which the core collection is well suited.

For clonal crops, the maintenance of the original population structure of a (vegetative) sample from a single field is more difficult and less justified than in seed crops. The tendency is to split conspicuously polymorphic samples into distinct clones. Indeed, the Centro Internacional de la Papa (CIP) restricts clonal maintenance of potato accessions to unique genotypes and removes duplicates to be maintained as seed collections (Huaman, 1984). This makes the genotype the unit of the clonal collection. The pattern of intra- and inter-field variation could be extracted from records in the database on the origin of each clone when needed.

Thus, in answer to the criticisms concerned with valid ways to sample variation, the overall conclusion is that stratified sampling and core designation is an effective way to deal with polymorphism within accessions. However, it should be stressed that experimental protocols must take account of the differences in variation structure with breeding system (allogamous, autogamous or clonal) or mating design (inbred lines or open-pollinated populations) in the collection.

HOW TO CHOOSE THE CORE COLLECTION

Other chapters in this volume describe specific examples and methods for the designation of a core collection. By way of introduction, let us set out a general scheme and note the problems that need a solution in each specific case. Figure 1 presents a flow diagram of the operations involved in developing a core collection.

Data assembly

We start with the collection in hand. The available passport (Tohme et al., *Chapter 3.1*) and characterisation (Rao and Rao, *Chapter 3.2*) data need to be assembled and updated as far as is practical. This step of data assembly raises the problems of how complete the data should be, how much and what kinds of data should be used, and how the various kinds of data should be combined

Figure 1 Procedure in developing a core collection

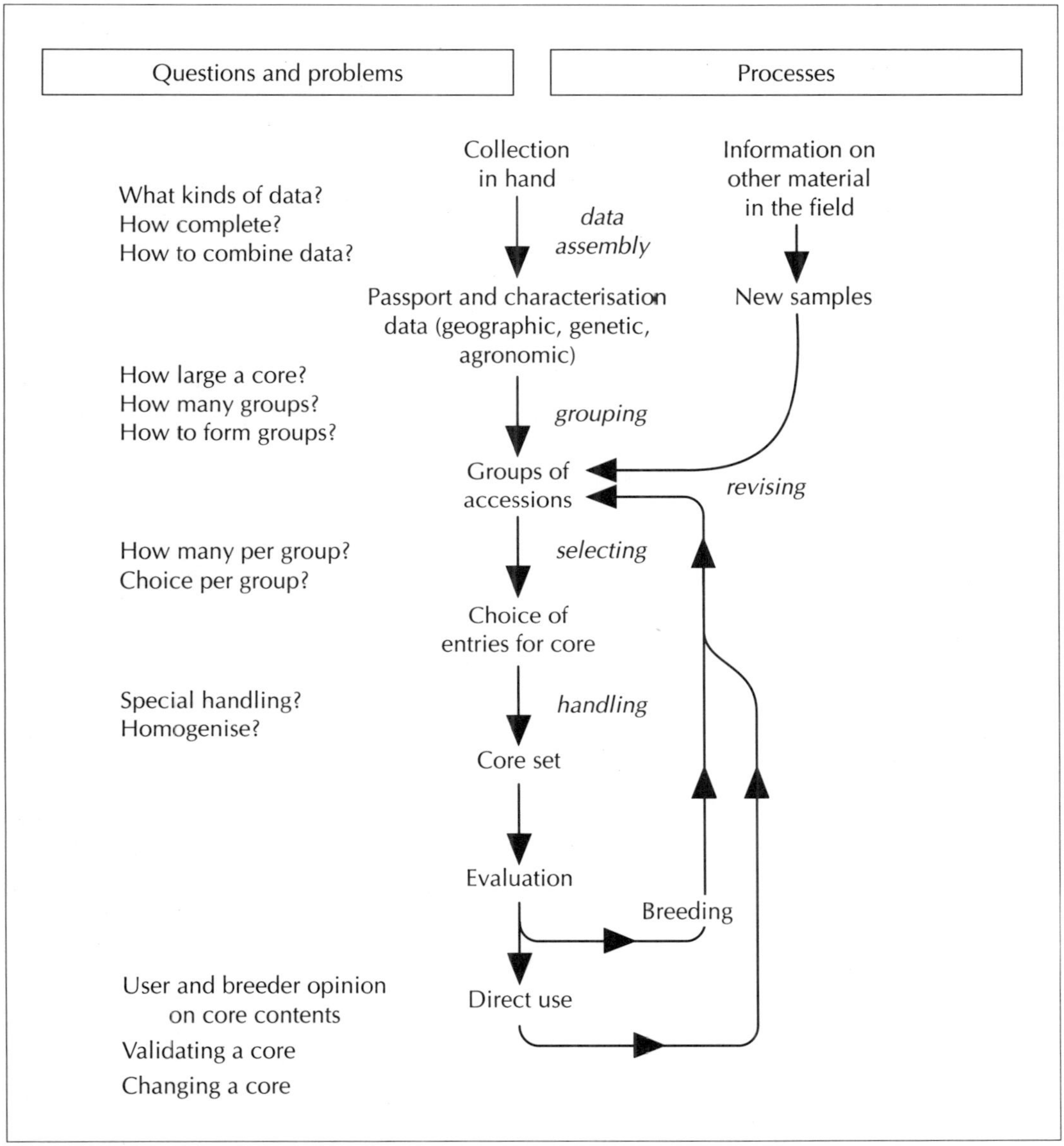

(Kresovich, *Chapter 3.5*). The most flexible approach is to organise the passport, characterisation and, in some cases, evaluation data for a hierarchical classification of the accessions into groups (Hamon et al., *Chapter 3.3*). The attributes are considered in descending order: taxonomy, geographic origin, ecological origin, genetic markers and agronomic data. At any level, accessions lacking data for an attribute can be assigned to a separate category as an unknown state. Often a set of similar attributes (such as climatic variables) may have no obvious hierarchical order. Such a set can be reduced to a few compound indexing variables using multivariate ordination techniques.

Grouping

The second step is to assign the accessions to groups, the members of which are likely to be genetically similar (Crossa et al., *Chapter 2.4*). If the hierarchical approach is adopted (Hintum, Chap*ter 2.1*), we employ a divisive procedure, splitting the collection into smaller and smaller groups within groups (Williams, 1971). The number of groups depends upon the size of the collection, the intended size of the core and the dissimilarity of the groups at the lowest level of sorting. With a total collection of N accessions, and a core size of 10%, the maximum number of non-empty groups would be 0.1 N. This number would be appropriate for small collections (N < about 100). For moderately sized collections (N about 1000), the number of groups may increase in proportion to the logarithm of N, so that more than one entry from a group would be in the core. If N = 1000, dividing the accessions into about 20 groups would imply an average of five entries from each group in the core. Again, the level to which the division of the collection into smaller groups proceeds should depend upon their distinctiveness.

Spagnoletti Zeuli and Qualset (1987) advocated another approach to grouping using, as an example, the multivariate analysis of eight quantitative spike characters in the United States Department of Agriculture (USDA) collection of durum wheat. The accessions were divided into groups according to their country of origin. The countries were then clustered using a canonical discriminant analysis of the spike data into those with phenotypically similar accessions. The clusters (rather than distinct countries) formed the basis of the stratified core sampling. The authors noted that some countries with contrasting agroecological conditions clustered together. Such a result might appear to argue that country of origin is not a reliable basis for stratification. However, their alternative procedure assumes that phenotypic similarity for a limited set of characters is a better indicator of genetic and ecological similarity than is country of origin. In another study, this time with a large set of data for barley, Peeters and Martinelli (1989) reached the contrary conclusion that country of origin is a reliable general indicator of diversity.

Selection

Having divided the collection into groups, the third step is to select entries from each group. The questions to be addressed here are the number to be chosen from each group and the method of choice within the group (Yonezawa et al., *Chapter 2.2*). Schoen and Brown (*Chapter 2.3*) discuss this issue in detail.

Using the Kimura and Crow (1964) theoretical equilibrium model of selectively neutral mutants with parameter (θ = 4 x effective population size x mutation rate), it is possible to prove a general theorem that the maximum number of alleles within the core will occur when each group is represented in proportion to its value of (θ). This is a very powerful conceptual result. However, its application in practice needs an estimate of the parameter (θ) for each group. Some approximate approaches are:

- First, suppose that we have no data on genetic variation. If redundancy or identity by descent do not vary among groups, the theorem implies representation should be in direct proportion to the size of each group. For example, Brown (1989b) for barley and *Glycine tomentella*, and Erskine and Muehlbauer (1991) for lentils, found that this strategy is better than taking a constant number from each group. If there are differences among the groups in redundancy, it usually occurs because redundancy tends to be present in larger groups. In such cases, representation in proportion to the logarithm of group size is more conservative (Brown, 1989b).

- Second, we may have estimates of genetic polymorphism among the accessions within a group based on a comparable sample of loci. The formula $\theta = h/(1-h)$ converts each single-locus estimate of gene diversity (h = heterozygosity in outbreeding species) in each group to an estimate of (θ). The number of entries from a group in the core is fixed in proportion to the average value of (θ) for that group.

- Third, comparative estimates of the amount of (additive) genetic variance for quantitative characters might be available. It is known from theory that the amount of quantitative genetic variance for a 'neutral' character (one that is not under selection) is proportional to the parameter (θ) at equilibrium. Thus the guide to numbers of entries from each group in the core is to represent them in proportion to the genetic variance in each group.

Of course, genetic data may be used in the second approach (Brown et al., 1990; Gepts, *Chapter 3.4*). For example, Perry et al. (1991) used canonical discriminant analysis of isozyme polymorphism in the USDA soybean collection to cluster country of origin into six groups in a way analogous to the durum wheat and barley studies mentioned earlier. However, they went on to use this genetic evidence of divergence to consider sampling rates. They suggested forming a core by selecting a fraction of the largest cluster (China, Korea and Japan) and related material (that together made up 85% of the whole collection) and all the accessions in the other four clusters (15% of the collection). They argued for such highly uneven representation because 'rare types' were present and redundancy was less in the favoured clusters, and because these were known centres of diversity. However, the effect of this partition on allele recovery was not tested.

Whether richness of diversity within a group (as outlined above) or degree of divergence from other groups should be the major criterion for determining representation in the core is a matter of debate. In general, however, caution is needed when suggested sampling proportions are highly uneven over groups within a species.

Handling

The final step in setting up the core concerns the handling of the entries. In many collections, such as those of cereals, the core entries remain within the whole collection. It may be sufficient merely to record in the database a signal that a certain accession is a core entry. In other cases (such as for clonal crops) a spatially separated plantation may be planned, or another conservation method (such as tissue culture, cryogenic set or DNA bank) may be employed. A second question is whether or not each core entry should be made homogeneous. For self-pollinating accessions, this would mean making a new pure-line from one inbred individual. For apomictic species, it would mean propagating one clone as the core entry. The reason for such a procedure is to achieve genotypic fixation that would provide a constant reference point. However, the cost is a substantial loss of genetic variation and the creation of a genetic stock rather than the representation of an accession in a collection (Brown, 1992).

With the core established, programmes of evaluation can begin. The evaluation could be for finding donors of various characters for use in breeders' crosses, or in biological projects, or for agronomic performance and direct use. This raises the question as to whether the breeder or other user of plant germplasm should influence what entries are in core collections. Ideally, the user of a core should send useful information back to the curator as to the core's validity and value (Galwey, *Chapter 4.3*).

The findings may well lead to the revision of group assignments or number of representatives per group (Jaradat, *Chapter 4.2*). Other major changes to the core would follow from new samples entering the collection. The new sampling may be planned from knowledge of what is still in the field and lacking from the collection, expressly targeted to fill gaps. Such new samples form new groups from which extra entries for the core would proceed.

CORE COLLECTIONS OF CLONAL CROPS

The conservation and use of the genetic resources of clonal crops offer particular challenges in South America. Table 1 summarises data from Lawrence et al. (1986) and other sources on the approximate sizes of several collections on this continent of three major clonal crops: cassava, potato and sweet potato. The data illustrate the number and range in size of clonal collections.

At first sight, a core collection seems to be difficult to justify and to apply in a clonal species. For example, a major theoretical argument for cores in seed crops is that a small number of samples is surprisingly efficient in retaining alleles at single loci (Brown, 1989a), and we can presume that the breeders would assemble alleles into genotypes at will in crossing programmes. The relative efficiency of few samples is attributable to the expectation that the number of alleles increases in proportion to the logarithm of the number of samples. In contrast, in clonal crops much more interest surrounds the whole genotype. Specific combinations of genes in highly heterozygous combinations can be worth preserving. In this case the number of genotypes (genets) preserved increases in direct proportion to the number of samples (assuming duplicates are eliminated), and there is no special sampling efficiency in small numbers.

Second, it appears from Table 1 and from data presented by Holden (1984) that collections of clonal crops are already an order of magnitude smaller than collections of seed crops. The largest of them are in the order of thousands of accessions, rather than the tens of thousands found in large cereal collections. If clonal collections by nature already have fewer accessions, do they need further rationalisation? Is there any scope for core collections in clonal species?

In fact, a combination of several factors suggests that gene banks for clonal crops, or for species with recalcitrant seeds, have much to gain from setting up core collections. These factors are:

- *Size:* Field gene banks are expensive to run and are exposed to damage from poor management, inappropriate environments, pathogens, herbivores, theft and loss of support (Morales et al., *Chapter 5.3*). Their expense and vulnerability increase with increasing numbers to maintain. Size of field gene banks is an increasingly pressing problem as national programmes assume responsibility for their indigenous germplasm, and core entries will need to be given first call on limited resources. Clearly, the elimination of proven duplicates is an obvious and important first step in rationalising clonal collections, but such reduction may be insufficient to meet the restrictions of size.

- *Conservation research:* Clonal species therefore stand in great need of new and alternative methods for conservation. An important option is the conservation of tissue culture *in vitro* under cryogenic or slow growing conditions (Roca et al., 1989), with minimised frequency of subculturing. These methods have several experimental variables that need optimising. Further, the optima are specific to species and often even to genotype. The ideal experimental material for developing and validating new *in vitro* methods and for monitoring the health of the conserved samples would be a restricted set of accessions representative of the genetic diversity within a collection (that is, a core). Thus the conservation of clonal species is technologically more complex than that of seed crops, and developmental research enhances the experimental role for core collections.

Table 1 Large collections in South America of three root crops

Country and institution[a]	Cassava[b]	Potato	Sweet potato
Argentina			
INTA, Balcarce		200 1362 (18)	
EEA, Obispo, Colombres		700	
Bolivia			
IBTA, Cochabamba		653 164 (29)	
Brazil			
CENARGEN, Brasília	800 79 (18)	387	
CNPH, Brasília		74 5 (1)	660 5 (1)
U, Brasília	2 469 (30)		
CNPMF, Cruz d. Almas	1500		198
EPACE, Fortaleza	474		
Chile			
UAC, Valdivia		579 38 (5)	
Colombia			
CIAT, Cali	5000 597 (30)		
ICA, Bogota		1062 120 (8)	
Ecuador			
INIAP, Quito		185	
Paraguay			
IAN, Cordillera	164	14	15
Peru			
CIP, Lima		3500 1500 (90)	5528 820 (59)
INIAA, La Molina	294 1 (1)		445
INIA, Puno		490	
UNSAAC, Cusco		3354	
UNA, La Molina		2500	
UNC, Huancayo		972	
UNSCH, Ayacucho		898 8	456 37 (16)
UNPRG, Lambayeque	167		365
FONAGRO, Chincha			400
Venezuela			
CIARCO, Araure	248		
UCV, Maracay	212		40

Note: a The full names of the institutions are given in the acronym list at the end of this volume
b The set of three numbers refers to the number of accessions of cultivated species, of the wild accessions and (in parenthesis) the total number of wild species

Source: Lawrence et al. (1986); CENARGEN and CIP (pers. comm.)

- *Conversion to seed:* Another option, that of seed storage, is available for those clonal species that are vegetatively propagated in agriculture but also produce seed. This group includes sugar cane, tuber crops, perennial herbs, vines, shrubs and trees (beverage, fruit and forest trees). In such species, one might question the growing of old clones in a field gene bank where they are subject to cryptic yield decline and worse. The retention of specific genotypes of sexual species in field collections can be justified as long as they have some prospect of future use in plantations. When old genotypes become outmoded, their primary future role is unlikely to be as cultivars, but rather as parents in breeders' crosses (as donors of genes and linked blocks). For this role, their sexual progeny (seed), produced by selfing or in a limited range of crosses, could store these genes and linkages and could later be grown and function as parents for the next generation of breeders. Thus, if seed conservation is an available option, more of such clones could be converted to seed. However, clonal maintenance has one advantage over seed storage in that it produces planting material for the evaluation of new traits. Therefore it is cheaper to evaluate, say, 500 clones representing a wide diversity, than 500 populations each of at least 10-20 genotypes. The core approach offers a way to choose which accessions to keep growing in the field. The seed collection could include both core entries and reserve accessions.

- *Evaluation:* A major role of clonal collections is to provide genotypes for direct use. This can be better done when genotypes have been evaluated. As noted earlier, the core approach is well suited to effective evaluation of a collection in that it allows an increased range of characters and their interactions to be tested over a representative set of the collection.

- *Introduction:* Linked with the above factor is the role of clonal collections in supplying genotypes for introduction to new areas or for new purposes. The testing of a core collection in the new area would be the first step of a two-stage strategy to find the best genotype for extension. Fewer entries in trials would allow more replication and use of resources in testing at more sites for the importance of genotype x environment interactions.

- *Combining ability:* The breeding of clonal species aims to capitalise on heterosis. Breeding advances come from finding combinations of parents with high specific combining ability. In sugar cane breeding, this has led to wide use of the 'proven cross' approach. The problem then is how to explore clonal collections for new parents or combinations that will improve upon the current cultivars. A set of exploratory crosses based on a subset of likely genotypes chosen from the core collection would be a logical beginning (Frankel and Brown, 1984).

- *Germplasm distribution:* Gene banks that have to distribute clonal material bear a greater burden in eliminating pathogens (especially viruses) from their shipments than do gene banks dispatching seed samples. A reduction in collection size is thus very important because of the high cost of producing and maintaining pathogen-free clonal material.

These factors represent many of the incentives for setting up core collections in clonal species. However, the formation of groups and the selection of core entries should give more emphasis to phenotypic characters expressed in the relevant environment than that given in seed crops. This is especially so in clonal species that rarely or never set seed, such as banana. The development of the core will need more integration of data (Kresovich et al., *Chapter 3.5*) than in seed collections. Also, the proportion of entries in the core should not be fixed at 10%. To fulfill the special roles in clonal collections discussed above, a higher or a lower proportion might be needed.

THE CORE CONCEPT AND *IN SITU* CONSERVATION

Another challenge prominent in the Keystone International Dialogue Series on Plant Genetic Resources (1991) is that of the need for programmes that conserve genetic resources *in situ*. Conservation *in situ* has long been the preferred method for the wild species related to crop plants. Its role in the conservation of cultivated species is more debatable (Marshall, 1989; Wilkes, 1989). The rationale of such programmes has proved so controversial that devising the best ways to implement them has received little attention.

Perhaps the core concept of stratified, representative sampling from a species has something to offer in the choice of populations or areas for conservation *in situ*. The need for selection of the crop populations to be maintained arises because of the costs or complexities of implementing such strategies. Brush (1991) discussed three examples (potato in the Andes of Peru, maize in Chiapas in Mexico and rice in northern Thailand) where farmers currently incorporate modern varieties into farming systems without discarding traditional varieties. He suggested that the conservation of landraces *in situ* could be achieved by encouraging farmers in areas of diversity to continue the diverse plantings that many already practise. His call for 'pilot projects' in areas of crop diversity requires improving the information base to locate farming regions where the maintenance of crop genetic resources is practised. Selection of areas is needed to target the improvement of marketing systems (by such means as better transportation), to conduct supporting agronomic research and to document genetic effectiveness. A limited number of such projects in areas selected to cover the agroecological range would constitute a core approach to *in situ* conservation.

In the case of wild crop relatives, the need for selection comes from limits on the area of a country that can be reserved and excluded from development. The benefits of selection are analogous to those for *ex situ* conservation. By restricting projects to modest proportions that are chosen to cover the genetic diversity of the species and its ecogeographic range, better use of limited resources will enable specific scientific goals (such as microevolutionary studies and the discovery of new resistances) to be achieved. A good example of such a project in wild cereal relatives is the Amiad project in *Triticum dicoccoides* and *Hordeum spontaneum* (Anikster and Noy-Meir, 1991). The same gene sampling theory indicates that modest projects would capture a substantial fraction of conceivable target alleles. Grandiose schemes including large acreages of farmland or large exclusive reservations are not justified.

CONCLUSION

The proposal to develop core collections has now been under discussion for several years and implemented in a few cases. The arguments for and against cores, the theory behind them and the methods to select them have matured. Core collections now stand at the crossroads as to whether they will be more widely adopted, modified or remain the special interest of a few students of germplasm. The views put forward in this volume are aimed at identifying and solving the problems that hinder the development of core collections. I believe that cores or similar strategies are needed to meet the current challenges in the conservation and use of genetic resources.

As national gene banks take on total responsibility for local genetic resources they are in danger of being swamped by large numbers of new deposits. This is especially the case for clonal crops. Some selection seems inescapable if gene banks are to avoid losing significant components of their collection. Far from being a threat to the integrity of collections, the concept of core collections is there to assist in dealing with a deluge of local material. Further, the development of a core will render a collection more workable to the user, a result that must help to ensure its conservation in the long term.

Acknowledgements

It is a privilege to acknowledge my debt to Sir Otto Frankel who first proposed core collections and who has continued to encourage their development. His constant questioning of any dogma or opinion for its scientific and practical validity has influenced my own writing on the subject. I also thank Dr Z. Huaman for providing data on potato collections and for many thoughtful comments on the draft of this chapter.

References

Abel, B.C. and Pollak, L.M. 1991. Rank comparisons of unadapted maize populations by testers and *per se* evaluation. *Crop Science* 31: 650-56.

Anikster, Y. and Noy-Meir, I. 1991. The wild-wheat field laboratory at Amiad. *Israel J. Botany* 40: 351-62.

Brown, A.H.D. 1989a. The case for core collections. In Brown, A.H.D., Frankel, O.H., Marshall, R.D. and Williams, J.T. (eds) *The Use of Plant Genetic Resources*. Cambridge, UK: Cambridge University Press.

Brown, A.H.D. 1989b. Core collections: A practical approach to genetic resources management. *Genome* 31: 818-24.

Brown, A.H.D. 1992. Genetic variation and resources in cultivated barley and wild *Hordeum*. *Barley Genetics* 6: 669-82.

Brown, A.H.D., Grace, J.P. and Speer, S.S. 1987. Designation of a 'core' collection of perennial *Glycine*. *Soybean Genetics Newsletter* 14: 59-70.

Brown, A.H.D., Burdon, J.J. and Grace, J.P. 1990. Genetic structure of *Glycine canescens*, a perennial relative of soybean. *Theoretical and Applied Genetics* 79: 729-36.

Brush, S.B. 1991. A farmer-based approach to conserving crop germplasm. *Economic Botany* 45: 153-65.

Burdon, J.J. and Speer, S.S. 1984. A set of differential *Glycine* hosts for the identification of races of *Phakopsora pachyrhizi* Syd. *Euphytica* 33: 891-96.

Chang, T.T. 1989. The case for large collections. In Brown, A.H.D., Frankel, O.H., Marshall, R.D. and Williams, J.T. (eds) *The Use of Plant Genetic Resources*. Cambridge, UK: Cambridge University Press.

Duvick, D.N. 1984. Genetic diversity in major farm crops on the farm and in reserve. *Economic Botany* 38: 161-78.

Erskine, W. and Muelbauer, F.J. 1991. Allozyme and morphological variability, outcrossing rate and core collection formation in lentil germplasm. *Theoretical and Applied Genetics* 83: 119-25.

Frankel, O.H. 1984. Genetic perspectives of germplasm conservation. In Arber, W., Llimensee, K., Peacock, W.J. and Starlinger, P. (eds) *Genetic Manipulation: Impact on Man and Society*. Cambridge, UK: Cambridge University Press.

Frankel, O.H. and Soulé, M.E. 1981. *Conservation and Evolution*. Cambridge, UK: Cambridge University Press.

Frankel, O.H. and Brown, A.H.D. 1984. Current plant genetic resources — a critical appraisal. In *Genetics: New Frontiers* (vol IV). New Delhi, India: Oxford and IBH Publishing.

Hawkes, J.G. 1992. Review of 'The Use of Plant Genetic Resources'. *Biological J. Linnaean Society* 45: 289-90.

Holden, J.H.W. 1984. The second ten years. In Holden, J.H.W. and Williams, J.T. (eds) *Crop Genetic Resources: Conservation and Evaluation*. London, UK: George Allen and Unwin.

Huaman, Z. 1984. The evaluation of potato germplasm at the International Potato Center. In Holden, J.H.W. and Williams, J.T. (eds) *Crop Genetic Resources: Conservation and Evaluation*. London, UK: George Allen and Unwin.

Keystone International Dialogue Series on Plant Genetic Resources. 1991. *Oslo Plenary Session. Final Consensus Report*. Washington DC, USA: Genetic Resources Communications Systems Inc.

Kimura, M. and Crow, J.F. 1964. The number of alleles that can be maintained in a finite population. *Genetics* 49: 725-38.

Lawrence, T., Toll, J. and van Sloten, D.H. 1986. *Directory of Germplasm Collections. 2. Root and Tuber Crops*. Rome, Italy: IBPGR.

Marshall, D.R. 1989. Crop genetic resources: Current and emerging issues. In Brown, A.H.D., Clegg, M.T., Kahler, A.L. and Weir, B.S. (eds) *Plant Population Genetics, Breeding and Genetic Resources*. Sunderland, UK: Sinauer.

Marshall, D.R. and Brown, A.H.D. 1975. Optimum sampling strategies in genetic conservation. In Frankel, O.H. and Hawkes, J.G. (eds) *Genetic Resources for Today and Tomorrow*. Cambridge, UK: Cambridge University Press.

Marshall, D.R. and Brown, A.H.D. 1981. Wheat genetic resources. In Evans, L.T. and Peacock, W.J. (eds) *Wheat Science — Today and Tomorrow*. Cambridge, UK: Cambridge University Press.

National Research Council. 1991. *Managing Global Genetic Resources: The US National Plant Germplasm System*. Washington DC, USA: National Academy Press.

Peeters, J.P. and Martinelli, J.A. 1989. Hierarchical cluster analysis as a tool to manage variation in germplasm collections. *Theoretical and Applied Genetics* 78: 42-48.

Perry, M.C., McIntosh, M.S. and Stoner, A.K. 1991. Geographical patterns of variation in the USDA soybean germplasm collection II. Allozyme frequencies. *Crop Science* 31: 1356-60.

Roca, W.M., Chavez, R., Martin, M.L., Arias, D.I., Mafla, G. and Reyes, R. 1989. *In vitro* methods of germplasm conservation. *Genome* 31: 813-17.

Schoen, D.J. and Brown, A.H.D. 1991. Intraspecific variation in population gene diversity and effective population size correlates with the mating system. *Proc. National Academy of Science, USA* 88: 4494-97.

Spagnoletti Zeuli, P.L. and Qualset, C.O. 1987. Geographical diversity for quantitative spike characters in a world collection of durum wheat. *Crop Science* 27: 235-41.

Vaughan, D.A. 1991. Choosing rice germplasm for evaluation. *Euphytica* 54: 147-54.

Wilkes, G. 1989. Germplasm preservation: Objectives and needs. In Knutson, L. and Stoner, A.K. (eds) *Biotic Diversity and Germplasm Preservation: Global Imperatives*. Amsterdam, Netherlands: Kluwer.

Williams, W.T. 1971. Principles of clustering. *Annual Review of Ecology and Systematics* 2: 303-26.

Part 2

METHODS OF DATA ANALYSIS FOR DEVELOPING CORE COLLECTIONS

Core Collections of Plant Genetic Resources
Edited by T. Hodgkin, A.H.D. Brown, Th.J.L. van Hintum and E.A.V. Morales

A Wiley-Sayce Publication

2.1

Hierarchical approaches to the analysis of genetic diversity in crop plants

TH.J.L. VAN HINTUM

Abstract

Genetic diversity is the result of evolution, including domestication and plant breeding. The processes of natural evolution resulted in a build-up of genetic diversity in natural populations. Domestication caused the further differentiation of a small part of the diversity of wild species, which became adapted to human requirements. Plant breeding manipulated genetic diversity of domesticated species and made it suitable for modern agricultural production systems. In the process of adapting crops to modern agricultural systems, the genetic diversity over vast areas became very small. Genetic diversity is not randomly distributed among plant populations, but has a structure that can generally be represented by a hierarchical model, a tree. A genetic diversity tree can be reconstructed by branching, based on knowledge of such factors as natural evolution, domestication, distribution and utilisation, and by clustering, based on a phenetic analysis of individuals. For the purpose of clustering, data obtained as close to the DNA as possible should be used (such as RFLPs and allozymes). Once a diversity tree is available, a core collection can be selected by deciding on the numbers of accessions per subgroup, and then choosing accessions with maximal diversity within each subgroup.

Genetic diversity is the raw material for plant breeding. For the efficient conservation and exploitation of genetic diversity it is important to know its nature and structure. This becomes even more important if the objective is to maximise the amount of useful genetic diversity in a germplasm collection, such as a core collection (Brown, 1989).

Among the methods for selecting a core collection, two types can be distinguished: those based on branching and those based on clustering. Branching methods are based on passport data combined with knowledge of, and assumptions about, the structure of the gene pool. Knowledge of all differentiating processes can be applied, including natural evolution, domestication, distribution and utilisation. No actual accessions are needed. Clustering methods are based on a phenetic analysis of characterisation data on accessions from which the core collection is to be selected. For example, in the case of the Barley Core Collection (Knüpffer and Hintum, *Chapter 4.1, this volume*) initial branching resulted in groups such as 'European cultivars', 'South Asian landraces' or 'wild species', which in turn will be divided into smaller groups such as 'Western European spring cultivars', 'Himalayan landraces' or '*Hordeum chilense*'.

Once branching is completed, the numbers of accessions to be chosen per group can be decided, depending upon the total size of the core collection and the relative importance of the groups. The structure within any group can be determined by clustering individual accessions from that group. The final choice of accessions can be made using the results of this analysis. For example, in the Barley Core Collection the selection of core accessions of the 'Himalayan landraces' group could be based on the results of the electrophoretical characterisation of a representative sample of this group (Konishi and Matsuura, 1991). If the phenetic structure of the group is known, accessions containing maximal genetic diversity should be chosen for inclusion in the core collection.

Both types of methods are based on assumptions about the structure of genetic diversity. In the case of branching it is assumed that information on the identity and origin of material allows predictions to be made about the genetic diversity in that material. This implies that passport data should be reliable. In the case of clustering it is assumed that the observed diversity in morphological, molecular or other marker systems represents the underlying total genetic diversity.

GENETIC DIVERSITY

The first modern system for classifying natural diversity into categories was devised by Linnaeus (1753), although it was highly artificial. The methodology of classical taxonomy proved useful for the classification of genera and species, but not very suitable for the classification of infraspecific crop diversity. This is illustrated by the problems concerning taxonomy and nomenclature of cultivated plants (Hawkes, 1986), which may be rooted in the opinion expressed by Linnaeus (1764) that 'the grouping of cultivated forms under species is the task of beginners in botany; a qualified botanist studies species and higher taxonomic levels'. This has not prevented many experimental taxonomists and agronomists since Linnaeus from studying crop diversity using a range of approaches, from the simple description and classification of phenotypes in a gene pool, through biogeography and ethnobotany, to the current molecular approaches. The results of these studies have partially clarified the complex structure of genetic diversity.

Genetic diversity can be defined as the extent to which heritable material differs within a group of plants. Genetic differentiation is the extent to which heritable material differs between groups of plants. The heritable material of a plant comprises its genomic and cytoplasmic DNA. Heritable material can differ both at the level of DNA sequences (alleles) and at the level of allele combinations (genotypes). The definition is concerned with all heritable material, not only with characters. The expression of a given character is merely the reflection of a relatively small amount of the total heritable material of a plant, often obscured by the environment.

Before discussing genetic diversity, genetic similarity should be briefly considered. Comparing the heritable material of plants, it can be seen that many sequences, representing a considerable part of the DNA, are shared by all plants. A comparison of the large subunit of the ribulose biphosphate carboxylase (*rbc*L) gene on the chloroplast DNA of totally different crops of tobacco and rice showed 93% sequence similarity (Sugiura, 1989). And although genomic DNA is known to be more diverse than cytoplasmic DNA, the genetic similarity between the species is evident. This similarity is the basis on which diversity is studied. Although similarities are numerous, for gene bank curators and plant breeders the differences in heritable material are what counts. To understand genetic diversity, it is important to know what caused differentiation and how these processes influenced the structure of genetic diversity.

Evolution

A first level of genetic differentiation is that among species. Evolutionary divergence resulted in an association of characters that makes it possible to distinguish well-defined groups. These groups can be described using a limited number of characters based on a minute part of the heritable material. It can be assumed that, because of reproductive isolation, a relatively large part of the rest of the heritable material is also unique to the group and that differences within a group will be smaller than those between groups. Often it will be possible to divide the genetic diversity in such a group into new smaller groups on the same basis.

This principle of subsequent divisions implies that species diversity has a basically hierarchical structure. Taxonomic nomenclature is based on this hierarchy. For example, Kentucky bluegrass can be classified into the following groups, each group being one of the subgroups of the previous one (Porter, 1959): kingdom Plantae, division Embryophyta, subdivision Phanaerogama, branch Angiospermae, class Monocotyledoneae, subclass Glumiflorae, order Poales, family Poaceae, subfamily Festucoideae, tribe Festuceae, genus *Poa*, section Pratenses, species *Poa pratensis*. This hierarchy will generally reflect the effects of evolution in time. Like most systems in nature, this structure is not always clear-cut. There are numerous natural factors, such as hybridisation and clinal variation, that can complicate the structure. In general, however, the hierarchical model will suffice.

Cladistics, also known as phylogenetic systematics, involves the study of this evolutionary hierarchy (Stuessy, 1990). In pre-defined groups, the so-called operational taxonomic units (OTU), a distinction is made between character states. Primitive 'wild type' states are distinguished from advanced 'new' states. On this basis, cladograms can be constructed, describing the sequence of branching of ancestral populations.

The reconstruction of evolutionary history resulted in the need to discriminate certain kinds of evolutionary trends. The most important ones are homology and homoplasy. Homology is resemblance as a result of inheritance from a common ancestry. Homoplasy is resemblance not resulting from such inheritance, and includes parallelism and convergence. Parallelism is the development of similar character states in separate lineages of common ancestry on the basis of characteristics of that ancestry. Convergence is the development of similar character states in different lineages but without a common direct ancestry. These trends are also important if we go further up in the tree of genetic differentiation.

Domestication

The second level of genetic differentiation results from infraspecific divergence as a result of domestication. The processes causing genetic change under domestication are basically the same as those occurring under evolution in nature, except that human selection, partly conscious, partly unconscious, is added. The major difference, apart from being centred around human activity, is that the rate of change under domestication is much higher than the relatively slower processes under natural evolution (Pickersgill, 1984).

Domestication usually causes a reduction of genetic diversity compared with the diversity of the species in the wild. This is attributable to the founder effect: only a small part of the wild gene pool is brought under cultivation (Ladizinsky, 1985). In general, this limited initial diversity hardly increases via introgression. Various isolating mechanisms prevent natural hybridisation between cultivated and wild plants, and if gene flow occurs this is more effective in the direction from the cultivated to the wild populations. Because of the generally recessive nature of cultivated characteristics (Ladizinsky, 1985; Lester, 1989) hybrids will show mainly wild characteristics, and will thus have a selective disadvantage in the human environment.

Landraces developed within domesticated species. Since landraces in a certain region can be expected to be adapted to that region and since exchange of heritable material between landraces grown in a certain region will be more frequent than between landraces grown in different regions, they can be assumed to share a certain genetic background, including specific adaptations to the area. In the geographical distribution of domesticated diversity it is possible to identify three primary spatial scales (Zimmerer and Douches, 1991). The largest scale (macro-geographic) was first described by Vavilov (1926) with his centres of origin, and later enlarged by Zhukovsky (1975) with his mega-centres of diversity (Zeven and de Wet, 1982). Several studies indicated that these often subcontinental centres divide into smaller subcentres (meso-geographical scale), with distinct diversity (Simmonds, 1976). Micro-geographical centres of crop diversity are often related to geographically distinct areas such as coasts and mountainous areas, as noted, for example, by Weltzien (1989).

Apart from this genetic differentiation in meso- and micro-geographical regions, within a region landrace groups (Zeven, 1986) can often be distinguished on the basis of name, morphology or usage, as reported, for example, by Elings (1991) and Zimmerer and Douches (1991). The formation of these landrace groups is the result mainly of human selection, which causes differentiation for specific applications. The resulting phenotypic differentiation can be excessive. An example is *Brassica oleracea*, where the domesticated phenotypes include cabbage, kale, Chinese kale, cauliflower, broccoli, Brussels sprouts and kohl rabi (IBPGR, 1981), an impressive range which indicates the diversity in the wild of *B. oleracea* from which the domesticated forms arose.

The extent to which the hierarchical structure applies to the genetic diversity within a domesticated crop species will generally be less than that at the species level. The association between characters within a crop species will generally be much smaller than between species since the reproductive isolation at the species level has occurred over a much longer period of time, resulting in a clearer delimitation of variation. However, the associations that exist in landraces can, like species diversity, be represented in a hierarchical structure.

Plant breeding

Modern plant breeding brings many of the above processes under human control and changes the rate radically. Some plant breeding techniques increase crop diversity (for example, induced mutations, hybridisation between previously incompatible populations, and introgression from previously isolated populations). Others cause a reduction of diversity (for example, the production of inbred lines and hybrid cultivars, and most of the breeding for adaptation to high-input agriculture). The replacement of relatively variable landraces by the resulting homogeneous cultivars is one of the causes of genetic erosion in many major and minor crops.

In the case of modern cultivars, specific adaptation to certain growing areas and usages can be expected, as in the case of landraces. A breeder will prefer to use material in his breeding programme that is already adapted to the specific market at which he is aiming. If 'exotic' material is used this is usually only for the introduction of specific characters. In the classical breeding techniques after a cross with less adapted material, the undesirable exotic background is reduced via recurrent backcrossing. Modern techniques promise to make it possible to transfer only the gene with the desired character.

Although the similarity of modern cultivars is striking, it is usually possible to distinguish types of similarly adapted material. The different types will not be totally distinct from each other; there will be a more or less continuous spectrum of cultivars between the extreme types. Even for this rather diffuse structure, however, the hierarchical model of differentiation can be used for the purpose of composing a core collection.

Structure of genetic diversity

Genetic diversity has a complex multi-dimensional structure. As a result of considerable association between characters in groups of plants, it is possible to describe the structure of genetic diversity to some extent by describing these groups and their relationships. In many cases this structure can be adequately represented in a hierarchical model.

Based on such a model, a genetic diversity tree can be constructed. Starting from the trunk of the tree, which is the entire domain of diversity studied (that is, the genus or a crop species), distinct groups such as species or major phylogenetic groups can be distinguished. Each group can be further divided, step by step, into smaller groups which have the highest genetic differentiation possible. This branching can continue until no more differentiating steps can be made. Material in a group at any level of differentiation can be sampled and characterised. If sufficient characterisation data on individual plants are available it is possible to construct a diversity tree using clustering techniques; individuals are grouped on the basis of observed similarity. It should be noted that gene bank accessions are often not genetically homogeneous since they correspond with wild populations, landraces or cultivars.

ANALYSING CROP DIVERSITY

A critical distinction in ways of analysing the genetic diversity in a group of plants lies in the data used for analysis. Clustering based on pedigree data, genetic markers, qualitative characters, or quantitative characters can be expected to differ because all data types require their own measures of genetic diversity, and can seem to present different aspects of genetic diversity.

Pedigree data

If pedigrees of the studied material are known, which is only rarely the case, it is possible to perform a pedigree analysis. The degree of co-ancestry of two individuals is usually quantified with the coefficient of parentage (r) as defined by Kempthorne (1969). The r between two individuals is the probability that a random allele at a random locus in one individual is identical by descent to a random allele at the same locus in the other individual (Cox et al., 1985). In the analysis, several assumptions have to be made (Martin et al., 1991); all of them are, to varying extents, open to dispute:

- a cultivar receives half its genes from each parent
- parents in crosses are homozygous and homogeneous
- ancestors for which no pedigree information is available are unrelated
- the r between a cultivar and a selection from that cultivar is 0.75

Pedigree analysis has been used to describe the genetic basis of crops (Knauft and Gorbet, 1989) and its development over time (Cox et al., 1986; Souza and Sorrells, 1989), but also, for example, to predict hybrid performance (Smith et al., 1990).

Genetic markers and qualitative characters

Genetic markers allow identification of the alleles for a certain gene of the plant without disturbing environmental interference. Genetic markers in diversity studies include morphological markers, storage proteins (for example, Hintum and Elings 1991), allozymes (for example, Brown and Weir, 1983) and restriction fragment length polymorphisms (RFLPs) (for example, Kesseli et al., 1991). Not all genetic markers are unambiguous; for example, allozyme expression may be tissue specific and an allozyme band may correspond to more than one DNA sequence. Qualitative characters, such as flower colour, waxiness and presence of anthocyanin, also show the genotype of the plant, but do not always allow a direct interpretation into alleles because of the often more complicated genetic basis. The increasingly popular RFLPs are analysed as qualitative characters (Graner et al., 1990; Miller and Tanksley, 1990; Smith et al., 1990; Kochert et al., 1991). Many morphological characters can also be considered to be qualitative. The analysis of this type of data is characterised by the fact that distinct classes of alleles or characteristics are studied.

Several measures of genetic diversity are available. Two concepts can be identified: the allelic richness (the number of distinct alleles in a sample); and the allelic evenness (the distribution of allelic frequencies) (Brown and Weir, 1983). Commonly used measures such as Nei's (1973) diversity index H or Shannon and Weaver's (1949) information index I combine these two concepts (Hennink and Zeven, 1991; Hintum and Elings, 1991). Genetic differentiation between populations can be quantified in several ways, including the use of G_{ST} (Nei, 1973), interpretable in genetic terms (Gregorius, 1987) or the use of measures such as the proportion of corresponding bands in the RFLP banding pattern.

Genetic diversity studies based on genetic markers and qualitative characters are used for many purposes. They have been used in taxonomic studies (Miller and Tanksley, 1990), to find the centre of diversity of a species (Lubbers et al., 1991), to trace the route of domestication (Konishi, 1988), to study the relationship between environment and diversity (Hintum and Elings, 1991) and to study a complete crop gene pool (Kesseli et al., 1991) or the diversity in a specific part of a gene pool (Cox et al., 1986). The distinct classes that result from descriptions based on genetic markers and qualitative characters also allow the study of association between characters or the multi-locus structure of groups (for example, Zhang et al., 1990).

Quantitative characters

The expression of quantitative characters is dependent upon both genotype and environment. They include morphometric characters and important agronomic characters such as yield, earliness and drought susceptibility. Since their genetic basis is complex and the influence of the environment is usually significant, and since similarity can be based on convergence, quantitative characters are generally not very suitable for diversity studies. They can be used for the identification of similar adaptations.

As a measure of diversity within populations, the common standard deviation (σ) and coefficient of variation (CV) can be used. Differentiation between populations is usually quantified using the difference in mean expression or analysis of variance components (for example, Spagnoletti Zeuli and Qualset, 1987).

Analysis

After calculating the similarity between individuals or between populations it is possible to classify the material, resulting in clusters of similar material using phenetic methods. Most common are hierarchical clustering techniques producing dendrograms (for example, Kesseli et al., 1991; *see also* Figure 3). An important factor in applying clustering techniques is the choice of the method for calculating the distance between clusters. Standard statistical software packages usually do not include genetic distance measures in their clustering modules, making it difficult to apply the proper method.

It is also possible to trace patterns using principal components analysis (PCA) (*see* Figure 1) or to check the validity of an existing classification using the related canonical variance analysis (CVA) (for example, Erskine and Muehlbauer, 1991).

Figures 1, 2 and 3 give three of the many different ways of graphically representing genetic diversity, and illustrate their relationship. The principal component plot (*see* Figure 1), the result of a fictional phenetic study, shows the structure of genetic diversity in a group of individuals in two dimensions. The two orthogonal axes, principal components, contain the maximal variance possible in the multi-dimensional space of descriptions. They are linear combinations of the descriptors used

Figure 1 Example of a principal component plot

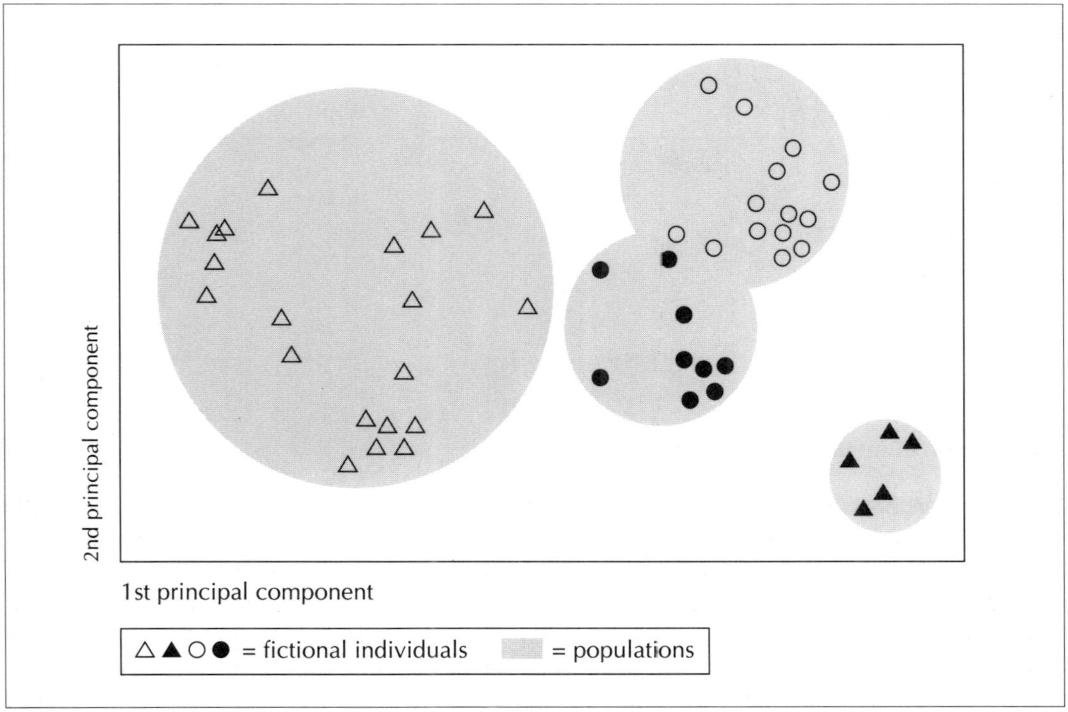

in the analysis. The three-dimensional diversity tree (*see* Figure 2) adds the dimension 'differentiation process' to the two presented in the principal component plot. It shows the differentiation process resulting in the patterns as shown in the principal component plot, corresponding with the upper surface in the three-dimensional diversity tree. The dendrogram (*see* Figure 3), which can result from a cladistic or a phenetic study or a combination of both, shows the distances between objects and clusters of objects, and can be seen as a side view of the three-dimensional diversity tree. The symbols used in the dendrogram correspond to those in the principal component plot.

Figure 2 Example of a three-dimensional diversity tree

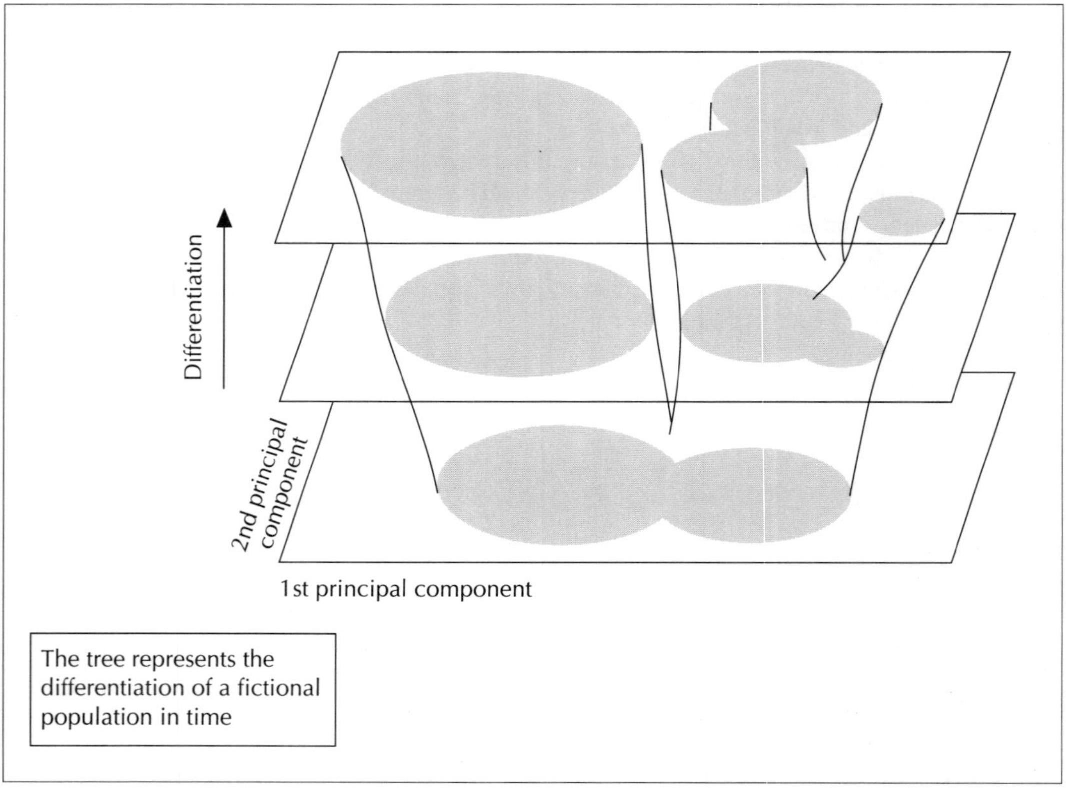

Comparison of methods

Several papers have compared the results of using the different types of data. Souza and Sorrells (1991a, b) studied North American oat cultivars, both old and modern, and found that grouping based on pedigrees was very similar to that based on qualitative data. Clusters formed on the basis of quantitative characters produced a good measure of similarity of environmental response but a biased measure of genetic relationship. This had also been reported by Wilson (1989) who found on the basis of allozyme markers that two morphometrically distinct Mexican *Cucurbita* species, one crop and one weed species, were genetically very similar, whereas the allozyme markers allowed for a level of

Figure 3 Example of a dendrogram

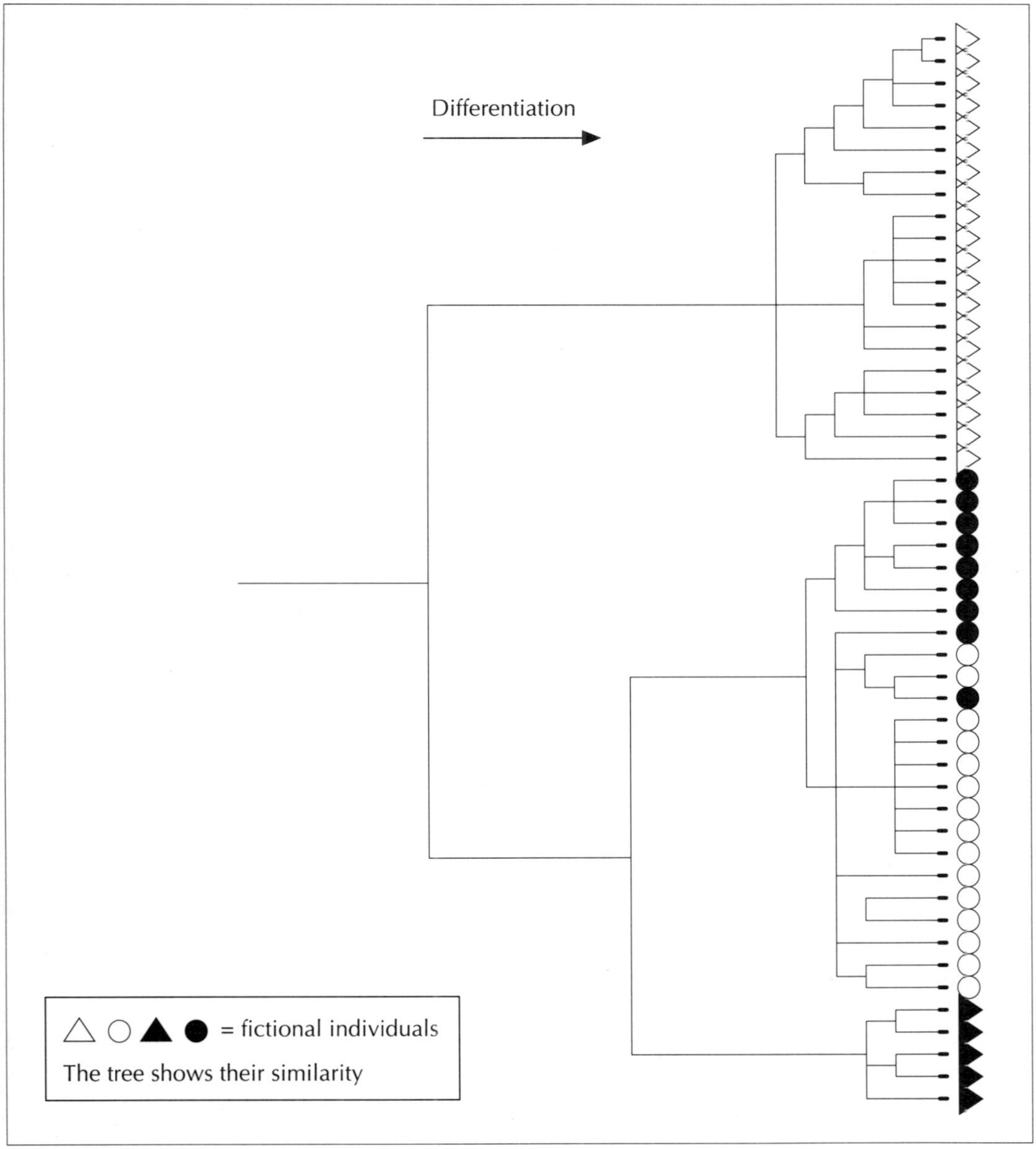

resolution that extended to infraspecific differentiation. Hintum and Elings (1991) studied storage protein diversity and phenotypic diversity in durum wheat landraces in relation to their geographical origin. They found that diversity measures based on quantitative characters had a much lower positive correlation with environmental factors than diversity measures based on the storage protein markers, implying that the markers were more suitable for measuring genetic diversity. Cox et al. (1985), in their study of old and modern soybean cultivars, found good correspondence between coefficient of parentage and similarity indexes based on allozymes and qualitative morphological markers. There was no indication that the allozyme data were better than the morphological data. Smith et al. (1990) studied the genetic distance among more or less related inbred lines of maize based on RFLPs and found a very high positive correlation with F1 yield components. This correlation was slightly higher than that of the coefficient of parentage, and much higher than those of all other measures based on allozymes, zeins or morphological characters.

These and many other studies confirm that:

- genetic diversity can best be quantified on the basis of data obtained as close to the DNA as possible
- the results of the analysis of pedigree data correspond to those of the analysis of genetic markers and qualitative characters
- the expression of morphological and agronomic characters indicates adaptation to environmental factors rather than genetic diversity and differentiation

Sometimes, no structure or only a very weakly defined structure can be found. This can be attributed to a lack of variation at the marker loci or to an insufficient resolving power of the combination of characters and analytical tools (for example, Gepts and Clegg, 1989). It could also be attributed to the chaotic nature of the genetic diversity that was studied, in cases where the hierarchical model did not apply. Sometimes, the results of different studies seem to be contradictory. This may stem from character incongruence or other problems in structuring continuous variation (Prentice, 1984). In some cases the structure is extremely clear, as in the case of an RFLP study conducted by Kesseli et al. (1991) on lettuce, where not only species clustered perfectly, but also infraspecific variation clustered in the way that had been expected.

CONCLUSION

Genetic diversity is not randomly distributed over plant populations. As a result of processes in natural evolution, domestication and modern plant breeding, genetic diversity has a structure that can generally be summarised in a hierarchical model, a tree.

The selection of accessions for a core collection is based on the structure of the genetic diversity to be represented by that core collection. To describe the structure, a genetic diversity tree can be constructed by branching (based on a knowledge of natural evolution, domestication, distribution and utilisation) or by clustering (based on a phenetic analysis of individual accessions). For clustering, data obtained as close to the DNA as possible should be used (that is, RFLPs and allozymes).

If a diversity tree is available, a core collection can be selected by deciding on the numbers of accessions per subgroup, and subsequently choosing accessions with maximal diversity within each group.

References

Brown, A.H.D. 1989. Core collections: A practical approach to genetic resources management. *Genome* 31: 818-24.

Brown, A.H.D. and Weir, B.S. 1983. Measuring genetic variability in plant populations. In Tanksley, S.D. and Orton, T.J. (eds) *Isozymes in Plant Genetics and Breeding, Part A*. Amsterdam, Netherlands: Elsevier.

Cox, T.S., Kiang, Y.T., Gorman M.B. and Rodgers, D.M. 1985. Relationship between coefficient of parentage and genetic similarity indices in the soybean. *Crop Science* 25: 529-32.

Cox, T.S., Murphy, J.P. and Rodgers, D.M. 1986. Changes in the genetic diversity in the red winter wheat region of the United States. *Proc. National Academy of Science (USA)* 83: 5583-86.

Elings, A. 1991. Durum wheat landraces from Syria. II. Patterns of variation. *Euphytica* 54: 231-43.

Erskine, W. and Muehlbauer, F.J. 1991. Allozyme and morphological variability, outcrossing rate and core collection formation in lentil germplasm. *Theoretical and Applied Genetics* 83: 119-25.

Gepts, P. and Clegg, M.T. 1989. Genetic diversity in pearl millet (*Pennisetum glaucum* [L.] R. Br.) at the DNA sequence level. *J. Heredity* 80: 203-08.

Graner, A., Siedler, H., Jahoor, A., Herrmann, R.G. and Wenzel, G. 1990. Assessment of the degree and the type of restriction fragment length polymorphism in barley. *Theoretical and Applied Genetics* 80: 826-32.

Gregorius, H.R. 1987. The relationship between the concepts of genetic diversity and differentiation. *Theoretical and Applied Genetics* 74: 397-401.

Hawkes, J.G. 1986. Problems of taxonomy and nomenclature in cultivated plants. *Acta Horticulturae* 182: 41-47.

Hennink, S. and Zeven, A.C. 1991. The interpretation of Nei and Shannon Weaver within population variation indices. *Euphytica* 51: 235-40.

Hintum, Th.J.L. van, and Elings, A., 1991. Assessment of glutenin and phenotypic diversity of Syrian durum wheat landraces in relation to their geographical origin. *Euphytica* 55: 209-15.

IBPGR. 1981. *Genetic Resources of Cruciferous Crops*. Rome, Italy: IBPGR.

Kempthorne, O. 1969. *An Introduction to Genetic Statistics*. Ames, USA: Iowa State University Press.

Kesseli, R., Ochoa, O. and Michelmore, R. 1991. Variation at RFLP loci in *Lactuca* spp. and origin of cultivated lettuce (*L. sativa*). *Genome* 34: 430-36.

Knauft, D.A., and Gorbet, D.W. 1989. Genetic diversity among peanut cultivars. *Crop Science* 29: 1417-22.

Kochert, G., Halward, T., Branch, W.D. and Simpson, C.E. 1991. RFLP variability in peanut (*Arachis hypogaea* L.) cultivars and wild species. *Theoretical and Applied Genetics* 81: 565-70.

Konishi, T. 1988. Genetic differentiation and geographical distribution of barley. *Crop Genetic Resources of East Asia*. Rome, Italy: IBPGR.

Konishi, T. and Matsuura, S. 1991. Geographical differentiation in isoenzyme genotypes of Himalayan barley. *Genome* 34: 704-09.

Ladizinsky, G. 1985. Founder effect in crop-plant evolution. *Economic Botany* 39: 191-99.

Lester, R.N. 1989. Evolution under domestication involving disturbance of genetic balance. *Euphytica* 44: 125-32.

Linnaeus, C. 1753. *Species Plantarum*. (1st edn). Stockholm, Sweden: Holmiae.

Linnaeus, C. 1764. Ordines naturales. In *Genera plantarum*. (6th edn). Stockholm, Sweden: Holmiae.

Lubbers, E.L., Gill, K.S., Cox, T.S. and Gill, B.S. 1991. Variation of molecular markers among geographically diverse accessions of *Triticum tauschii*. *Genome* 34: 354-61.

Martin, J.M., Blake, T.K. and Hockett, E.A. 1991. Diversity among North American spring barley cultivars based on coefficients of parentage. *Crop Science* 31: 1131-37.

Miller, J.C. and Tanksley, S.D. 1990. RFLP analysis of phylogenetic relationships and genetic variation in the genus *Lycopersicon*. *Theoretical and Applied Genetics* 80: 437-48.

Nei, M. 1973. Analysis of gene diversity in subdivided populations. *Proc. National Academy of Science (USA)* 70: 3321-23.

Pickersgill, B. 1984. Evolution of hierarchical variation patterns under domestication and their taxonomic treatment. In Styles, B.T. (ed) *Infraspecific Classification of Wild and Cultivated Plants*. Oxford, UK: Clarendon.

Porter, C.L. 1959. *Taxonomy of Flowering Plants*. San Francisco, USA: W.H. Freeman and Co.

Prentice, H.C. 1984. Continuous variation and classification. In Styles, B.T. (ed) *Infraspecific Classification of Wild and Cultivated Plants*. Oxford, UK: Clarendon.

Shannon, C.E. and Weaver, W. 1949. *The Mathematical Theory of Communication*. Urbana, Chicago, USA: University of Illinois.

Simmonds, N.W. (ed) 1976. *Evolution of Crop Plants*. London, UK: Longman.

Smith, O.S., Smith, J.S.C., Bowen, S.L., Tenborg, R.A. and Wall, S.J. 1990. Similarities among a group of elite maize inbreds as measured by pedigree, F_1 grain yield, grain yield, heterosis, and RFLPs. *Theoretical and Applied Genetics* 80: 833-40.

Souza, E. and Sorrells, M.E. 1989. Pedigree analysis of North American oat cultivars released from 1951 to 1985. *Crop Science* 29: 595-601.

Souza, E. and Sorrells, M.E. 1991a. Relationships among 70 North American oat germplasms. 1. Cluster analysis using quantitative characters. *Crop Science* 31: 599-605.

Souza, E. and Sorrells, M.E. 1991b. Relationships among 70 North American oat germplasms. 2. Cluster analysis using qualitative characters. *Crop Science* 31: 605-12.

Spagnoletti Zeuli, P.L. and Qualset, C.O. 1987. Geographical diversity for quantitative spike characters in a world collection of durum wheat. *Crop Science* 27: 235-41.

Stuessy, T.F. 1990. *Plant Taxonomy: The Systematic Evaluation of Comparative Data*. New York, USA: Columbia University Press.

Sugiura, M. 1989. The chloroplast chromosomes in land plants. *Annual Review of Cell Biology* 5: 51-70.

Vavilov, N.I. 1926. Studies on the origin of cultivated plants. *Bulletin of Applied Botany 26*. Leningrad, USSR: Institute of Applied Botany and Plant Breeding.

Weltzien, E. 1989. Differentiation among barley landrace populations from the Near East. *Euphytica* 43: 29-39.

Wilson, H.D. 1989. Discordant patterns of allozyme and morphological variation in Mexican *Cucurbita*. *Systematic Botany* 14: 612-23.

Zeven, A.C. 1986. Landrace groups of bread wheat (*Triticum aestivum* L. em. Thell.). *Acta Horiculturae* 182: 365-76.

Zeven, A.C. and Wet, J.M.J. de. 1982. *Dictionary of Cultivated Plants and Their Regions of Diversity, Excluding Most Ornamentals, Forest Trees and Lower Plants*. Wageningen, Netherlands: PUDOC.

Zhang, Q.F., Sagai Maroof, M.A. and Allard, R.W. 1990. Worldwide pattern of multilocus structure in barley determined by discrete log-linear multivariate analysis. *Theoretical and Applied Genetics* 80: 121-28.

Zhukovsky, P.M. 1975. *World Gene Pools for Plant Breeding: Mega-Genecentres and Endemic Micro-Genecentres*. Leningrad, USSR: USSR Academy of Sciences.

Zimmerer, K.S. and Douches, D.S. 1991. Geographical approaches to crop conservation: The partitioning of genetic diversity in Andean potatoes. *Economic Botany* 45: 176-89.

Core Collections of Plant Genetic Resources
Edited by T. Hodgkin, A.H.D. Brown, Th.J.L. van Hintum and E.A.V. Morales

A Wiley-Sayce Publication

2.2

Sampling strategies for use in stratified germplasm collections

K. YONEZAWA, T. NOMURA and H. MORISHIMA

Abstract

Sample size and stratification strategy in sampling core entries from a stratified or structured collection were investigated on the basis of a theoretical model that takes account of both the amount of genetic diversity retained and its maintenance. Calculations using the model showed that the optimum sample fraction depends upon various genetic and resources parameters, primarily on the degree of genetic redundancy among accessions comprising the whole collection and the amount of resources available for the maintenance (rejuvenation) of the core entries. The optimum sample fraction was large, with a lower redundancy among accessions or a lower initial allelic diversity within accessions, indicating that a larger sample fraction is better in species with a higher selfing rate. A 20-30% sample was estimated to be appropriate in the situations where accessions in the collection are neither very heavily nor very lightly redundant ($0.9 > Dr > 0.2$, in terms of the degree of the redundancy defined in this chapter), and the core entries are maintained for a duration of about 10 cycles of rejuvenation, with plants of the order of 10^3 being grown for one cycle of rejuvenation. Five stratification strategies were compared, using the calculations of 14 hypothetical and four real collections composed of several groups. It was concluded that, in general, the optimal strategy is a proportional one, where the sample size for each group is in proportion to the number of accessions in that group. When the genetic diversity is known in advance, a strategy where the stratification is made in proportion to the range of genetic diversity is best. Sampling procedures for collections where accessions are hierarchically structured among groups or single accessions were assessed. To retain both the pattern and the range of genetic diversity in the whole collection, procedures using the uniqueness of accessions were developed.

Plant germplasm collections have grown fairly large in size (for example, Yeatman et al., 1984). Developing procedures for reducing the size of a collection to a manageable and accessible level (a core size) is becoming one of the more important issues in the management and utilisation of plant germplasm collections (Frankel, 1984; Brown, 1989a; Marshall, 1990). Two basic issues to be addressed are the optimum sample size (that is, how large a fraction of a whole collection should be sampled) and the strategy for stratified sampling when the collection is grouped or structured in some way.

Discussion on these issues was initiated by Brown (1989a, b). He recommended a sample size of about 10%, which should be stratified in log proportion (strategy L, defined below) or absolute proportion (strategy P) to the number of accessions in groups when the collection is composed of several groups. This conclusion was based on a theoretical model which, as will be explained later, may not always hold in actual collections.

In this chapter, we report on investigations into the optimum sample size and its allocation strategy using another theoretical model where not only the range of genetic diversity retained by the sampling was considered but also the resources required for the maintenance of the genetic diversity initially retained. The accessions in a collection may be hierarchically structured among groups or single accessions. Sampling procedures for structured collections are also discussed.

OPTIMUM SAMPLE FRACTION

Definitions and formulations

Sample size (that is, the number of accessions to be sampled) is the first issue to be addressed when forming a core collection from a whole collection. As noted above, a sample fraction of about 10% has been proposed by Brown (1989a) as the optimum sample size. This estimation was based on the sampling theory put forward by Ewens (1972) where the number of alleles (n_r in Brown's notation) expected to be retained from a population of effective size N_e, with N_r individuals being randomly chosen, was discussed. Applying Ewens' theory, Brown (1989a) showed that the fraction (or more precisely, the lower 95% limit of it) of the alleles retained from a population increased rapidly as the sample fraction N_r/N_e increased to 0.1, but rather slowly after N_r/N_e surpassed 0.1.

The estimation based on Ewens' theory could not be directly applied in many cases of sampling core entries from a collection. Sampling N_r plants from a population of effective size N_e should not be regarded as genetically equivalent to sampling N_r core entries from N_e accessions. The accessions in a collection are the genetic units which have adapted to or evolved in different habitat conditions in isolation from others, and have been collected and maintained separately from others. In addition, an accession may not be of a single genotype but a mixture of plants with different genotypes. The accessions in a collection therefore could not generally be regarded as equivalent to individuals interbreeding in one population. Ewens' theory also assumed neutrality of alleles and a state of equilibrium in allelic frequencies. The neutrality principle may not hold for many genes controlling the adaptive traits of wild species and economic traits of cultivars because these traits are the products of long-term natural and artificial selection. It also has been suggested (for example, by Morishima, 1991) that some, if not many, isozymes are distributed highly associated with adaptive characters, and thus the neutrality principle may not apply even for isozyme variations. The assumption of an equilibrium state also may not be valid since accessions in a collection do not interbreed with each another.

In this chapter, the optimum sample size is discussed using another theoretical model. Suppose that the whole collection is composed of M accessions which, as shown in Table 1, are classified into K classes of allelic composition at a locus. Based on this definition of the collection, each accession belongs to only one K class, and all the accessions belonging to the same class have an identical or almost identical allelic composition. To measure the degree of genetic redundancy in this collection, the following quantity is introduced:

$$Dr \quad = \quad 1 - K/M \tag{1}$$

Dr equals 0 when all the accessions have a different allelic composition, approaching unity (precisely, 1 - 1/*M*) as *K* gets smaller. With m_s accessions being randomly sampled from this collection, the retention of genetic diversity, referred to as *RT*, is:

$$RT \quad = \quad E[k]/K \tag{2}$$

In this formula, $E[k]$ indicates the expected number of the allelic composition classes retained in the m_s accessions. This was used, although in different context, by Brown (1989a), being formulated as:

$$E\,[k] \quad = \quad K - \sum_{j=1}^{K} Q_j \tag{3}$$

where:

$$Q_j \quad = \quad \begin{cases} \prod_{x=0}^{m_s-1} \left\{ 1 - \frac{mj}{M-x} \right\} & \text{...when } m_s < M - m_j \\ 0 & \text{...otherwise} \end{cases} \tag{4}$$

Retention (*RT*) reaches unity when all the *K* classes are retained, taking the minimum value of 1/*K* when only one class is retained. The sample size m_s is related to the sample fraction *p* thus:

$$m_s \;=\; M \cdot p \qquad \text{or} \qquad p \;=\; m_s/M \tag{5}$$

The optimum sample fraction cannot be derived using the quantity *RT* alone since *RT* increases steadily towards unity, as does *p* towards unity or m_s towards *M*. To define the optimum sample fraction, a quantity to measure degree of maintenance of the allelic composition within accessions (that is, the proportion with which the initial state of allelic composition within entries is maintained for a certain duration of the core collection) was introduced. Using this quantity, denoted by *GM*, the overall efficiency of the core collection, *EF*, is quantified as:

$$EF \;=\; RT \cdot GM \tag{6}$$

which may be taken as the total fraction of allelic diversity that is expected to be sampled and maintained in the core collection. With the input of the total amount of resources (such as manpower and facility resources) for the maintenance of the core collection being fixed, *GM* decreases, while *RT* increases, as *p* increases. The maintenance of the allelic composition of single entries will be jeopardised if more entries with lower resource input per entry are maintained. The sample fraction that maximises *EF* with a fixed total amount of resources is defined as the optimum fraction.

There are a number of different ways of formulating *GM*. One of the simplest ways is:

$$GM = \frac{1 - \theta_t}{1 - \theta_o} \tag{7}$$

where variable θ_t indicates the probability that any two alleles randomly chosen from a single entry are identical by descent, the subscript t being the number of rejuvenations for maintenance (θ_o defines the value at initial state). Therefore, θ_t measures the degree of allelic homogeneity within entries; it increases as the allelic diversity within entries is reduced as generation advances, reaching unity when the entries become genetically fixed (containing only a single type of allele). The quantity $1 - \theta_t$, being standardised by its initial value $1 - \theta_o$, therefore measures degree of maintenance of the initial allelic composition within entries.

When an entry of monoecious species is rejuvenated, with N plants being grown each generation, the probability θ_t is calculated by the recurrence relationship:

$$\theta_t = \frac{1}{N} \cdot \frac{1 + f_{t-1}}{2} + \{1 - \frac{1}{N}\} \cdot \theta_{t-1}$$

$$f_t = s \cdot \frac{1 + f_{t-1}}{2} + (1 - s) \cdot \theta_{t-1} \tag{8}$$

where s indicates the selfing rate of each plant, f_t being the probability of two homologous alleles of a plant being identical by descent at generation t, a quantity that measures the excess degree of homozygosity over the Hardy-Weinberg ratio. A relationship $f_o = s / (2 - s)$, which is expected to hold in a natural plant population at genetic equilibrium, is assumed in the numerical calculations presented in the next section. Absence of selection and mutation of alleles was assumed in formula (8), which is not a problem in the present discussion since entries in this case are maintained under managed conditions with a small size; population genetics theory shows that, with a small population size, the random drift determines the genetic fate of a population (for example, Crow and Kimura, 1970).

Both θ_o and f_o are equal to unity for a genetically fixed entry that from the beginning ($t = 0$) contains only a single homozygous genotype. The quantity *GM* for this entry should be defined as unity since no reduction in allelic composition occurs in this entry. In reality, *GM* would take different values with different entries, but as a first step it is assumed to be uniform throughout all the m_s entries sampled.

Lastly, cost parameters are incorporated into the model. In maintaining the core entries, most of the resources will be allocated to the procedures for rejuvenation, such as field preparation, seeding, growing and harvesting. When a total amount of resources, R, is allocated for the maintenance of the core collection, the following relationship may be assumed:

$$R = m_s \cdot r \cdot N \cdot t$$

and therefore

$$N = (R / r) / (t \cdot m_s) \tag{9}$$

where r denotes the resource input per plant for one cycle of rejuvenation, t being the number of rejuvenations (the duration of the core collection). By definition, R / r, which is symbolised by α in formula (10), represents the aggregate number of plants treated for the maintenance of the m_s entries through t cycles of rejuvenation, and then the quantity α / t, symbolised by β, indicates the plant number treated in a single generation. Using these symbols, formula (9) is written as:

$$N = \alpha / (t \cdot m_s) \quad \text{or} \quad \beta / m_s \tag{10}$$

EF is now related to the cost parameters by substituting the right side of formula (10) for the plant number N in formula (8).

Numerical computations

The retention of genetic diversity (*RT*), where $M = 500$, $\alpha = 10^4$ and $t = 10$, was computed for three cases (a, b and c, as defined in Figure 1) differing in *Dr*. The figure shows that, with a high degree of redundancy ($Dr = 0.9$), *RT* increases very sharply with an increase in p, reaching a plateau at a relatively small p value ($0.2 \sim 0.3$). With a moderate or small *Dr* value, however, *RT* increases steadily towards unity at a rather constant increasing rate. The optimum fraction, therefore, cannot be deduced from the retention alone. Calculations for the case of $M = 200$ (not presented here) gave essentially the same results, indicating that *RT* with a given p depends primarily upon the *Dr* in the collection.

The retention is influenced not only by *Dr* but also by the distribution of accessions on the allelic composition classes (*see* Table 1). *RT* with a given value of *Dr* is reduced as the number of classes that

Figure 1 **The retention of genetic diversity (*RT*) in three cases differing in degree of genetic redundancy (*Dr*)**

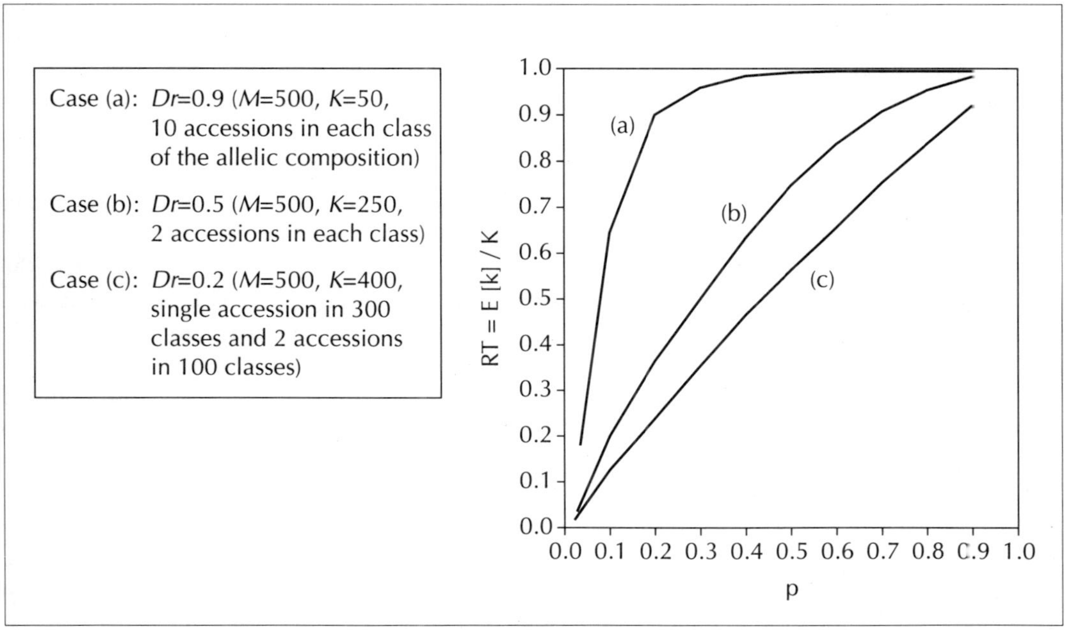

Table 1 Genetic composition of accessions contained in a collection

	Class of allelic composition				Total
	1	**2**	**...**	**K**	
Number of accessions	m_1	m_2	...	m_k	M $(= \Sigma m_i)$

contain only one accession increases. However, the calculations (not presented) showed that this influence is not important unless *Dr* is as high as or higher than 0.9.

The overall efficiency (*EF*) was calculated for a number of combinations of the constituent parameters. The results for the three cases (a, b and c) are shown in Figure 2. The four curves (i to iv) in each of the three cases show the influences of the mating system (s) and the initial allelic homogeneity (θ_o) within entries.

A comparison of the three cases in Figure 2 shows that the optimum sample fraction (as expected) tends to be large as the *Dr* decreases from 0.9 to 0.2. This increase, however, is not very large; the optimum sample fraction stays within the range of 0.2 ~ 0.3, although this goes up to about 0.5 when predominantly outcrossing (s = 0.1) entries with a high initial co-ancestry (θ_o = 0.9) are to be maintained (curve iii).

With a given degree of the redundancy, curves (iii) and (iv) show higher retentions and larger optimum fractions than (i) and (ii), indicating that allelic diversity within outcrossing entries is easier to maintain than that within selfing entries. Where entries with a high selfing rate are rejuvenated by bulk seed handling, as assumed in the formulation of θ_t and f_t, rare alleles will soon drop off because of the random drift. Comparisons between curves (i) and (ii) and between (iii) and (iv) show that the optimum sample fraction increases with an increase in the initial allelic homogeneity (θ_o). In an entry with low initial allelic diversity, genetic fixation proceeds rapidly even when a fairly large number of plants are grown per generation. In this situation, keeping more entries, with lower resource input (fewer plant numbers for rejuvenation) being allocated for individual entries, is more rewarding than keeping fewer entries with high resource input for each entry.

Calculations for the case of M = 200 (not presented) produced similar trends, except that the optimum sample fraction was inflated compared to that for M = 500, rather slightly for(i), (ii) and (iv) but appreciably for (iii).

The optimum sample fraction is also influenced by the resource parameters α and ß (the plant number in total and per generation) and the duration of the core collection (t). The effects of these parameters are seen in Tables 2 and 3, respectively. Table 2 shows that, with a given duration of the core collection (t = 10), the optimum fraction is markedly influenced by α (and ß), increasing with increasing values of these parameters. The optimum fraction is extremely small when the plant number is as small as $\alpha = 10^3$ (and $\beta = 10^2$), suggesting that the plant number allocated to an entry per generation — that is, $\beta/m_s = \beta/(p \cdot M)$ — should not be too small. In other words, there is a critical lower limit of entry size for keeping the allelic diversity within entries. In Table 2, the critical value may be said to be about 10; the plant number per entry per generation is not much reduced from about 10 by the plant number α (and ß) being reduced from 10^4 (and 10^3) to 10^3 (and 10^2).

The influence of t with a fixed plant number per generation is shown in Table 3. The optimum sample fraction decreases with an increase in t, although not so drastically as it does with a decrease in α (and ß), as noted above; this indicates that a larger number of plants per entry is needed to retain the allelic diversity for a longer duration of the core collection.

Figure 2 The overall efficiency (*EF*) of a core collection in different sample fractions (*p*)

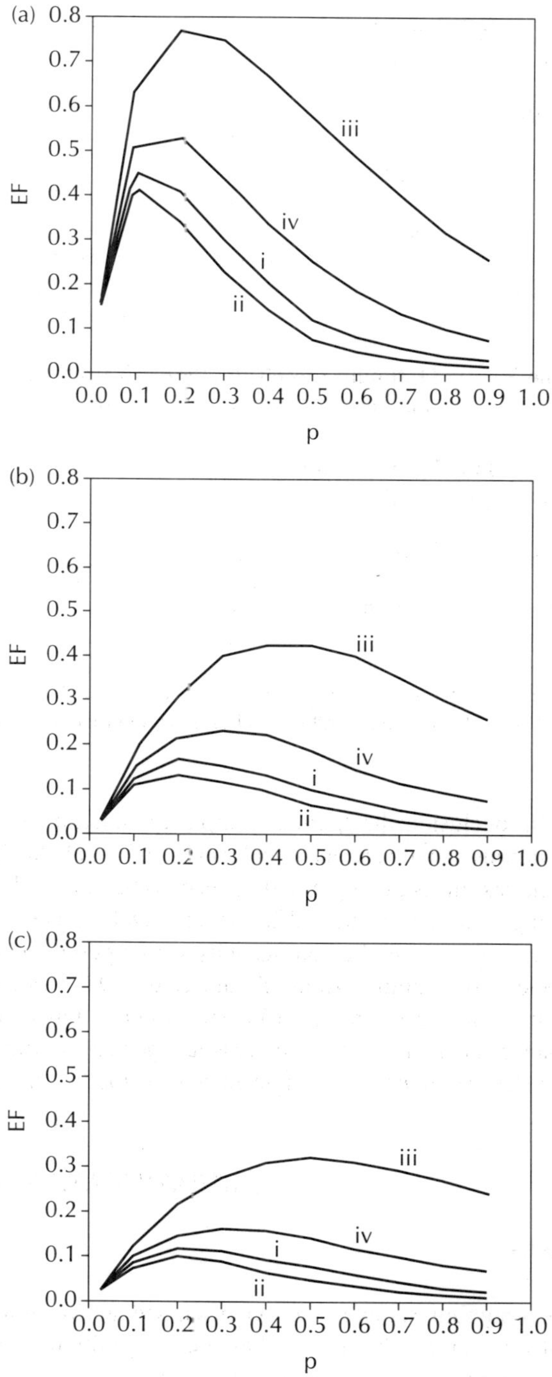

The definitions of cases (a), (b) and (c) are given in Figure 1

Curve (i) $s = 0.9$, $\theta_0 = 0.9$
Curve (ii) $s = 0.9$, $\theta_0 = 0.1$
Curve (iii) $s = 0.1$, $\theta_0 = 0.9$
Curve (iv) $s = 0.1$, $\theta_0 = 0.1$

Table 2 **Optimum sample fraction with different amounts of resources in total (α) and per generation (ß)**[a]

Initial genetic state			α, ß		
s	f_0	θ_0	$10^3, 10^2$	$10^4, 10^3$	$10^5, 10^4$
0.9	0.82	0.9	0.02 (10)[b]	0.2 (10)	0.8 (25)
		0.1	0.02 (10)	0.2 (10)	0.8 (25)
0.1	0.05	0.9	0.06 (4)	0.4 (5)	0.9 (23)
		0.1	0.04 (5)	0.3 (7)	0.9 (23)

Note: a $M = 500$; $K = 250$; two accessions in each class; $t = 10$
b Figures in parenthesis indicate plant number per entry per generation

Table 3 **Optimum sample fraction with different maintenance durations (t) for a fixed plant number per generation ($ß = 10^4$)**[a]

Initial genetic state			t		
s	f_0	θ_0	10	20	50
0.9	0.83	0.9	0.8 (25)[b]	0.6 (34)	0.4 (50)
		0.1	0.8 (25)	0.6 (34)	0.3 (67)
0.1	0.05	0.9	0.9 (23)	0.8 (25)	0.6 (34)
		0.1	0.9 (23)	0.8 (25)	0.5 (40)

Note: a $M = 500$; $K = 250$; two accessions in each class
b Figures in parenthesis indicate plant number per entry per generation

In conclusion, the optimum sample fraction cannot be uniquely specified or delimited. It is affected by various genetic and technical factors, of which the degree of redundancy and amount of available resources (measured by α and ß) seem to be most influential. It has been widely acknowledged that selfing or predominantly selfing species tend to have a high degree of genetic differentiation between populations and allelic homogeneity within populations, compared with predominantly outcrossing species (for example, Hamrick and Godt, 1990; Morishima et al., 1992). It is therefore logical to assume that accessions of selfing species have a much lower genetic redundancy and a higher initial co-ancestry than those of outcrossing species. A strategy of maintaining more entries, with fewer resources being used for single entries, would be efficient for species with higher selfing rates.

STRATIFICATION STRATEGY

Model

The optimum sample fraction discussed above assumed completely random sampling from a collection or that the classes of allelic composition as defined in Table 1 are not identified at the time of sampling. The accessions in a collection may be divided into a number of groups according to passport data. The efficiency of sampling in this situation will be much improved if an appropriate stratified sampling approach is adopted.

Five sampling strategies (four stratified strategies and one completely random strategy, as a check) are discussed here:

- *Random strategy (R):* Entries are completely randomly sampled from a collection, groups being ignored
- *Constant strategy (C):* An equal number of entries is sampled from all groups, irrespective of the number of accessions (group size) in each group
- *Proportional strategy (P):* Entries are sampled in proportion to group size
- *Logarithmic strategy (L):* Entries are sampled in proportion to the logarithm of group size
- *Genetic diversity-dependent strategy (G):* Entries are sampled in proportion to the amount of genetic diversity in the groups

Strategies C, P and L have been compared by Brown (1989b). Strategy G is addressed for the first time here. To quantify the efficiency of these strategies, the parameters of the genetic composition of the collection are established, using the following variables:

m_{ij} = number of accessions in group i ($i = 1, 2, ..., g$), belonging to the allelic composition class j ($j = 1, 2, ..., K$)

$m_{i\cdot}$ = total accessions in group i ($= \sum_j m_{ij}$)

$m_{\cdot j}$ = total accessions for class j of allelic composition ($= \sum_i m_{ij}$)

M = grand total number of accessions ($= \sum_{ij} m_{ij}$)

k_i = number of allelic composition classes contained in group i

K = total number of allelic composition classes contained in the whole collection ($\leq \sum_i k_i$)

m_{is} = number of entries sampled from group i, allocated by any of the five sampling strategies

$m_{\cdot s}$ = total accessions sampled ($= \sum_i m_{is} = M \cdot p$)

The structure of this collection is set out in Table 4. The classes of allelic composition in this table are the same as those defined in Table 1, indicating that each accession in the collection belongs to only one class in only one group.

Table 4 **Establishing the parameters of the genetic composition of a collection**

Group (i)	— **Class of allelic composition** (j) — 1	2	...	K	**Group size** ($m_{i.}$)	**Number of non-empty classes** (k_i)
1	m_{11}	m_{12}	...	m_{1k}	$m_{1.}$	k_1
2	m_{21}	m_{22}	...	m_{2k}	$m_{2.}$	k_2
.	.			.	.	.
.	.	.	(m_{ij})	.	.	.
.	.	.		.	.	.
g	m_{g1}	m_{g2}	...	m_{gk}	$m_{g.}$	k_g
Sum over groups ($m._j$)	$m._1$	$m._2$	...	$m._k$	M	$k.$

Not all groups may have unique classes of allelic composition (that is, some classes may be shared by more than one group). To measure the degree of overlapping between groups of the allelic composition classes, the following quantity is introduced:

$$Do = 1 - K / \sum_i k_i \qquad (11)$$

Do takes zero in the absence of the overlapping ($K = \sum k_i$), approaching unity (precisely, 1 - 1/g) as the overlapping increases.

If $m._s$ from the total M accessions are sampled, m_{is} accessions being allocated to group i, the retention of the classes of allelic composition (RT_s) would be represented by:

$$RT_s = E\,[k]_s / K$$
$$= 1 - \sum_{j=i}^{K} \{\prod_{t=i}^{g} Q_j(i)\} / K \qquad (12)$$

where

$$Q_j(i) = \begin{cases} 1 & \text{...when } m_{ij} = 0 \\ \prod_{x=0}^{m_{is}-1} \{1 - \frac{m_{ij}}{m_i.-x}\} & \text{...when } m_{is} < m_i.- m_{ij} \; (m_{ij} \neq 0) \\ 0 & \text{...otherwise } (m_{ij} \neq 0) \end{cases}$$

The retention for strategy R with sample size m_s is obtained by substituting $m._j$ and $m._s$ for m_j and m_s in Q_j of formula (4), respectively.

The allocation to group i (m_{is}) in the four strategies C, P, L and G are represented by:

$$m_{isC} = [m._s / g]$$
$$m_{isP} = [m_i. \cdot p]$$
$$m_{isL} = [m._s \cdot \log m_i. / \log (\prod_{i=1}^{g} m_i.)] \qquad (13)$$
$$m_{isG} = [m._s \cdot k_i / \sum_{i=1}^{g} k_i]$$

Numerical computations

The genetic structure of a collection is usually unknown in advance. Fourteen hypothetical collections, chosen to represent some typical situations, were investigated to determine some general trends in the relative efficiencies of the sampling strategies. One of these collections is shown in Table 5, and characteristic features of all 14 collections are summarised in Table 6. The results of the calculations are presented in Table 7. A sample fraction of 0.2 was assumed for all the calculations.

Table 5 **Genetic composition of one of the 14 hypothetical collections used for numerical computations**

Group	Class of allelic composition (j)															
(i)	1	2	3	4	5	6	7	8	9	10	11	12	13	K	m_i	k_i
1	10	10	10	10											50	5
2			8	8	8	8	8								40	5
3					6	6	6	6							30	5
4							5	5	5	5	5				20	5
5									2	2	2	2	2		10	5
$m_{\cdot j}$	10	10	18	18	24	14	19	11	13	7	7	2	2		150(M)	25($k.$)

Table 6 **Summary of features of 14 hypothetical collections used for numerical computations (g = 5 for all cases)**

Collection	M	K	Dr	Change in[a] $m_{i\cdot}$	k_i
1	150	13	0.48	Changeable (10 to 50)[b]	Constant (5)
2		21	0.16	Changeable (10 to 50)	Constant (5)
3		21	0.16	Highly changeable (5 to 75)	Constant (5)
4		26	0.13	Changeable (10 to 50)	Changeable (2 to 10)
5		56	0.07	Highly changeable (5 to 75)	Highly changeable (2 to 30)
6	300	13	0.48	Highly changeable 910 to 150)	Constant (5)
7		21	0.16	Changeable (20 to 100)	Constant (5)
8		21	0.16	Highly changeable (10 to 150)	Constant (5)
9		26	0.13	Constant (60)	Changeable (2 to 10)
10		26	0.13	Changeable (20 to 100)	Changeable (2 to 10
11		26	0.13	Highly changeable (10 to 150)	Changeable (2 to 10)
12		56	0.07	Constant (60)	Highly changeable (2 to 30)
13		56	0.07	Changeable (20 to 100)	Highly changeable (2 to 30)
14		56	0.07	Highly changeable (10 to 150)	Highly changeable (2 to 30)

Note: a Changeable = changing in an equal difference between two neighbouring groups
Highly changeable = changing in an equal ratio between two neighbouring groups
b Figures in parenthesis show the range

Table 7 **Retention of genetic diversity (RT_S) expected to be sampled using the five sampling strategies ($p = 0.2$)**

Collection	Strategy R	C	P	L	G
1	.8286 (5)[a]	.9193 (1)	.8398 (4)	.8950 (3)	.9193 (1)
2	.7228 (5)	.8242 (1)	.7430 (4)	.8042 (3)	.8242 (1)
3	.6305 (5)	.8741 (1)	.6437 (4)	.8051 (3)	.8741 (1)
4	.7122 (3)	.6545 (5)	.7330 (1)	.6907 (4)	.7330 (1)
5	.4586 (3)	.3753 (5)	.4671 (2)	.4263 (4)	.4677 (1)
6	.8532 (5)	.9899 (2)	.8616 (4)	.9962 (1)	.9899 (2)
7	.8806 (5)	.9652 (1)	.8951 (4)	.9638 (3)	.9652 (1)
8	.7772 (5)	.9741 (2)	.7919 (4)	.9790 (1)	.9741 (2)
9	.8481 (5)	.8604 (2)	.8604 (2)	.8604 (2)	.9321 (1)
10	.9111 (3)	.8520 (5)	.9240 (1)	.8775 (4)	.9240 (1)
11	.8876 (4)	.8541 (5)	.9048 (2)	.8998 (3)	.9292 (1)
12	.5353 (5)	.5408 (2)	.5308 (2)	.5408 (2)	.7432 (1)
13	.6435 (3)	.5305 (5)	.6512 (2)	.5640 (4)	.7088 (1)
14	.6909 (3)	.5301 (5)	.7009 (1)	.5935 (4)	.7009 (1)
Average rank	4.21	3.00	2.64	2.93	1.14
Frequency of:					
best rank	0	4	2	2	12
worst rank	8	6	0	0	0
worse than R	—	6	0	5	0

Note: a Figures in parenthesis show rank in superiority

Two collections in pairs of (3) and (8), (4) and (10), and (5) and (14) differed in the total number of accessions (M), with other parameters being the same. Two in pairs of (1) and (2), and (6) and (8) differed in the number of the allelic composition classes (K) and consequently in the degree of overlapping (Do). Comparisons in Table 7 between two collections in these pairs showed that while the absolute value of RT_s increased as M and Do increased, the rank in the superiority of the five strategies was little modified.

Strategy R, as expected, was in no case the best strategy, being worst in most (8 out of 14) collections. Strategy C was best (as good as G) in collections (1), (2), (3) and (7), where all the groups had the same range (k_i = constant) of genetic diversity, although they differed in size ($m_i.$). However, this strategy gave lower RT_s than strategy R in collections (4), (5), (10), (11), (13) and (14) where k_i varied among the groups (*see* Table 7) or was highly changeable. Strategy P achieved the best or nearly the best ranks in collections (4), (5), (10) and (14), in all of which $m_i.$ changed in parallel with k_i within groups. Strategy L was the most efficient in collections (6) and (8), where $m_i.$ was highly variable but k_i was almost constant. It ranked lower than strategy R in five collections — (4), (5), (10), (13) and (14) — where strategy C was poorest. Strategy G almost always achieved the best rank, although it gave slightly lower RT_s than strategy L in collections (6) and (8). As shown in the lower part of Table 7, the average superiority of the five strategies for the 14 collections was in the following order: G, P, L, C and R.

Calculations were conducted on four real collections: two collections of *Glycine tomentella* (reproduced from Brown, 1989b) and two of *Oryza sativa* (Oka, 1953, 1954). The characteristic features of these collections are presented in Table 8, and the results of the calculations are summarised in Table 9. Trends similar to those in the 14 hypothetical collections were observed. The average superiority of the five strategies was in the order of G, P, L, R and C, the difference between the last two strategies being insignificant.

From these investigations, we conclude that a stratified strategy weighted by either the group size or range in genetic diversity is superior to random or constant strategies. Of the three weighted strategies (P, L and G), strategy G was nearly always the best. The amount of genetic diversity of the groups, however, could not be known in many practical collection projects. Strategy P is recommended in this situation. Strategy L, conceived on the neutrality principle put forward by Brown (1989b) and thought to be as good as or better than P, does not appear to have a wide application.

Table 8 **Characteristic features of four real collections (two of *Glycine tomentella* and two of *Oryza sativa*) used for numerical computations**

Collection	g	M	K	Do	Range of m_i	Range of k_i	Species	References
1	7	270	6	0.647	5~125	2~4	*G. tomentella* (4n)	Brown (1989b) (Table 1)
2	5	125	5	0.375	8~60	1~3	*G. tomentella* (2n)	Brown (1989b) (Table 1)
3	13	147	15	0.722	5~16	2~7	*O. sativa* (2n)	Oka (1953) (Table 8)
4	10	83	11	0.633	2~12	2~5	*O. sativa* (2n)	Oka (1954) (Table 6)[a]

Note: a Some groups were pooled

Table 9 **Retention of genetic diversity (RT_S) from the four real collections described in Table 8, expected to be sampled using the five sampling strategies**

Collection	Strategy R	C	P	L	G
1	.8478 (5)[a]	.9829 (3)	.8532 (4)	.9873 (2)	.9938 (1)
2	.9522 (2)	.8515 (5)	.9597 (1)	.8967 (4)	.9281 (3)
3	.5345 (5)	.5453 (4)	.5557 (2)	.5498 (3)	.5716 (1)
4	.6569 (4)	.6423 (5)	.6885 (2)	.6740 (3)	.6942 (1)
Average rank	4.00	4.25	2.25	3.00	1.50

Note: a Figures in parenthesis show rank in superiority

SAMPLING FROM A HIERARCHICALLY STRUCTURED COLLECTION

Structure among groups

The groups in a collection may be hierarchically related, as illustrated in Figure 3. The hierarchy dendrogram is based on the genetic distances between groups which are calculated from the trait scores averaged within each group. In this structure, accessions within a group are treated equally (that is, sampling within the group is made without giving any priority to any single accessions).

Figure 3 Hierarchical relationship of six groups in a hypothetical collection

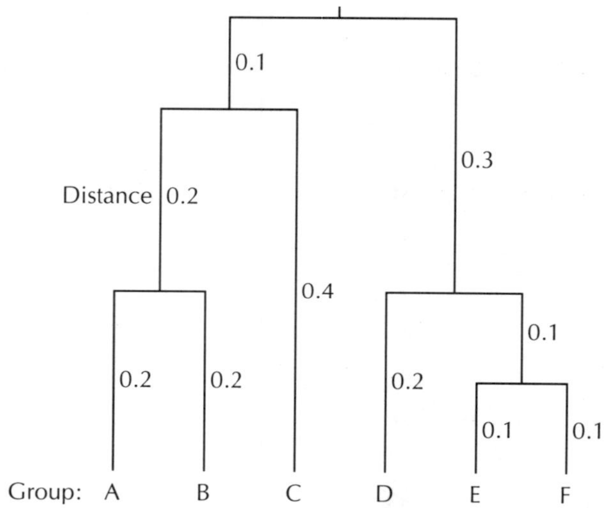

Source: Crozier (1992)

Since accessions belonging to more closely related groups tend to have a higher degree of genetic similarity or overlapping, the sample size allocation weighted by the group size alone (strategy P) may not be the most appropriate. A modification system using the uniqueness value proposed by Crozier (1992) is described here. The uniqueness value was introduced to measure the degree of genetic remoteness of a species from others, based on calculations which take account of both the genetic distances and phylogenic topology among the species. Crozier (1992) defined the uniqueness value of a species as 'the probability of this species being unique in character state, which equals the probability of this species being different in state from its node in the phylogenic network plus the product of the chance of this species not being different from the node and the probabilities of all other species being different from this node'.

For the simplest example of three species (1, 2 and 3) being connected to a common node, H, the uniqueness of species 1 is calculated as $d_{1H} + (1 - d_{1H}) .d_{2H} .d_{3H}$, where d_{1H} etc. indicate the genetic distance of species 1 etc from the node H. For phylogenic trees with more species, the calculation 'involves proceeding down the tree in stepwise fashion' (Crozier, 1992). The absolute (u_i) and relative

(ru_i) values of the uniqueness of the six groups (A to F) composing the phylogenic tree of Figure 3 are presented in the third and fourth columns of Table 10, respectively. With the sizes of the groups being as presented in the second column of Table 10, the sample sizes allocated by strategy P (m_{isP}) are as presented in the fifth column of the table. These sample sizes are, in turn, are modified by the uniqueness values thus:

$$m_{isU} = m_s \cdot \frac{m_{isP} \cdot u_i}{\sum_i (m_{isP} \cdot u_i)} \quad \text{or} \quad m_s \cdot \frac{m_{isP} \cdot ru_i}{\sum_i (m_{isP} \cdot ru_i)} \tag{14}$$

The sample sizes thus obtained are presented in the last column of Table 10. By this modification, the sample size for group C with the highest uniqueness value of 0.47 was considerably inflated, rising from 16 to 34. Conversely, the sizes for the two relatively closely connected groups E and F were reduced.

Table 10 Sample sizes modified by the uniqueness value of six hierarchically related groups in a hypothetical collection

Group	Number of accessions (m_i)	Uniqueness (u_i)	Relative[a] uniqueness (ru_i)	Sample size[b] with allocation P (m_{fsP})	Sample size modified by uniqueness (m_{fsU})
A	20	0.26	1.30	4	5 (4.59)
B	40	0.26	1.30	8	9 (9.17)
C	80	0.47	2.38	16	34 (33.59)
D	160	0.25	1.25	32	35 (35.28)
E	320	0.20	1.00	64	56 (56.45)
F	640	0.20	1.00	128	113 (112.91)
Total	1260 (M)	1.64	8.23	252 (M_s)	252 (M_s)

Note: a Relative to the smallest value of 0.2
b Assuming a sample fraction of $p = 0.2$

Structure among accessions

A collection may be structured from the level of single accessions (for example, Souza and Sorrells, 1991a, b; Zeven and Hintum, 1992). In this situation, the uniqueness value is calculated for each single accession, and a step-by-step allocation from the node of the highest rank to those of lower ranks in the phylogenic tree must be used. A system for this allocation is described here and is illustrated in Figure 4 (the lower part of the hierarchy in the figure has been abbreviated because of space constraints).

In the first step, the total sample size ($m_s = M \cdot p = 800 \text{ x } 0.2 = 160$) is calculated at the highest node of the dendrogram. These 160 accessions are then allocated to the three branches from this node, being weighted in two steps: first, by the total number of accessions belonging to each branch ($m_{i\cdot}$) and,

Figure 4 **Dendrogram of a hypothetical collection that is hierarchically structured from the level of single accessions**

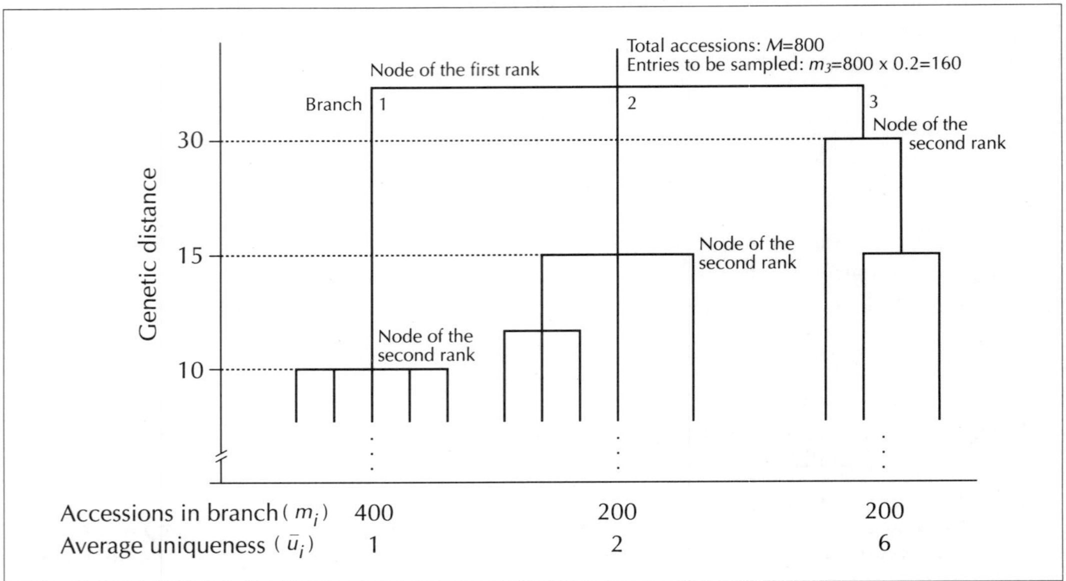

second, by the average uniqueness value through accessions within each group ($\bar{u}_i$). The sample sizes in the first step (allocation by strategy P) are obtained thus:

Branch 1: $160 \times \frac{400}{800} = 80$

Branches 2 and 3: $160 \times \frac{200}{800} = 40$

which, in the second step, are modified by the average uniqueness thus:

Branch 1: $160 \times \frac{1 \times 80}{1 \times 80 + 2 \times 40 + 6 \times 40} = 32$

Branch 2: $160 \times \frac{2 \times 40}{1 \times 80 + 2 \times 40 + 6 \times 40} = 32$

Branch 3: $160 \times \frac{6 \times 40}{1 \times 80 + 2 \times 40 + 6 \times 40} = 96$

In some cases, it may be better to make the calculation in the second step using the genetic diversity within each branch, which is most simply measured by the maximum genetic distance within each group (d_i) (that is, 10, 15 and 30 in the case of Figure 4). The sample sizes in this weighting are:

Branch 1: $$160 \times \frac{10 \times 80}{10 \times 80 + 15 \times 40 + 30 \times 40} = 49\ (49.23)$$

Branch 2: $$160 \times \frac{15 \times 40}{10 \times 80 + 15 \times 40 + 30 \times 40} = 37\ (36.92)$$

Branch 3: $$160 \times \frac{30 \times 40}{10 \times 80 + 15 \times 40 + 30 \times 40} = 74\ (73.84)$$

The above two-step allocation is repeated at each node of lower rank down to a node where the number of accessions allocated is equal to, or less than, that of branches. In the former case, one accession should be allocated to each of the branches so that the widest range of genetic diversity is covered. In the latter case, priority should be given to branches containing a larger numbers of accessions, with one accession being allocated to each of the branches thus chosen; this will give the largest representation and the widest coverage of the genetic diversity in the node.

In the last step of choosing one accession from within a branch, an accession with the lowest uniqueness should be preferred, as this accession has the largest number of related accessions within the branch and is therefore most representative of the branch.

To summarise, the rules for allocating sample sizes to branches within a node are:

Rule 1: Allocate accessions to all branches at a node, giving weights in two steps (first, by the number of accessions contained in the branch and, second, by the average uniqueness value of the group).

Rule 2: When the accessions allocated to a node are as many as or more than the accessions for all branches belonging to this node, choose all of these, the excess (of the number allocated over the number available) being allocated to another node of nearest distance.

Rule 3: When the number of accessions allocated to a node is exactly the same as the number of branches, allocate one accession to each branch.

Rule 4: When accessions allocated to a node are fewer than the number of branches, give priority to branches containing more accessions, allocating one accession to each of the branches thus chosen.

Rule 5: When sampling a single accession from within a branch, choose one that has the lowest uniqueness (highest representativeness of the branch).

CONCLUSION

Brown (1989a, b) proposed that a fraction of about 10% is an appropriate sample size for sampling core entries from a whole collection. He based this proposal on the theory that had been put forward by Ewens (1972), where the retention of allelic diversity sampled from a finite population was formulated on the assumption that the population is at genetic equilibrium for neutral alleles. The calculations using this theory led to the result that at least 70% of the existent alleles could be drawn with 95% certainty if 10% or more of the plants were sampled from the population (Brown, 1989a). In the work reported in this chapter, the sample size investigation was based on a different model, with two components: the amount of genetic diversity expected from sampling a certain fraction of a whole collection [$E[k] / K$ in formula (2)], and the maintenance of allelic diversity within single accessions (*PM*). It was calculated from this model that the optimum sample fraction depends largely upon the degree of genetic redundancy among accessions [*Dr* in formula (1)], the total amount of resources available for the maintenance of core entries (quantified by the total plant number, α, treated for the maintenance), and the duration (number of rejuvenations) of the core collection. While the optimum sample size could not be uniquely determined, a fraction in the range of 20 ~ 30% was estimated to be the best or nearly the best under conditions where $0.2 < Dr < 0.9$, $\alpha \doteqdot 10^4$ and $t \doteqdot 10$.

The terms constituting the overall efficiency of the collection (*EF*) may be quantified in different ways according to the types of genetic factors to be retained and maintained. For instance, where particular kinds of alleles are to be maintained rather than the allelic compositions assumed above, *PM* should be quantified in terms of the probability that these particular alleles are maintained, not in terms of the co-ancestry θ_t as adopted in this chapter. Accessions in some cases may be characterised by traits controlled by genes at multiple loci. The success of maintenance in this situation must be measured considering combinations of allelic states at multiple loci. The optimum sample size tends to decrease with a larger amount of resources being required for maintaining the genetic factors that characterise the core entries.

The quality of maintenance of single entries may be important in some situations, rather than *EF* as defined in this chapter. Occasionally, the procedures of management for individual entries may be predetermined for some reason. In this case, the optimum sample fraction cannot be defined since the resource input for individual entries is predetermined. The number of accessions in the core collection is then determined by the amount of resources available (primarily, manpower and facilities).

Core entries in some plant species may be maintained without rejuvenation in the form of seeds, vegetative organs or plants. The optimum sample fraction in these cases must be defined differently and quantified in terms of parameters describing the persistence of viability of seeds, organs or plants.

A stratified sampling strategy will be efficient when the collection is subdivided into a number of groups of accessions. A comparison of five stratification strategies led to a rather simple conclusion. Among the five strategies compared, strategy G (where accessions are sampled in proportion to the genetic diversity within groups) was almost always the best. When the absolute or relative amounts of genetic diversity within groups are not known, as would be the case in many collections, strategy P (where accessions are sampled in proportion to group size) is recommended. Allocation by strategy P should be appropriately modified if any additional information other than group size is available; for instance, a group which containing more accessions from similar habitat conditions may be less weighted than other groups. Strategy L (where accessions are sampled in proportion to the logarithm of group size), considered by Brown (1989b) to be as or more efficient than strategy P, was found to be rather of limited use.

Lastly, we proposed various sampling procedures for hierarchically structured collections. Procedures where strategy P is modified using the uniqueness value (Crozier, 1992) were described. The uniqueness value of a group or accession (*see* Figures 3 and 4) measures the degree of genetic remoteness from others, being calculated from the topology of dendrogram and genetic distances among groups or accessions. Both the pattern and the range of genetic diversity in a collection could be well conserved in core entries by using the sampling procedures which we have described in this chapter.

References

Brown, A.H.D. 1989a. The case for core collections. In Brown, A.H.D., Frankel, O.H., Marshall, D.R. and Williams, J.T. (eds) *The Use of Plant Genetic Resources.* Cambridge, UK: Cambridge University Press.

Brown, A.H.D. 1989b. Core collections: A practical approach to genetic resources management. *Genome* 31: 818-24.

Crow, J.F. and Kimura, M. 1970. *An Introduction to Population Genetics Theory*. New York, USA: Harper and Row.

Crozier, R.H. 1992. Genetic diversity and the agony of choice. *Biological Conservation* 61: 11-15.

Ewens, W.J. 1972. The sampling theory of selectively neutral alleles. *Theoretical Population Biology* 3: 87-112.

Frankel, O.H. 1984. Genetic perspectives of germplasm conservation. In Arber, W., Llimensee, K., Peacock, W.J. and Starlinger, P. (eds) *Genetic Manipulation: Impact on Man and Society*. Cambridge, UK: Cambridge University Press.

Hamrick, J.L. and Godt, M.J.W. 1990. Allozyme diversity in plant species. In Brown, A.H.D., Clegg, M.H., Kahler, A.L.and Weir, B.S. (eds) *Plant Population Genetics, Breeding, and Genetic Resources.* Sunderland, Massachusetts, USA: Sinauer.

Loveless, M.D. and Hamrick, J.L. 1984. Ecological determinants of genetic structure in plant populations. *Annual Review of Ecology and Systematics* 15: 65-95.

Marshall, D.R. 1990. Crop genetic resources: Current and emerging issues. In Brown, A.H.D., Clegg, M.H., Kahler, A.L.and Weir, B.S. (eds) *Plant Population Genetics, Breeding, and Genetic Resources.* Sunderland, Massachusetts, USA: Sinauer.

Morisima, H. 1991. Association between *Pox-1* variation and seed productivity potential in wild rice. In *Rice Genetics*. Manila, Philippines: IRRI.

Morishima, H., Sano, Y. and Oka, H.I. 1992. Evolutionary studies in cultivated rice and its wild relatives. *Oxford Surveys in Evolutionary Biology* 8: 135-84.

Oka, H. 1953. Phylogenetic differentiation of the cultivated rice plant. I. Variation of various characters and character combinations among rice varieties. *Japanese J. Breeding* 3: 33-43.

Oka, H. 1954. Varietal variation of the responses to day-length and temperature and the number of days of growth period. *Japanese J. Breeding* 4: 92-100.

Souza, E. and Sorrells, M.E. 1991a. Relationships among 70 North American oat germplasms. I. Cluster analysis using quantitative characters. *Crop Science* 31: 599-605.

Souza, E. and Sorrells, M.E. 1991b. Relationships among 70 North American oat germplasms. II. Cluster analysis using qualitative characters. *Crop Science* 31: 605-12.

Yeatman, C.W., Kafton, D. and Wilkes, G. (eds). 1984. *Plant Genetic Resources, A Conservation Imperative.* Boulder, Colorado, USA: Westview Press.

Zeven, A.C. and Hintum, Th.J.L. van. 1992. Classification of landraces and improved cultivars of hexaploid wheats (*Triticum aestivum, T. compactum* and *T. spelta*) grown in the USA and described in 1922. *Euphytica* 59: 33-47.

Core Collections of Plant Genetic Resources
Edited by T. Hodgkin, A.H.D. Brown, Th.J.L. van Hintum and E.A.V. Morales

A Wiley-Sayce Publication

2.3

Maximising genetic diversity in core collections of wild relatives of crop species

D.J. Schoen and A.H.D. Brown

Abstract

The conservation of genetic diversity in wild relatives of crop species presents challenges beyond those faced in the case of domesticated species. In particular, there are many potentially useful accessions and natural populations, yet it is possible to conserve only a representative sample of these. The core collection may thus have special relevance in the conservation of germplasm of wild crop relatives. Unequal diversity and differentiation among accessions create problems for constructing diverse and representative core collections.

In this chapter, two new core collection strategies (termed the 'H' and 'M' strategies) are developed to address these problems. Both utilise stratified sampling and marker gene data to select allelically rich accessions from different geographical or ecological groups. In addition, the M strategy attempts to reduce redundancy in the core collection by pinpointing sets of well-differentiated accessions. The H, M, and several other core collection strategies were compared by simulating the sampling of accessions for the construction of core collections in nine wild crop relatives. Allelic richness in these simulated core collections was calculated in all cases. The results show the importance of stratified sampling, demonstrate that genetic marker data help to maximise allelic richness in the core, and confirm that genetically diverse and representative germplasm collections of wild crop relatives can be established using the core collection approach.

The conservation of wild relatives of crop species is an important part of efforts to maintain crop genetic diversity (Harlan, 1984; Chapman, 1989; Ladizinsky, 1989), but the inclusion of wild relatives in gene banks presents a number of technical and logistical problems. First, there are many more species and populations of wild relatives than of crops, and it is practical to collect and maintain only a small fraction of these. Second, because of their often wide-ranging geographical distributions, the exploration and collection of wild relatives may involve significant costs. Third, reproductive biological characteristics such as poor seed set and germination make it difficult and costly to maintain some wild relatives species. Fourth, many breeders use wild relatives only when the variability they need is absent from elite lines. The challenge, therefore, is to make the variability of wild crop relatives more accessible to plant breeders but at a minimum cost. By emphasising representative and better-documented germplasm samples, the establishment of core collections (Frankel, 1984; Frankel and Brown, 1984) of wild crop relatives should help ameliorate the problems outlined above.

This chapter is concerned with the methodology for constructing core collections of wild relatives of crop species. We begin by reviewing some population genetic factors that contribute to the organisation of genetic variability in natural plant populations and that can complicate sampling efforts designed to conserve genetic variation. In particular, it is suggested that unequal levels of genetic diversity and differentiation among accessions may present problems in sampling wild relatives of crop species. Unequal diversity may be especially prevalent in inbreeders (Schoen and Brown, 1991), while unequal degrees of differentiation are likely to be ubiquitous. In view of these complications, two new sampling strategies, termed the 'H' and 'M' strategies, for constructing core collections are developed here. These differ from previously described sampling strategies in that information gained from screening accessions at genetic marker loci is incorporated directly into the sampling protocol. After introducing the H and M strategies, their effectiveness with respect to allele retention in the core collection is evaluated. This is done by simulating the construction of core collections using published data on geographical range and allozyme variation of wild relatives of crop species. The results obtained with the H and M strategies are compared with those from several other strategies for constructing core collections. Employing genetic marker data to guide sampling can lead to significant gains in the overall allelic richness of core collections. Moreover, the results show under what circumstances such marker-assisted methods will be the most effective. While only wild relatives of crop species are examined in this study, we expect that the results will hold in general.

PLANT POPULATION GENETICS AND ALLELIC RICHNESS

It is generally accepted that combinations of alleles play an important role in adaptation but, as noted elsewhere, such combinations are likely either to behave as supergenes (and hence, be inherited like single alleles) or to be difficult to conserve in the face of recombination pressure (Brown and Briggs, 1991). Maximising the numbers of kinds of distinct alleles (allelic richness) is, therefore, the most appropriate theoretical objective in constructing core collections. The conservation of a diverse and representative collection of alleles should ensure that plant breeders have at their disposal the genetic resources required to breed crops to respond to future changes in the physical and biotic environment. Hence, in what follows the primary concern will be with the effects of various population genetic factors on allelic richness and, later, with methods for maximising allelic richness in core collections.

The way in which allelic diversity occurs both among and within the accessions of a given wild relative of a crop species will influence the success of different core collection strategies. Patterns of allelic richness in wild relatives are ultimately the products of mutation, selection, migration, genetic drift and non-random mating. Some of these factors, however, are more influential than others. For instance, populations of most plant species are unlikely to be numerically stable over long periods of time. Such demographic instability may be especially pronounced in natural populations of wild relatives, many of which are weedy colonists of disturbed sites. Demographic instability as manifested by periodic large reductions in the size of existing populations (population bottlenecks) or establishment of new populations from one or a few individuals (founding events) allow the effects of genetic drift and inbreeding to come to the fore. One of the more pronounced consequences of population bottlenecks is the reduction in the number of alleles maintained in the population (Nei et al., 1975). Theory suggests that allelic richness is more sensitive to the effects of bottlenecks than other measures of genetic variation, such as gene diversity and heterozygosity (Nei et al., 1975; Maruyama and Fuerst, 1985). Leberg (1992) has recently published evidence based on the experimental manipulation of founding population size that supports this point. Rare alleles are especially prone to loss following a bottleneck (Sirkkomaa, 1983), but when populations remain small over many generations, the effects of population bottlenecks are cumulative and there is an increasing probability that common alleles will also be lost. Barrett and Kohn (1991) have reviewed some of the empirical evidence pertaining to the effects of population bottlenecks on plant population genetic diversity. Low allelic richness was a common denominator of studies that focused on taxa which experienced a population bottleneck or founding event as a result of population size reductions following glaciation, introductions into new regions and colonisation of islands.

Many crops and their relatives are autogamous. Baker (1955) argued that in autogamous species, long-distance dispersal and establishment of populations from single propagules may be relatively common (that is, in selfers as opposed to outcrossers, a single propagule is sufficient to found a new sexually reproducing population). A genetic correlate is that selfing species may often consist both of populations that have recently experienced a founding event and of older and more stable (source) populations. This dichotomy gives rise to unequal allelic richness levels among populations and, in the case of a wild relative of a crop species, renders some accessions more valuable to the core collection than others. Such variation complicates the task of maximising diversity in the core. Variation in allelic richness among the populations of a wild relative, particularly in autogamous species, is evident from allozymic surveys of natural populations (for example, Rick et al., 1977; Brown, 1979; Nevo et al., 1979; Schoen and Brown, 1991).

In addition to interpopulation variation in levels of genetic diversity, populations of wild relatives of crop species are also expected to vary in their degree of differentiation from one another. There are several reasons for this. First, spatially varying selection in addition to or reinforced by inbreeding and linkage disequilibrium leads to different amounts of interpopulation divergence. The pattern of selective differentiation may often be complex, as, for example, in the patchy spatial relationships observed between plant parasites and their hosts (Burdon et al., 1989). Such underlying complexity makes it difficult to obtain a representative sample of useful variation in wild relatives of crop species (for example, a diverse set of rust resistance genes). Second, natural populations differ in timing of divergence from a common ancestor. In the case of neutral or near-neutral variation, populations isolated from one another by longer periods of time will have diverged more as the result of the accumulation of mutations compared with populations isolated for less time. The overall situation might be summarised as in Figure 1. In this hypothetical example, populations differ with respect to the effects of genetic drift and the time since divergence from a common ancestor. The problem is that a collector or gene bank manager interested in establishing a diverse and representative collection is unlikely to be aware of these underlying genetic patterns. For example, the genetically most diverse accessions in Figure 1 are D and F, but a diverse and more representative set of accessions might be obtained by choosing A and G instead. In the section that follows we present some possible solutions to problems such as this one.

MARKER GENES AND MAXIMISING ALLELIC RICHNESS IN CORE COLLECTIONS

Genetic markers, such as allozyme, restriction fragment length polymorphism (RFLP) and random amplified polymorphic DNA (RAPD) markers, have been successfully employed in many applied

Figure 1 **Uneven levels of diversity and differentiation among accessions in a hypothetical wild relative of a crop species**

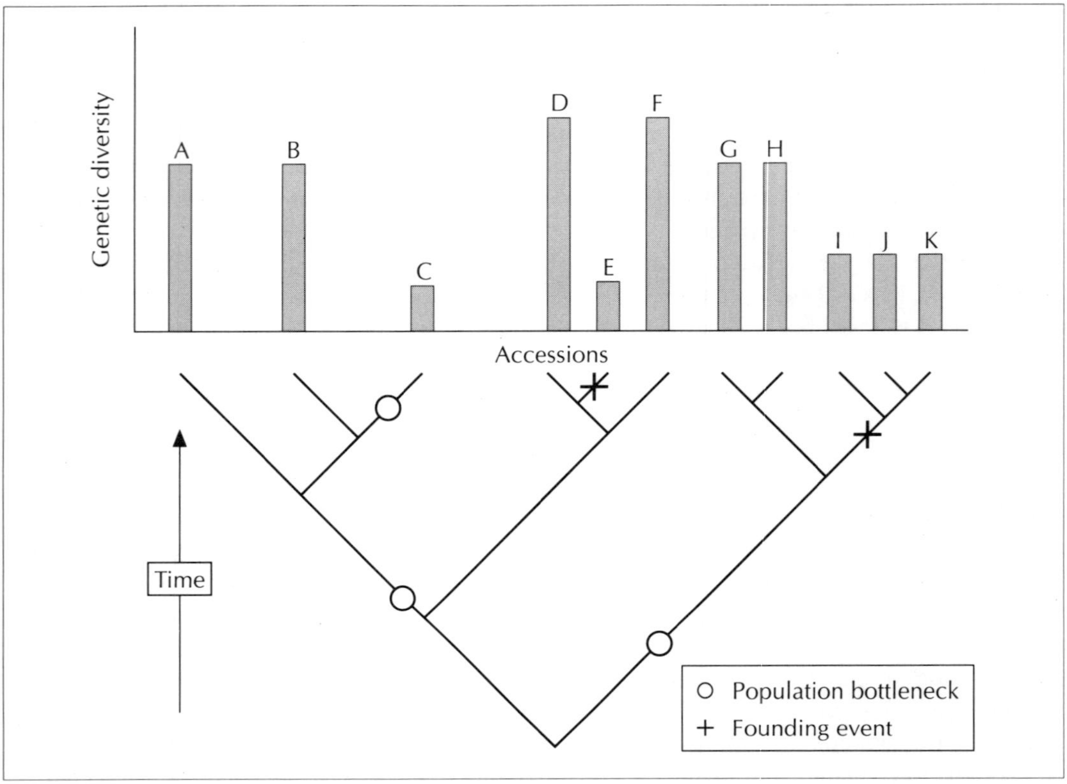

aspects of biology and medicine, including disease resistance gene mapping, dissection of quantitative traits (Landry and Michelmore, 1987; Tanksley et al., 1989), marker-assisted selection (Stuber et al., 1980) and studies of mating systems and inbreeding levels in wild and captive populations (Ritland, 1983). Information gained from surveys of marker loci (allozymes) might also be profitably applied to the task of assessing genetic resources (Brown, 1978) and assembling core collections of crop species and wild relatives (Brown, 1989b; Perry et al., 1991). However, although the utility of allozymes has been recognised, an explicit procedure for incorporating such information directly into a core collection strategy is lacking. Two methods for doing this are presented here. Both methods can be viewed as extensions of previous procedures. For instance, setting up a core sample typically involves the selection of approximately 10% of the total collection, a level of sampling that in theory is expected to retain roughly 70% of the alleles in the whole collection (Brown, 1989a). This sampling intensity was used in the methods developed in this chapter. In addition, division of the overall collection into a number of non-overlapping ecogeographical regions or groups, followed by sampling within the groups, has been recommended as a practice to help ensure inclusion in the core of all ecotypes (Frankel and Brown, 1984; Brown, 1989a, b). This practice was also followed. The main departure from previous methods of core sampling, therefore, pertains to the use of information gained from marker loci to guide in sampling accessions.

H strategy

Real populations can have a diverse array of genetic structures, depending upon the predominance of the different population genetic factors as well as upon historical accident. As a prelude to sampling for germplasm conservation, one may choose to learn about this structure or, alternatively, one may base the sampling of a species on a theoretical model that approximates the structure of population genetic variation. The infinite neutral alleles model (Kimura and Crow, 1964) as applied to a large number of isolated (island) populations has proved useful in this regard. This model assumes that every allele that arises through mutation is unique and selectively neutral, and that populations are reproductively isolated from one another. The adoption of the neutral model is not meant to imply that the genetic variation of interest to conservationists is selectively neutral. The neutral model is useful because it specifies an exact distribution of allelic frequencies, with sampling characteristics that are well defined (Ewens, 1972). The allelic frequency distributions specified by the neutral model lie between models that assume heterotic selection (many alleles each with relatively high frequency) and those that assume mutation-selection balance (fewer alleles, one common and the rest rare) (Marshall and Brown, 1975). In fact, when selection is weak, allele distributions specified by the neutral alleles model yield distributions of alleles that are experimentally indistinguishable from models invoking selection. Clearly, it is important to establish whether the distribution of alleles of interest to conservationists is well approximated by neutral theory, but as a starting point the neutral model provides a convenient 'intermediate' model of allelic distribution (Marshall and Brown, 1975, Brown and Briggs, 1991).

To see how the neutral model can be used as a guide to constructing the core collection, consider two isolated (geographical) groups of accessions (groups 1 and 2) and two polymorphic loci (loci A and B). Following the definition of core collections, it is assumed that there is some finite amount of resources, such that the overall number of accessions collected for the core is a constant. Based on Ewens' (1972) sampling theory for the neutral model, and its integral approximation (Brown, 1989b), the expected number of distinct alleles sampled from the two groups of accessions (at both loci) is:

$$K \cong \theta_{11} \ln [\theta_{11} + n_1] + \theta_{21} \ln [\theta_{21} + n_1] + \theta_{12} + \ln [\theta_{12} + n_2] + \theta_{22} \ln [\theta_{22} + n_2] - \Sigma_{ij} \theta_{ij} \ln \theta_{ij} + \text{constant} \qquad (1)$$

where:

θ_{ij} = an estimate of $\theta_{ij} = 4 N_{e(j)} \upsilon_{(i)}$

$N_{e(j)}$ = effective population size for group j

$\upsilon_{(i)}$ = the mutation rate for the ith locus

To find the optimal n_1 and n_2 (that is, the values of n_1 and n_2 that maximise K, assuming $\theta_{ij} << n_1$ and n_2), it is sufficient to find the value of n_1 such that:

$$dK/dn_1 \cong [\theta_{11}/n_1] + [\theta_{21}/n_1] - [\theta_{12}/n_2] - [\theta_{22}/n_2] = 0$$

or (2)

$$[(\theta_{11} + \theta_{21})/n_1] - [(\theta_{12} + \theta_{22})/n_2] = 0$$

This leads to the solution $n_1/n_2 = (\theta_{11} + \theta_{21}) / (\theta_{12} + \theta_{22})$. In other words, the two groups should be sampled in proportion to the ratio of the sums (or average) of the estimates of θ obtained from the marker loci. The above result may be generalised to any number of loci ($i = 1, \ldots, I$). Moreover, using the method of Lagrange's multipliers, it can be shown that with $j > 2$ ($j = 1, \ldots, J$) groups, the allocation of sampling effort (number of accessions selected per group) should be in direct proportion to the ratio of the sums (across the i loci) of estimates of θ_{ij} for the J groups:

$$n_1 : n_2 : n_3 : \ldots : n_g = (\Sigma_i \, \theta_{i1}) : (\Sigma_i \, \theta_{i2}) : (\Sigma_i \, \theta_{i3}) : \ldots : (\Sigma_i \, \theta_{iJ}) \qquad (3)$$

This method is referred to as the H strategy (after Nei's gene diversity index, described below).

Individual estimates of the θ_{ij} may be obtained in several ways. One approach, based on sampling theory for the neutral model, is to estimate the θ_{ij} directly from the number of alleles observed per locus (Ewens, 1972; Chakraborty and Neel, 1989). An alternative is to use the relationship:

$$\theta_{ij} = h_{ij} / (1 - h_{ij}) \qquad (4)$$

where h_{ij} is Nei's gene diversity index, defined as one minus the sum of the squared allelic frequencies at the ith locus in the jth group (Nei, 1973); the h_{ij} should be estimated on the same set of loci in each population. Estimation of θ_{ij} using equation (4) is affected less by variation in sample size (not always reported in published studies) than estimation based on observed numbers of alleles. Once the sums over loci of the θ_{ij} are obtained for each group, they are averaged, and accessions are sampled at random from within the groups in the proportions specified by equation (3). In principle, the H strategy could be extended to information gained from quantitative genetic variation. For example, in a population that is in equilibrium between mutation and genetic drift, the additive component of quantitative genetic variation will be proportional to θ (Lande and Barrowclough, 1987). This suggests that estimates of quantitative genetic variance gained for a number of metric traits in each group of accessions could be used to determine the proportions in equation (3).

M strategy

The H and other core collection strategies (Brown, 1989b) specify how sampling efforts should be divided up among the different groups to be sampled (that is, how many accessions should be randomly sampled per group). The strategy developed here differs from these approaches in that it pinpoints the *individual accessions* from within each geographic group to be selected as entries in the core, so that sampling of accessions within groups is no longer done randomly. The approach is to use a number of genetic marker loci to identify accessions having both high allelic richness and pairwise differentiation (low redundancy). It is assumed that variation at the marker loci is representative of the variation at loci of interest in genetic conservation (hereafter referred to as 'target loci') or, in other words, that maximisation of marker allelic diversity in the core will maximise target allelic diversity. This strategy is termed the M (maximisation) strategy. The key assumption of similarity between marker and target locus variation is based on two postulates: first, that genetic diversity levels at neutral or near-neutral loci are correlated; and, second, that the pattern of genetic differentiation among sets of accessions as observed with marker loci is similar to that in target loci. Before outlining the operational details of the M strategy, the theoretical basis and empirical evidence for these postulates is reviewed.

To begin, consider the variance associated with the average population value of Nei's diversity index, $h_{.j} = \Sigma_i \, h_{ij} / I$. The variance of this average [$\mathrm{var}_j\,(h_{.j})$], also called the population diversity component (PDV), has two components (Brown and Schoen, 1994). The first, gene diversity variance (GDV), arises from variation within populations between the single-locus gene diversities measured at different loci, whereas the second component, gene diversity covariance (GDC), arises from the covariance between pairs of single-locus diversities within populations (*see* Appendix 1). When populations that exhibit high (or low) levels of diversity at one locus tend also to exhibit high (or low) levels of diversity at other loci, both GDC/GDV and the correlation of gene diversity, R, will be positive. This is especially likely when the separate populations of a species differ in the magnitude and/or recency of past population bottlenecks, or show variation in the strength of directional selection. In populations that have experienced a recent or severe bottleneck or that are under strong directional selection, one expects uniformly low h_{ij} values as compared with uniformly higher h_{ij} values in larger and demographically more stable populations or those where directional selection pressures are relaxed. Such variation translates directly into significant GDC or positive R. In fact, significantly positive GDC/GDV and R is commonly found among allozyme loci in wild relatives of crop species, particularly in autogamous species (Brown and Schoen, 1993). Such correlation of gene diversity is expected to apply not only to marker loci but also to other loci. In one case, Barrett and Husband (1989) compared a large source population of the aquatic plant *Eichhornia paniculata* with a smaller, disjunct and derived population. They found correlated reductions in both allozymic variability (percentage loci polymorphic, allelic richness) and in the heritability levels of 15 quantitative traits.

The second postulate outlined states that accessions that are well differentiated at marker loci will also tend to be well differentiated at other loci. Such a correlation may arise for a number of reasons. First, migration among populations and regions, as well as establishment of new populations, takes place via whole genomes, so pairs of accessions with a recently shared evolutionary history should also share a large number of alleles. Second, local selection may act on suites of traits to foster adaptation, and many of these traits may be genetically correlated (Lande, 1982). Third, in autogamous species, high levels of inbreeding and the accompanying slowdown in decay of linkage disequilibrium will help to enforce associations between marker loci and other loci. Examples suggestive of associations between marker locus genotypes and other genotypes are found in *Avena barbata*, where there is an association between allozyme genotypes and microhabitat (Hamrick, 1975), in wild relatives of maize where a correlation of allozyme genotype with altitudinal position has been found (Bretting et al., 1990) and in the legume *Amphicarpaea bracteata*, where correlations between allozyme and disease resistance genotypes have been reported (Parker, 1988).

To make the M strategy operational a linear programming approach is followed. In other words, the optimal solution to the problem is sought, subject to one or more sets of constraints. In the case of the M strategy, the aim is to minimise loss of marker alleles with respect to those present in the entire collection, subject to the following constraints: that a fixed proportion of the total number of accessions (10%) be entered into the core collection, and that the core include at least one accession per group.

A formal representation of the problem is as follows. Let r_j = the total number of accessions present in group j ($j = 1, ..., J$; $\Sigma_j r_j = r.$), n_j = the actual number of accessions selected for entry into the core from group j ($\Sigma_j n_j = n.$), $\{A_{aj}\}$ = set of accessions chosen as the core collection (that is, the set contains several accessions a selected from group j), and $Q_i\{A_{aj}\}$ = the probability that allele i is *not* included in the core when the set of accessions $\{A_{aj}\}$ is sampled. The problem can now be stated in the notation of linear programming as:

Minimise: $\Sigma_i Q_i\{A_{aj}\}$

Subject to: $\Sigma_j n_j = \text{n.} = 0.10\,(r.)$

and (5)

$n_j \geq 1$ (for all $j = 1, ..., J$)

Once a particular set of accessions $\{A_{aj}\}$ is specified, then $\Sigma_i Q_i\{A_{aj}\}$ is determined; note, however, that there are many ways of selecting individual accessions from the different J groups. For example, with $J = 2$ groups, the first with 10 and the second with 30 accessions (that is, $r. = 40$ accessions), a core collection consisting of $n = 4$ accessions may be selected such that $(n_1, n_2) = (3,1)$, (2,2) or (1,3). The total number of possible combinations of groups and accessions (total number of unique sets A_{aj}) is:

$$Cm = \binom{30}{3}\binom{10}{1} + \binom{30}{2}\binom{10}{2} + \binom{30}{1}\binom{10}{3} = 63\,775 \qquad (6)$$

In general, the number of combinations of groups and accessions for J groups each with n_j accessions is:

$$Cm = \sum_{\mathbf{n}_j} \binom{r_1}{n_1}\binom{r_2}{n_2} \cdots \binom{r_J}{n_J} \qquad (7)$$

where the $\mathbf{n}_j$ (a vector) indexes all the ways of taking n_j accessions per group from each of the J groups (for instance, the summation is over the vectors (3,1), (2,2) and (1,3) in the above example).

There is no known analytical solution to the programming problem as defined above. It is, however, possible to solve the problem by searching through all Cm combinations of groups and accessions that satisfy the constraints outlined in equation (5), and evaluating the quantity $\Sigma_i Q_i\{A_{aj}\}$ for each combination. To do this, a computer program (MSEARCH) was written that searched through all possible sets $\{A_{aj}\}$, calculated the quantity $\Sigma_i Q_i\{A_{aj}\}$, and stored the one(s) that minimised this quantity. Such sets of accessions are referred to below as 'M sets'. Note that since Cm is typically very large, it is possible that more than one M set will be found for a given species. In such cases, the effectiveness of the M strategy was evaluated by calculating target allele retention averaged across all the M sets.

COMPARISON OF CORE COLLECTION STRATEGIES

Methods

The effectiveness of the H and M strategies was examined by simulating the construction of core collections for a number of wild relatives of crop species. The species studied were selected based on the availability of data on geographical origin and allozyme variation for a large number of accessions per taxon (> 20 in most cases). Data were obtained directly from the literature or through the generosity of individual researchers (*see* Table 1). The species included wild relatives of several domesticated grasses (*Hordeum spontaneum*, *Sorghum bicolor* ssp. *arundinaceum*, *Zea mays* ssp. *parviglumis* and *Z. mays* ssp. *mexicana*), vegetables (*Capsicum annuum*, *C. frutescens*, *Lycopersicon pennellii* and *L. pimpinellifolium*), beans (*Phaseolus vulgaris*), potatoes (*Solanum berthaultii* and *S. tarijense*) and cotton (*Gossypium davidsonii*). In a few of these cases, two taxa that were closely related to one another were pooled and treated as a single taxon (for example, *Zea mays* ssp. *parviglumis* and *Z. mays* ssp. *mexicana*; *Capsicum annuum* and *C. frutescens*; *Solanum berthaultii* and *S. tarijense*). While such a practice might not be followed when constructing real core collections, it was necessary to do so in a few cases here in order to ensure an adequate sample size.

To serve as a benchmark for evaluating the H and M strategies, several other core collection strategies were examined. Like the H and M strategies, these other strategies also invoke stratified sampling from designated geographic groups, but they differ from the H and M strategies in that genetic marker data are not used to guide sampling. They involve the sampling of a *constant* number of

Table 1 **Wild relatives of crop species included in the study of sampling strategies for core collections**

Species	Origin	Regions for stratified sampling	Source
Hordeum spontaneum	Israel	(1) Inland — mesic (2) Inland — xeric (3) Coastal	Nevo et al. (1979)
Sorghum bicolor ssp. *arundinaceum*	Africa	(1) South Africa (2) East Africa (3) West Africa	Morden et al. (1990) J.F. Doebley (pers. comm.)
Zea mays ssp. *parviglumis/mexicana*	Mexico	(1) North (2) South	Doebley et al. (1984) J.F. Doebley (pers. comm.)
Lycopersicon pimpinellifolium	Peru	(1) Piara, Lambayeque (2) La Libertad, Ancash (3) Lima	Rick et al. (1977)
Solanum pennellii	Peru	(1) Lima (2) Ica	Rick and Tanksley (1981)
Capsicum annuum/frutescens	Mexico	(1) Campeche, Tabasco to Jalisco, Michoacan (2) Nuevo Leon, Tamaulipas (3) Sonora	Loaiza-Figueroa et al. (1989), K. Ritland (pers. comm.)
Phaseolus vulgaris	Central America, South America	1) Mexico, Guatemala, Costa Rica (2) Columbia, Peru (3) Argentina	Koenig and Gepts (1989)
Solanum berthaultii/tarijense	Bolivia, Argentina	(1) North (2) South	D. Douches and D. Spooner (pers. comm.)
Gossypium davidsonii	Baja California, Mexico	(1) Cabo San Lucas (2) La Paz	Wendel and Percival (1990)

accessions per group for entry into the core (C strategy); the sampling of accessions in *proportion* to the number available per group (P strategy); and the sampling of accessions in proportion to the *logarithm* of the number available per group (L strategy). The rationale for these strategies has been discussed elsewhere (Brown, 1989b). Also examined was the simplest strategy, which involves *random* sampling of accessions for inclusion in the core without regard to either geographic origin or marker locus data (R strategy).

Accessions were assigned to two or three non-overlapping geographic groups encompassing the entire range of the accessions available for the taxon under consideration (*see* Table 1). In some cases,

Table 2 Gene diversity in wild relatives of crop species

Species	Total loci[a]	Total alleles	Marker loci	Marker alleles	Target loci	Target alleles	Average gene diversity ($h_{..}$) T[b]	Average gene diversity ($h_{..}$) M	R_{ST}[c] T	R_{ST}[c] M
Hordeum spontaneum	25	101	15	50	10	51	0.11	0.10	0.99	1.08
Sorghum bicolor ssp. *arundinaceum*	25	96	14	54	11	42	0.04	0.05	3.95	4.23
Zea mays ssp. *parviglumis/ mexicana*	22	154	10	81	12	73	0.22	0.25	0.47	0.51
Lycopersicon pimpinellifolium	10	37	5	25	5	12	0.17	0.38	0.78	0.43
Solanum pennellii	15	75	8	38	7	37	0.26	0.19	0.48	0.62
Capsicum annuum/ frutescens	19	61	10	34	9	27	0.03	0.03	10.04	9.68
Phaseolus vulgaris	8	23	4	11	4	12	—[d]	—	—	—
Solanum berthaultii/ tarijense	10	27	5	13	5	14	0.14	0.20	1.05	0.91
Gossypium davidsonii	13	26	7	13	6	13	0.09	0.08	2.99	2.98

Note: a Polymorphic loci only
b T = total; M = markers
c Ratio of inter- to intra-population diversity
d Allele frequency data not available

the designated groups represented different known physiographic or ecological habitats; for example, for *H. spontaneum*, accessions were assigned to coastal, inland-mesic and inland-xeric groups (Nevo et al., 1979). In other cases where detailed ecological data were unavailable (for example, *L. pimpinellifolium* and *Z. mays*), the overall range was divided arbitrarily into two or three geographical groups of roughly equal area on the basis of position in a north-south or east-west cline, or according to obvious discontinuities in spatial distribution. These groups are not intended to represent the 'best' division of the taxon, and were done only to demonstate and compare the various core collection strategies; it is likely that a worker with a more detailed knowledge of the taxonomic and ecological diversity of these species would, perhaps, divide up the accessions differently.

The data on allozyme variation (numbers and identities of alleles) in each taxon were split into two parts. Roughly half the loci and associated alleles were considered as *markers*, and used only to assist in sampling via the H and M strategies, whereas the remaining loci and their alleles were considered to be *target* alleles whose inclusion in the core was assessed and compared among all the various sampling strategies (R, C, P, L, H and M). Data for 8-25 polymorphic loci in total were available per taxon. Average gene diversity for the different taxa ($h_{..}$) varied from 0.03 to 0.26 (mean = 0.13) for all loci, and from 0.03 to 0.38 for marker loci (*see* Table 2). The ratio of inter- to intra-population gene diversity (R_{ST}), a measure of geographic differentiation, ranged from 0.48 to 10.04 for all loci, and from 0.43 to 9.68 for markers (*see* Table 2). Alleles limited in their frequency of occurrence to one or two accessions often comprised a significant fraction of the variability in the nine taxa investigated — up to 52% in the two subspecies of *Z. mays* (*see* Table 3). The retention of this class of alleles is expected to be most influenced by variation in sampling strategy (Brown, 1989b).

Also of interest is the large degree of variation in gene diversity found among different accessions within each species (*see* Figure 2). The discovery of such variation accords with the notion that populations of these wild relatives of crop species have experienced variation in the magnitude or recency of genetic drift and directional selection.

The unit of sampling for all strategies was the individual accession (that is, an individual accession was either included or excluded as an entry in the core). The expected allele retention in the core and its variance were determined using the procedure outlined by Brown (1989b), as well as by simulating the random sampling of accessions (*see* Appendix 2). The different core collection strategies were also

Table 3 **Distribution of target alleles in wild relatives of crop species**

Species	Number of target alleles	Number of localised[a] target alleles (%)	
Hordeum spontaneum	51	17	(33%)
Sorghum bicolor ssp. *arundinaceum*	42	17	(40%)
Zea mays ssp. *parviglumis/mexicana*	73	36	(52%)
Lycopersicon pimpinellifolium	12	3	(25%)
Solanum pennellii	37	10	(27%)
Capsicum annuum/frutescens	27	7	(26%)
Phaseolus vulgaris	12	2	(17%)
Solanum berthaultii/tarijense	14	2	(14%)
Gossypium davidsonii	13	2	(15%)

Note: a Occurring in no more than two accessions per taxon

Figure 2 **Distribution of average population gene diversity ($h._j$) in wild relatives of crop species**

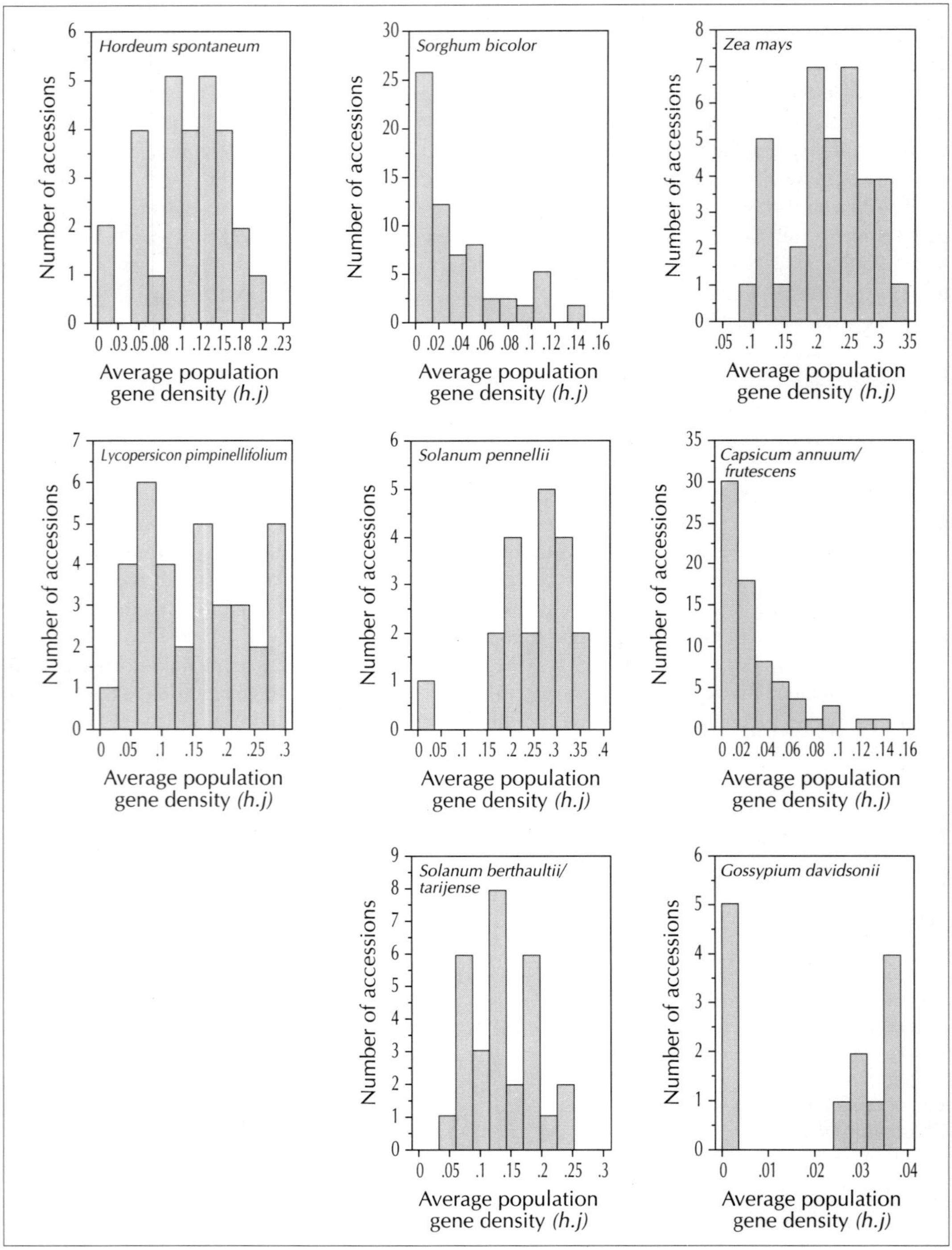

evaluated with respect to the *maximum* target allele retention possible given the sampling effort used (that is, the number of accessions in the core). This was done by applying the MSEARCH programme to the set of target (as opposed to marker) alleles. The maximum possible target allele retention provides a benchmark for comparing allele retention for each strategy with the best possible result.

Table 4 lists several characteristics of the simulated core samples. In a few instances it was necessary to sample more than 10% of the overall collection so that sufficient numbers of accessions were available for comparing the different sampling strategies. Moreover, several sampling strategies often coincided; for example, when the ratio of numbers of accessions per group closely matched those

Table 4 Information on sampling strategies for core collections of wild relatives of crop species

Species	No. of acc.[a] in overall collection[a]	No. of geog.[b] regions for stratified sampling	No. (%) of acc. selected for inclusion in core	No. of acc. selected[c,d]/region for strategy P	L	H	No. of possible combinations of geog. regions and acc. (M)[e]	No. of M sets (M)
Hordeum spontaneum	28	3	6 (21%)	3,2,1	2,2,2	2,2,2	271 791	1
Sorghum bicolor ssp. *arundinaceum*	68	3	6 (9%)	2,3,1	2,3,1	2,3,1	48 050 772	16
Zea mays ssp. *parviglumis/mexicana*	37	2	4 (11%)	2,2	2,2	2,2	59 109	1
Lycopersicon pimpinellifolium	35	3	6 (17%)	2,3,1	2,2,2	3,2,1	1 197 000	66
Solanum pennellii	21	2	4 (19%)	2,2	2,2	3,1	5 455	12
Capsicum annuum/frutescens	66	3	6 (9%)	2,1,3	2,2,2	3,2,1	60 031 686	372
Phaseolus vulgaris	70	3	6 (9%)	3,1,2	3,1,2	—[f]	55 778 000	>10 000
Solanum berthaultii/tarijense	29	2	4 (14%)	2,2	2,2	2,2	21 385	13
Gossypium davidsonii	13	2	4 (31%)	1,3	2,2	2,2	640	155

Note:
a Accessions
b Geographical
c For the C strategy, equal numbers of accessions were sampled per region; the numbers of accessions sampled per region for the R and M strategy are variable (*see text*)
d Order of accessions per region follows order of regions listed in Table 1
e The number of combinations searched for the M strategy (at least one accession per region included in the core)
f Allele frequency data not available

Figure 3 **Expected target allele retention in simulated core collections of nine wild relatives of crop species**

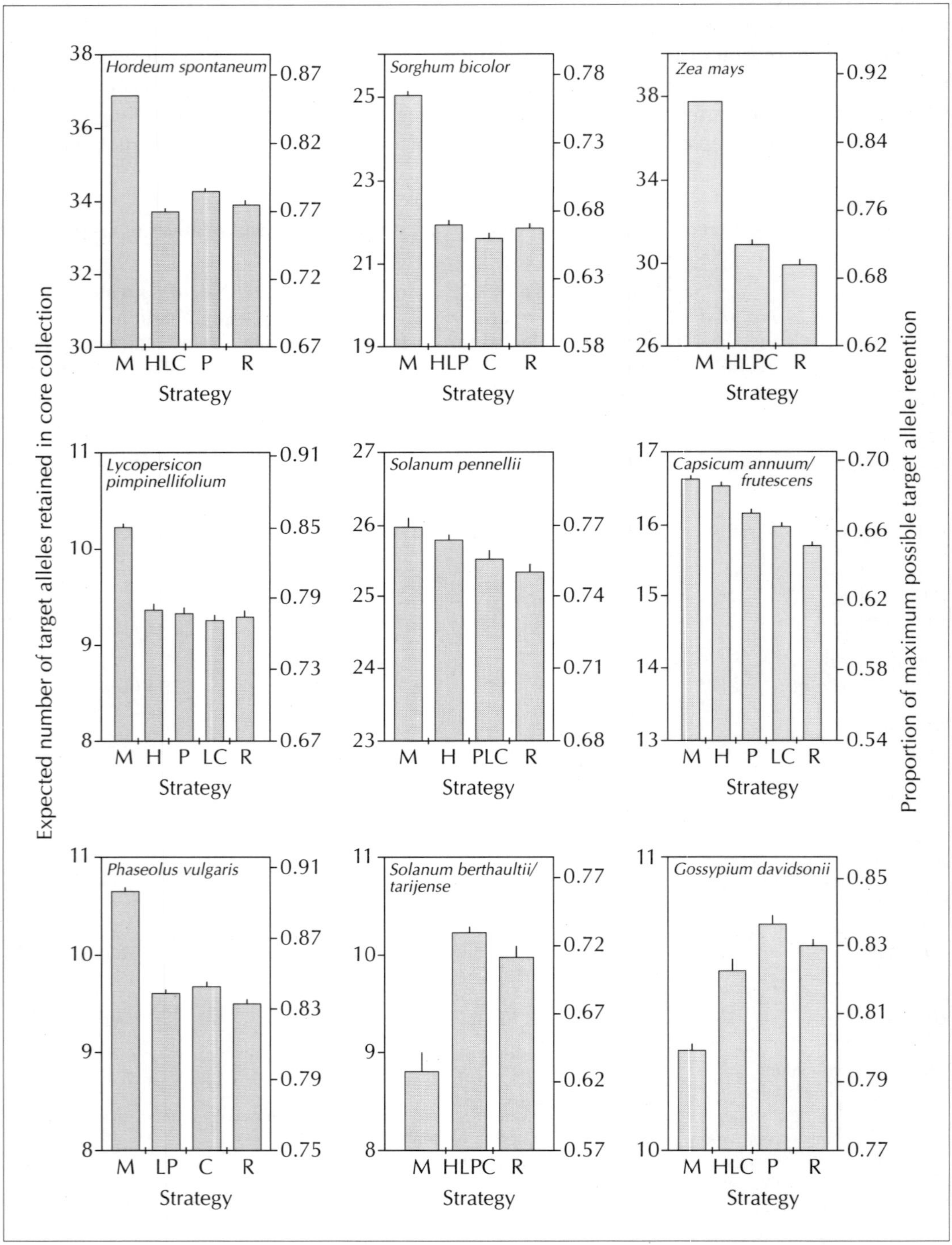

of the estimates of effective population size, the P, L and H strategies were often identical (*see* Table 3). For the M strategy, the penultimate column of Table 4 shows the *Cm* possible combinations of geographic groups and accessions when *r* accessions are selected for entry into the core (all $r_j > 0$), whereas the last column gives the actual number of M sets found. In most cases, the number of M sets represented only a very small proportion (<< 1%) of the all the *Cm* combinations, and in two instances (*H. spontaneum* and *Z. mays*) only a single M set was found.

Results

Overall, the simulated core collections captured approximately 70% of the available target alleles (*see* Figure 3). There was, however, some variation among species. In the species with many localised target alleles (principally the three grass taxa), core collections captured on average 40-65% of the all target alleles, whereas in the species with fewer localised target alleles, 70-80% of all target alleles were captured in the core collection (*see* Table 3 and Figure 3). Less interspecific variation was observed when allele retention was compared in terms of the proportion of the maximum *possible* allele retention given the sampling intensity used, as opposed to retention of all target alleles (*see* Figure 3).

It is notable that the form of the empirical relationship relating fraction of all target alleles retained to sampling intensity (*see* Figure 4) parallels the general theoretical expectation for the neutral allele

Figure 4 **Proportion of all available target alleles captured in the core collection (average of the six strategies M, H, L, P, C and R) versus the proportion of all accessions from the entire collection included as entries in the core collection**

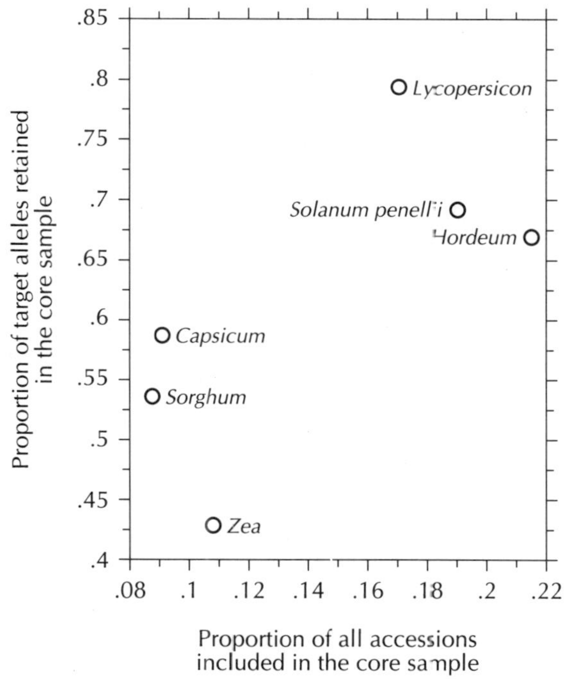

Figure 5 **Allele retention as a function of core collection strategy (M, H, L, P, C and R) and frequency of occurrence of target allele among all accessions**

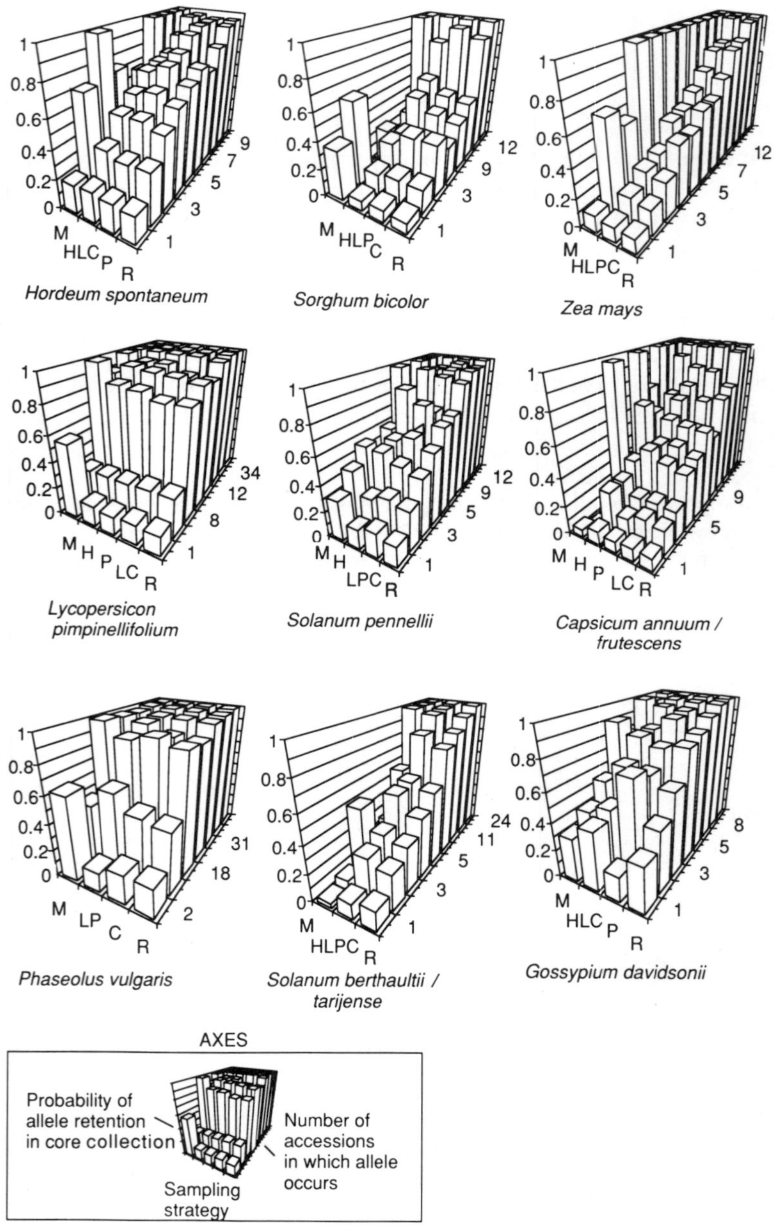

model with many loci (that is, diminishing returns in target allele retention with continued investment in sampling) (Brown, 1989a). The overall average ranking of the six strategies (highest rank = 1, lowest rank = 6) for expected allele retention is: M (2.1) > H (3.0) > P (3.1) > L (3.7) > C (4.1) > R (4.7) (*see* Figure 3). Allele retention as a function of both sampling strategy and frequency of occurrence of the target allele among the *r* accessions is shown in Figure 5. To perform this analysis, all alleles with a given frequency of occurrence were pooled in a single class. From these results it is apparent that stratified sampling (strategies M, H, P, L and C) was more efficient than simple random sampling (strategy R) and that the use of genetic markers to guide core sampling (strategies M and H) further increases allele retention.

Of particular interest was the significant improvement in performance of the M strategy over the other strategies. This improvement was particularly pronounced for alleles whose distribution is not widespread (*see* Figure 5). Such localised alleles may be important sources of disease resistance and adaptation to specific environmental conditions (Marshall and Brown, 1975). In all but two of the species studied, the M strategy gave the highest target allele retention. The exceptions (*S. b. tarijense* and *G. davidsonii*) are revealing in that they were the only taxa among the nine that did not exhibit positive covariation of gene diversity among either the marker loci or all loci (high GDC/GDV and R) (*see* Table 5). When these species are excluded from consideration, the ranking of strategies becomes

Table 5 **Association of gene diversity within populations as determined by variance component analysis and correlation diversity[a]**

Species	Association of gene diversity (GDC/GDV) All loci	Association of gene diversity (GDC/GDV) Marker loci	Average correlation of gene diversity (R) All loci	Average correlation of gene diversity (R) Marker loci
Hordeum spontaneum	0.09 ***[b]	0.05 ***	0.05	0.02
Sorghum bicolor ssp. *arundinaceum*	0.08 ***	0.07 ***	0.07	0.06
Zea mays ssp. *parviglumis/mexicana*	0.09 ***	0.13 ***	0.08	0.14
Lycopersicon pimpinellifolium	0.18 ***	0.10 **	0.20	0.14
Solanum pennellii	0.10 ***	0.07 *	0.11	0.07
Capsicum annuum/ frutescens	0.08 ***	0.05 ***	0.06	0.12
Phaseolus vulgaris	—[c]	—	—	—
Solanum berthaultii/ tarijense	0 ns[d]	0.01 ns	0	0
Gossypium davidsonii	0 ns	-0.04 ns	-0.11	-0.32

Note: a See text for details
b * P < 0.05, ** P < 0.01, *** P < 0.001; *see* Brown and Schoen (1994) for details of significance testing
c Gene frequency data not available
d Not significant

M (1.0) > H (2.9) > P (3.5) > L (3.8) > C (4.4) > R (5.0). Thus, the M strategy works best in all cases where it is expected to do so, and the successful application of both the M and H strategy were predictable from the analysis of association of gene diversity in the marker locus subset.

Additional analysis of the results from *H. spontaneum* and *Z. mays* helps to confirm the theoretical rationale for the H and M strategies. In both cases only a single M set was found. This makes it easy to compare the genetic characteristics of the accessions selected by the M strategy against those present in the entire collection. In both taxa, the accessions comprising the M set had higher gene diversity and average pairwise genetic distances compared with all accessions. It is interesting to note that in one case, *H. spontaneum*, the M set contained an accession which is monomorphic at all marker loci. Further examination of this accession revealed that it was highly differentiated from the others, and this high level of differentiation is the basis for its entry into the core. The result underscores the difference between the H and M strategies.

Clearly, it would also have been of significant interest in this study to examine how marker-assisted strategies of germplasm conservation compare when examining the retention of genes coding for traits such as disease resistance and water or temperature stress. By using allozyme loci in place of such target genes to assess the effectiveness of the different core collection strategies, we do not mean to imply that such loci are representative of the loci of interest in gene conservation. Rather, this step was taken out of necessity, as there are no comprehensive data sets for 'useful' variation. It should be noted, however, that at least some of the variation of potential interest to genetic conservation is likely to be selectively neutral at present, and so its distributional and sampling properties may be approximated by use of allozymes. Moreover, in inbreeders, variation at neutral and selected loci is likely to be correlated because of a slowdown in the decay rate of linkage disequilibrium.

CONCLUSION

In the face of unequal diversity and differentiation, the allelic richness of core collections can be maximised by attention to several practical steps. The simplest step involves the use of passport and other data to create groups of accessions for stratified sampling of the entire collection. This ensures broad coverage and, as the simulation results show, increases the allelic richness in the core. Stratified sampling almost always increases the gene diversity of the core and does not demand much in terms of labour spent to characterise accessions.

The discussion has also shown that information gained from marker loci can be useful in maximising core diversity once groups for stratified sampling have been defined. In particular, collecting more accessions from groups of high marker gene diversity (H strategy) or targeting particular accessions that are both high in allelic richness and well differentiated (M strategy) can offer important advantages over other sampling strategies and help alleviate difficulties brought about by unequal levels of diversity and differentiation. As new technology for detecting marker locus variation develops, particularly the screening of accessions for RFLPs and RAPDs (Williams et al., 1990), it should be possible to assess variation levels at many more loci than with allozymes alone. Because such loci are likely to be randomly distributed throughout the genome, they should provide a more representative picture of patterns of variation.

Despite the apparent usefulness of genetic markers, the question arises as to whether the potential gains made in using genetic marker data to construct the core sample merit the effort required. Such an endeavour requires resources that might justifiably be spent on other tasks associated with germplasm conservation, such as additional exploration and phenotypic evaluation of accessions. Are the gains achieved by a strategy such as the M strategy worth the extra effort? Unfortunately, there is no clear answer to this question. Much depends upon the nature of the variation and breeding system of the species in question. For example, if an increase in allelic richness of the core sample leads to the retention of alleles important in disease resistance, the extra richness may be well worth the investment of effort, particularly in the case of an important crop pest. With *in situ* conservation, where there are limitations on the number of natural habitats that can be preserved, the maximisation of allelic richness in the core through adoption of the H or M strategy may be especially compelling.

While we have concentrated on core samples of wild relatives, the principles behind the methods developed and discussed above are general and should be of use in constructing core samples of domesticated materials (such as crop varieties and landraces). Indeed, there is a compelling need to apply such procedures and rationalise the continuing growth of germplasm collections. The results of the work presented above show that accessions of wild relatives of crop species differ widely in gene diversity and therefore are not of equal value to conservation. This, together with redundancy in the genetic composition of accessions, provides a strong impetus for seeking to construct core collections that will serve as useful focal points of well-evaluated germplasm for breeders to exploit.

Appendix

1 *Correlation of gene diversity among loci*

The population diversity component (PDV) has two components: gene diversity variance (GDV) and gene diversity covariance (GDC) (Brown and Schoen,1994), or:

$$\text{PDV} = \text{GDV/I} + \text{GDC (I-1)/I}$$

$$\text{Var}_j(h_{\cdot j}) = \{\Sigma_i \text{ var}_j(h_{ij}) + 2\,\Sigma_i \Sigma_{k>i} \text{ cov}_j\,[h_{ij}, h_{kj}]\,\}/\,\text{I}^2 \qquad \text{(A1)}$$

The ratio of GDC/GDV is akin to the pairwise correlation of gene diversity for separate loci within populations, averaged over all pairs of loci:

$$\text{R} = \Sigma_i \Sigma_{k>i}\ \{\text{cov}_j(h_{ij}, h_{kj})/\sqrt{\text{var}_j(h_{ij})\ \text{var}_j(h_{kj})}\ \}\,/\{\text{I(I-1)}/2\} \qquad \text{(A2)}$$

2 *Calculation of expected allele retention and its variance*

For stratified samples, the expected number of alleles retained in the core whose entries are randomly selected from within each geographic group (that is, according to the C, P, L or H strategies) is:

$$\text{E}(K) = \sum_i [1 - \prod_{k=1}^{g} \text{Q}_i(k)]$$

with

$$\text{Q}_i(k) = \prod_{m=0}^{n_k - 1} (r_k - s_{ik} - m)/(r_k - m) \qquad \text{...when } n_k < r_k - s_{ik} \qquad \text{(A3)}$$

$$= 0 \qquad \text{...otherwise}$$

where:

s_{ik} = the number of accessions from group k having allele i

Note that in the case of the M strategy, the assumption of the accession as the unit of sampling means that $Q_i\{A_{jk}\}$, which is analogous to $Q_i(k)$ in formula (A3), takes on a value of either 0 or 1.

The variance of expected allele retention in the case of the R, C, P, L and H strategies was estimated via a computer algorithm which simulated sampling of accessions without replacement from the entire collection. The algorithm was used to choose accessions randomly from each group in accordance with the r_k specified by the C, P, L and H strategies, or the r (R strategy). For each set of accessions sampled, the algorithm determined whether each target allele in the overall collection was retained in the core collection. Since there are typically many possible ways of selecting the accessions, the algorithm was iterated until 300 unique sets of accessions were selected. The variance of target allele retention was then calculated. The analytical formula (A3) used for calculating E(K) and the random sampling algorithm gave virtually identical estimates of average target allele retention.

For the M strategy, there is no analytical formula that can be used to calculate E(K). Instead, this was done by averaging target allele retention over all M sets found using the MSEARCH programme. Variance of expected allele retention for the M strategy was also calculated over all M sets.

References

Baker, H.G. 1955. Self-compatibility and establishment after 'long-distance' dispersal. *Evolution* 9: 347-49.

Barrett, S.C.H. and Husband, B.C. 1989. The genetics of plant migration and colonisation. In Brown, A.H.D., Clegg, M.T., Kahler, A.L. and Weir, B.S. (eds) *Plant Population Genetics, Breeding and Genetic Resources*. Sunderland, Massachusetts, USA: Sinauer Associates.

Barrett, S.C.H. and Kohn, J.R. 1991. Genetic and evolutionary consequences of small population size in plants: Implications for conservation. In Falk, D.A. and Holsinger, K.E. (eds) *Conservation of Rare Plants: Biology and Genetics*. Oxford, UK: Oxford University Press.

Bretting, P.K., Goodman, M.M. and Stuber, C.W. 1990. Isozymatic variation in Guatemalan races of maize. *American J. Botany* 77: 211-25.

Brown, A.H.D. 1978. Isozymes, plant population genetic structure and genetic conservation. *Theoretical and Applied Genetics* 52: 145-57.

Brown, A.H.D. 1989a. The case for core collections. In Brown, A.H.D., Frankel, O.H., Marshall, D.R. and Williams, J.T. (eds) *The Use of Plant Genetic Resources*. Cambridge, UK: Cambridge University Press.

Brown, A.H.D. 1989b. Core collections: A practical approach to genetic resources management. *Genome* 31: 818-24.

Brown, A.H.D. and Briggs, J.D. 1991. Sampling strategies for genetic variation in *ex situ* collections of endangered plant species. In Falk, D.A. and Holsinger, K.E. (eds) *Conservation of Rare Plants: Biology and Genetics*. Oxford, UK: Oxford University Press.

Brown, A.H.D. and Schoen, D.J. 1994. A revised measure of association of gene diversity values. *Hereditas* 120: 77-79.

Burdon, J.J., Jarosz, A.M. and Kirby, G.C. 1989. Pattern and patchiness in plant-pathogen interaction — causes and consequences. *Annual Review of Ecology and Systematics* 20: 119-36.

Chakraborty, R. and Neel, J.V. 1989. Description and validation of a method for simultaneous estimation of effective population size and mutation rate from human population data. *Proc. National Academy of Sciences, USA* 86: 9407-11.

Chapman, C.G.D. 1989. Collection strategies for the wild relatives of field crops. In Brown, A.H.D., Frankel, O.H., Marshall, D. R. and Williams, J.T. (eds) *The Use of Plant Genetic Resources*. Cambridge, UK: Cambridge University Press.

Doebley, J.F., Goodman, M.M. and Stuber, C.W. 1984. Isoenzymatic variation in *Zea* (Gramineae). *Systematic Botany* 9: 203-18.

Ewens, W.J. 1972. The sampling theory of selectively neutral alleles. *Theoretical Population Biology* 3: 87-112.

Frankel, O.H. 1984. Genetic perspectives of germplasm conservation. In Arber, W.K., Llimensee, K., Peacock, W.J. and Starlinger, P. (eds) *Genetic Manipulation: Impact on Man and Society*. Cambridge, UK: Cambridge University Press.

Frankel, O.H. and Brown, A.H.D. 1984. Current plant genetic resources — a critical appraisal. In *Genetics: New Frontiers* (vol. 4). New Delhi, India: Oxford and IBH Publishing Co.

Hamrick, J.L. and Allard, R.W. 1975. Correlations between quantitative characters and enzyme genotypes in *Avena barbata*. *Evolution* 29: 438-42.

Harlan, J.R. 1984. Evaluation of wild relatives of crop plants. In Holden, J.H.W. and Williams, J.T. (eds) *Crop Genetic Resources: Conservation and Evaluation*. London, UK: Allen and Unwin.

Kimura, M. and Crow, J.F. 1964. The number of alleles that can be maintained in a finite population. *Genetics* 49: 725-38.

Koenig, R. and Gepts, P. 1989. Allozyme diversity in wild *Phaseolus vulgaris*: Further evidence for two major centers of genetic diversity. *Theoretical and Applied Genetics* 78: 809-17.

Ladizinsky, G. 1989. Ecological and genetic considerations in collecting and using wild relatives. In Brown, A.H.D., Frankel, O.H., Marshall, D.R.,and Williams, J.T. (eds) *The Use of Plant Genetic Resources*. Cambridge, UK: Cambridge University Press.

Lande, R. 1982. A quantitative genetic theory of life history evolution. *Ecology* 63: 607-15.

Lande, R. and Barrowclough, G.F. 1987. Effective population size, genetic variation, and their use in population management. In Soulé, M.E. (ed) *Viable Populations for Conservation*. Cambridge, UK: Cambridge University Press.

Landry, B.S. and Michelmore, R.W. 1987. Methods and applications of restriction fragment length polymorphism analysis to plants. In Bruening, G., Harada, J., Kosuge, T. and Hollaender, A. (eds) *Tailoring Genes for Crop Improvement*. New York, USA: Plenum Press.

Leberg, P.L. 1992. Effects of population bottlenecks on genetic diversity as measured by allozyme electrophoresis. *Evolution* 46: 477-94.

Loaiza-Figueroa, F., Ritland, K., Laborde Cancino, J.A. and Tanksley, S.D. 1989. Patterns of genetic variation in the genus *Capsicum* (Solanaceae) in Mexico. *Plant Systematics and Evolution* 165: 159-88.

Marshall, D.R. and Brown, A.H.D. 1975. Optimum sampling strategies in genetic conservation. In Frankel, O.H. and Hawkes, J.G. (eds) *Crop Genetic Resources for Today and Tomorrow*. Cambridge, UK: Cambridge University Press.

Maruyama, T. and Fuerst, P.A. 1985. Population bottlenecks and nonequilibrium models in population genetics. II. Number of alleles in a small population that was formed by a recent bottleneck. *Genetics* 111: 675-89.

Morden. C.W., Doebley, J. and Schertz, K.F. 1990. Allozyme variation among the spontaneous species of *Sorghum* section Sorghum (Poaceae). *Theoretical and Applied Genetics* 80: 296-304.

Nei, M. 1973. Analysis of gene diversity in subdivided populations. *Proc. National Academy of Sciences, USA* 70: 3321-23.

Nei, M., Maruyama, T. and Chakraborty, R. 1975. The bottleneck effect and genetic variability in populations. *Evolution* 29: 1-10.

Nevo, E., Zohary, D., Brown, A.H.D. and Haber, M. 1979. Genetic diversity and environmental associations of wild barley, *Hordeum spontaneum*, in Israel. *Evolution* 33: 815-33.

Parker, M.A. 1988. Disequilibrium between disease resistance variants and allozyme loci in an annual legume. *Evolution* 42: 239-47.

Perry, M.C., McIntosh, M.S. and Stoner, A.K. 1991. Geographical patterns of variation in the USDA soybean germplasm collection. II. Allozyme frequencies. *Crop Science* 31: 1356-60.

Rick, C.M. and Tanksley, S.D. 1981. Genetic variation in *Solanum pennellii*: Comparisons with two other sympatric tomato species. *Plant Systematics and Evolution* 139: 11-45.

Rick, C.M., Fobes, J.F. and Holle, M. 1977. Genetic variation in *Lycopersicon pimpinellifolium*: Impact of genetic variation in floral characters. *Plant Systematics and Evolution* 127: 139-70.

Ritland, K. 1983. Estimation of mating systems. In Tanksley, S.D. and Orton, T.J. (eds) *Isozymes in Plant Genetics and Breeding*. Amsterdam, Netherlands: Elsevier.

Schoen, D.J. and Brown, A.H.D. 1991. Intraspecific variation in population gene diversity and effective population size correlates with the mating system in plants. *Proc. National Academy of Sciences, USA* 88: 4494-97.

Sirkkomaa, S. 1983. Calculations on the decrease of genetic variation due to the founder effect. *Hereditas* 99: 11-20.

Stuber, C.W., Moll, R.H., Goodman, M.M. and Wendel, J.F. 1980. Allozyme frequency changes associated with selection for increased grain yield in maize (*Zea mays*). *Genetics* 95: 225-36.

Tanksley, S.D., Young, N.D., Paterson, A.H. and Bonierbale, M.W. 1989. RFLP mapping in plant breeding: New tools for an old science. *Biotechnology* 7: 257-64.

Wendel, J.F. and Percival, A.E. 1990. Molecular divergence in the Galapagos Islands-Baja California species pair, *Gossypium klotzschianum* and *G. davidsonii* (Malvaceae). *Plant Systematics and Evolution* 171: 99-115.

Williams, J.G.K., Kubelik, A.R., Livak, K.J., Rafalski, J.A. and Tingey, S.V. 1990. DNA polymorphisms amplified by arbitrary primers are useful as genetic markers. *Nucleic Acids Research* 18: 7213-18.

Core Collections of Plant Genetic Resources
Edited by T. Hodgkin, A.H.D. Brown, Th.J.L. van Hintum and E.A.V. Morales

A Wiley-Sayce Publication

2.4

The use of multivariate methods in developing a core collection

J. Crossa, I.H. DeLacy and S. Taba

Abstract

The concept of core collections was introduced for the purpose of increasing the efficiency of germplasm description and exploitation. Multivariate methods facilitate the interpretation and description of complex two-way (accession x attribute) data. The objectives of this study were to review and describe multivariate procedures applied in genetic conservation, study patterns of phenotypic diversity of 175 maize accessions of the Tuxpeño race complex, and help the germplasm bank curator decide how and which accessions should be selected for the core collection. Numerical measures of association between each pair of accessions are necessary for both classification (cluster analysis) and ordination (principal components analysis). A hierarchical agglomerative clustering procedure with incremental sum of squares as the fusion strategy and squared Euclidean distance as the dissimilarity measure was used for classifying the Tuxpeño accessions. These accessions come from dry, wet and mixed ecologies. The results of cluster analysis confirmed those obtained from principal components analysis. This procedure was carried out in a series of steps where randomly selected accessions, within groups from cluster analysis and confirmed with ordination analysis, were included in the next step for further analysis. At each step, about half or fewer of the accessions were advanced to the next step. The process ended when 48 of the original 175 accessions were retained to form the core collection. A final check of the selection procedure was made using canonical variate analysis, a multi-group generalisation of discriminant analysis.

Frankel (1984) introduced the concept of core collections to increase the efficiency of germplasm description and use by ensuring maximum genetic diversity while reducing the size of a collection. Frankel and Brown (1984) and Brown (1989) described how a core subset could be assembled using information on the origin and agronomic and morphological characters of the accessions.

Two of the more important issues that a germplasm bank curator should address when forming core collections are to determine:

- the optimum number of accessions for retaining most of the alleles present in a given collection
- how to select accessions for the core collection

The first question was addressed by Brown (1989), who recommended including only 10% of the total collection and a maximum of 3000 accessions per species. The second question relates to the issues of how to identify diverse subgroups of accessions and how to sample and select from them. Brown (1989) suggested a stratified random sampling strategy where the core members would be selected from groups formed by some method from the entire collection. He suggested, as an example, that the collection could be subdivided into non-overlapping groups based on racial or ecogeographical criteria.

Suppose we have a data set consisting of measurements of several attributes (morphological or agronomic characters) for various accessions. This two-way table of data can be conceptualised as placing the accessions into a position in a multi-dimensional space defined by the attributes, where each dimension is determined by an attribute. The coordinates of the position of each accession are determined from the value of each attribute. Those accessions with similar values for each attribute would be close to each other in this space and have a similar pattern over the attributes. Such a multivariate data set is complex in structure and multivariate statistical techniques facilitate the interpretation and description of the relationships among the patterns of the accessions.

Classification (grouping of entities with similar patterns) and ordination (description of spatial relationships among entities) methods are two major multivariate techniques commonly used in such areas as numerical taxonomy, genetic analysis, plant breeding and biotechnology to describe and analyse multivariate data sets. Pattern analysis, which is the combined use of cluster analysis and ordination techniques (Williams, 1976), provides a powerful tool for dealing with the complexity of examining large data sets. A number of authors have recommended pattern analysis for the analysis of multi-attribute data, including Byth et al. (1976), Williams (1976), DeLacy and Cooper (1990) and DeLacy et al. (1990) for agricultural data, and Digby and Kempton (1987) for ecological data. These authors have provided many examples of the use of these techniques.

The use of numerical cluster analysis to classify accessions from germplasm collections according to their degree of similarity was assessed by Peeters and Martinelli (1989). Canonical discriminant analysis and cluster analysis were used to separate geographical patterns of variation in soybean germplasm based on morphological traits (Perry and McIntosh, 1991) and allozyme frequencies (Perry et al., 1991). Cluster analysis, principal components analysis (PCA) and canonical discriminant analysis were used on morphological and agronomic data to study patterns of genetic diversity in cultivated common bean (Singh et al., 1991). Classification and ordination techniques were used to assess genetic diversity among and within maize races using agronomic and morphological traits (Goodman and Bird, 1977; Bird and Goodman, 1978). Of the above, canonical discriminant analysis is a classification technique and PCA is an ordination technique.

The objectives of this chapter are to review and describe relevant multivariate methods and illustrate their use in forming a breeding core collection, to study phenotypic diversity patterns in accessions from a maize Tuxpeño collection, and to help the gene bank curator decide which accessions should be selected for the core collection.

CLASSIFICATION METHODS

The objective of these methods is to classify items (accessions) into groups which have similar values for all attributes. The data are simplified and summarised by determining more homogeneous subsets of accessions. We will describe two classification methods: cluster analysis and discriminant analysis.

Cluster analysis

Clustering is the partitioning of a set of objects into groups so that objects within a group are similar and objects in different groups are dissimilar. The aim of any numerical cluster analysis is to find discontinuity in the data by means of a cluster strategy that groups the objects. Cluster analysis is efficient in grouping objects with similar characteristics.

Clustering methods can be either hierarchical or non-hierarchical. In hierarchical methods, the individuals (accessions or attributes) are organised into a tree or hierarchy where individuals or groups are fused one at a time to individuals or groups with the most similar patterns for all attributes. In non-hierarchical systems, the individuals are organised into a set number of groups in the 'best' possible manner. The hierarchical methods can be used to form a fixed number of groups by truncating the hierarchy at a fixed level. In general, the 'best' grouping at any level is incompatible with a hierarchy, since the members in a group at a fixed level are constrained by their membership at a lower level in the hierarchy. Hierarchical methods optimise a path through the individuals as distinct from optimising a grouping at any level. However, the path methods have the advantage that a significant amount of information among the individuals resides in the hierarchy.

Agglomerative hierarchical clustering methods

All agglomerative methods start with a dissimilarity matrix and fuse the two individuals which have the smallest dissimilarity between them to form a group with two members. Next, the group-individual dissimilarity between this new group and all the remaining individuals is calculated. This set of dissimilarities is then added to the matrix of dissimilarities among the remaining individuals to form a new dissimilarity matrix that is one row and column smaller than the original. The procedure is repeated and another fusion made. When two or more groups are present, group-group dissimilarities must be calculated. The procedure ends when all the individuals are in one group.

The method for calculating the group-individual and group-group dissimilarity is called the clustering strategy. A number of strategies for agglomerative clustering have been proposed and their properties investigated. The methods discussed here were termed SAHN (sequential agglomerative hierarchical combinatorial strategies) by Sneath and Sokal (1973).

All the SAHN methods can be described in relation to flexible sorting formulae. This provides the new group-group dissimilarity between a new group just formed from two old groups and some other group. This makes the methods combinatorial, which is expressed in terms of the original dissimilarities before fusion and four user-defined parameters that determine the nature of the strategy. If we write the dissimilarity between two groups I and J as D(I,J), then the distance D(K,L) between a new group K (formed from old groups I and J) from another group L is given as:

$$D(K,L) = A_1{*}D(I,L) + A_2{*}D(J,L) + B{*}D(I,J) + G{*}abs[D(I,L)-D(J,L)]$$

The new distance D(K,L) is a function of the old distances between groups I and J and L. The four user-defined parameters are A_1, A_2, B and G; 'abs' means absolute value. Monotonicity is guaranteed if the following relationships hold among the four parameters: $A_1 = A_2 = (1-B)/2$ and $G = 0$.

Six strategies based on this general formula have been proposed and used:

1 *Single linkage (nearest neighbour):* Set $A_1 = A_2 = 0.5$, $B = 0$ and $G = -0.5$. This strategy defines the group-group dissimilarity as the smallest of all the pairwise dissimilarities between members of one group and another. It is highly space contracting, monotonic and not recommended.

2 *Maximum linkage (furthest neighbour):* Set $A_1 = A_2 = 0.5$, $B = 0$ and $G = 0.5$. This strategy defines the group-group dissimilarity as the largest of all the pairwise dissimilarities among the two groups. It is highly space dilating, monotonic and not recommended.

3 *Flexible sorting:* There are three forms of Lance and Williams' (1967) general method.

a) User flexible sorting: Set A_1, A_2, B and G to any constants. The methods vary from monotonic to major reversals and from highly space dilating to extremely space contracting, depending upon the values of the constants. It is recommended only for investigating the properties of the strategies.

b) Unweighted flexible sorting (flexible WPGMA): Set $A_1 = A_2 = (1\text{-}B)/2$, $G = 0$ and specify B. WPGMA stands for 'weighted pair group arithmetic averaging' (Sneath and Sokal, 1973). This was Lance and Williams' original definition. The method is monotonic and varies from highly space dilating ($B = -1$) through space conserving ($B = 0$) to highly space contracting ($B = 1$). The method changes the weights of objects inversely with group size. It is generally believed that the next version of the flexible sorting strategy is preferable.

c) Weighted flexible sorting (flexible UPGMA, flexible group average): Set $A_1 = (1\text{-}B)*N(I)/[N(I) + N(J)]$, $A_2 = (1\text{-}B)*N(J)/[N(I) + N(J)]$, $G = 0$ and specify B. N(I) is the number of members in group I. UPGMA stands for 'unweighted pair group arithmetic averaging' (Sneath and Sokall, 1973). The objects are weighted equally throughout the fusion process. If $N(I) = N(J)$, this is equivalent to the dissimilarity in 3b. The method is monotonic and varies from highly space dilating (with $B = -1$) to highly space contracting ($B = 1$). The recommended group average clustering strategy is $B = 0$ if the purpose of the analysis is to search for true discontinuities in the data. If the purpose is to describe the group structure, B should be set at -0.25 to give a strong space dilating strategy. The group-group dissimilarity for group average is the unweighted average of all pairwise dissimilarities between the members of one group and all the members of the other group.

4 *Median (WPGMC):* Set $A_1 = A_2 = 0.5$, $B = -0.25$ and $G = 0$. WPGMC stands for 'weighted pair group centroid' (Sneath and Sokal, 1973). It is the centroid equivalent of WPGMA, is not monotonic and is not recommended.

5 *Centroid (UPGMC):* Set $A_1 = N(I)/[N(I) + N(J)]$, $A_2 = N(J)/[N(I) + N(J]$, $B = A_1*A_2$ and $G = 0$. UPGMC stands for 'unweighted pair group centroid' (Sneath and Sokal, 1973). It is the centroid equivalent of UPGMA, is not monotonic and is not recommended.

6 *Ward's method (ISS):* Set $A_1 = [N(I) + N(L)]/[N(I) + N(J) + N(L)]$, $A_2 = [N(J) + N(L)]/[N(I) + N(J) + N(L)]$, $B = N(L)/[N(I) + N(J) + N(L)]$ and $G = 0$, provided always that the dissimilarity measure is the squared Euclidean distance (SED). The incremental sum of squares (ISS) then fuses the two groups which increase the within-group sum of squares the least. It is monotonic, strongly clustering and is the preferred method when the attribute types are all qualitative because the SED measure can be used. Otherwise, the flexible group average with B set at -0.25 mimics the clustering intensity of ISS. The advantage of this strategy is that it is a generalised form of the analysis of variance and enables the properties of the classifications to be integrated into standard theory (DeLacy and Cooper, 1990).

Non-hierarchical clustering methods

All non-hierarchical methods operate by 'guessing' a grouping and then employing some method or algorithm to improve the classification. The process is repeated (iterated) until no further improvement in the classification occurs. The algorithm requires both a criterion for evaluating the improvement in the grouping when individuals are transferred from one group to another and a procedure for determining how individuals should be reallocated to groups to improve the classification.

There are many ways of providing the initial grouping. Most methods provide a set of seeds and individuals chosen by some method from the original population; the remaining individuals are then allocated to their nearest seed. Any practical method must provide a mechanism to create a new group if individuals are found which are too unlike the initial seeds.

Discriminant analysis

Linear discriminant analysis is perhaps the most widely used method of classification (Fisher, 1936). The formal purpose of discriminant analysis is to assign distinct sets of items (accessions) to one or several groups or classes based on a set of measurements (attributes). The method allocates new items to previously defined groups by finding mathematical discriminant functions (linear combinations of the original variables) that minimise the chance of misclassification.

For two populations with means X_1 and X_2 and pooled variance S_x, Fisher's sample discriminant function is: $D = l'X$, where $l' = (X_1 - X_2)' S_x^{-1}$. Thus $D = l_{i1}X_1 + l_{i2}X_2 + \ldots\ l_{ik}X_k$. The l's are weighting coefficients, and the X's are the values of the k discriminant variables. An observation X_0 is allocated to population 1 if $D_0 = l'X_0 > 1/2(X_1 - X_2)' S_x^{-1} (X_1 + X_2)$. Otherwise, X_0 is allocated to population 2.

Fisher's discriminant functions and the allocation rule can be generalised to more than two clustered populations.

ORDINATION METHODS

Ordination methods summarise multidimensional data by producing a low-dimensional space in which similar items (accessions) are close together and dissimilar items are far apart. Field data have high dimensions because of the large number of accessions and attributes, whereas the final results must be of lower dimensions. Spatial representation of the items in two or three dimensions will reflect their relationship in higher dimensions with minimum distortion.

Ordination techniques are used in forming a core collection in order to study diversity of traits among accessions, complement and confirm groupings obtained through classification, establish whether further classification (subdivision) is needed based on variability shown in lower dimensions, and choose accessions that have the combined traits of interest.

Principal components analysis

The most common ordination technique is PCA. This technique involves using a similarity matrix among items (accessions) given by either the covariance matrix or correlation matrix. The main functions of PCA are to explain variance with a linear combination of the original variables, project points from high dimensions to fewer dimensions (data reduction), and facilitate interpretation of the existing patterns among entities.

In PCA, successive components are constructed to be uncorrelated with the previous ones, and often most of the variation can be summarised with only a few principal components.

PROPERTIES OF THE DATA AND PROXIMITY MEASURES

Classification and ordination procedures require a measure of association among individuals which is (mostly, but not always) calculated from measurements of a number of attributes on each individual. This type of data is common for germplasm banks, since two-way tables are generated by passport data on the accessions and characterisation and evaluation data on the whole or parts of the collection.

The effective use of classification and ordination methods requires an understanding of the properties of the forms of data collected, types of data collected, transformation and standardisation of the data, and measures of association.

Forms of data

As stated above, most data consist of a two-way table of measurements of a number of attributes on a set of accessions with an occasional symmetrical association matrix among the accessions. These tables consist of several different sets of passport, characterisation and evaluation data on different but overlapping subsets of the accessions in the collections. For example, for evaluation data the accessions may be evaluated for an attribute (yield) at several sites giving an accession by site matrix.

Types of data

An examination of a typical accession by the attribute two-way table will show that the data for each attribute are often of different types. Data can be qualitative (either binary or multinominal) or quantitative. Binary data occur when an attribute is scored as having two states, such as being present or absent, red or white. Multinominal attributes can take more than two states and can be unordered, such as a series of colours, or ordered, such as a five-point rating for a disease score. Quantitative measures are counts, percentages or measurements. Data tables can consist of measurements which are of one type or various mixtures of types. When the attributes are of one type, the analyses are more tractable, but passport and characterisation data are unlikely to be uniform.

Transformation and standardisation of data

The information contributed to a measure of the differences of a number of accessions is affected by the scale of the measurement. A common procedure to eliminate scale differences, which also has the effect of making all attributes equally important to the analyses, is to 'standardise' each attribute in the normal way (the mean is subtracted from each value and then divided by the standard deviation). This is a two-step procedure whereby the attribute is first centered to zero mean and then scaled to unit variance. In this particular instance, which step is performed first is irrelevant, but in general this is not the case. Standardising each attribute after row centering produces a different result than that from standardising before row centering. Scaling by range or mean are also common. Careful consideration has to be given to centering and scaling and how they affect the analyses.

Measures of association or proximity measures

Numerical measures of likeness between each pair of accessions form a symmetrical square matrix and are necessary for both classification and ordination techniques. They are often called proximity measures because they can be considered to provide a measure of distance or closeness in the multi-dimensional space referred to above. There are two fundamental types of proximity measures: similarity measures, such as the Pearson correlation coefficient, which are larger for accessions that are more similar for a set of attributes, and dissimilarity measures, such as the Euclidean distance, which are larger the more the accessions differ.

The first point to note is that similarity and dissimilarity measures always occur in pairs, referred to as companion or complementary (dis)similarity measures, that exploit the same information among the accessions (Gower, 1966). This is significant because many classification procedures require dissimilarity measures, while ordination procedures require similarity measures. Since one can always be calculated from the other, complementary classification and ordination (that is, a pattern analysis) can be achieved. While it is always possible to calculate the complementary proximity measure, what these measures are is not always obvious. DeLacy et al. (1990) list some of the complements: the unstandardised squared Euclidean distance is the complement of the covariance matrix, the standardised squared distance is the complement of the product moment correlation matrix, and the squared Euclidean distance on ranked data is the complement of the rank correlation matrix.

The second point is that there is a general form of the association measure. This form, although it is formulated in the traditional manner for qualitative attributes, if used for binary or multi-state attributes calculates a measure which can be recognised as a traditional measure defined for these attribute types. The standard form defines a measure between two accessions for one attribute. A weighted average is then produced over all attributes, although many common measures are the result of an unweighted average. The squared Euclidean distance between any two accessions is the unweighted average over all attributes of the squared pairwise difference between the two accessions. If this measure is applied to binary attributes, then the complement of the simple matching coefficient, a traditional similarity measure for binary data, is produced. The averaging process enables missing data to be accommodated. If one and/or another of the scores for an attribute are missing for a comparison between two accessions, that difference is excluded from the average.

The third point is that association measures among the attributes can be found by simply transposing the original data matrix. An association measure between two attributes derived from scores of all the accessions provides a measure of how (dis)similar the attributes were in discriminating among the accessions. Classification and ordination analyses of attributes describe relationships among all the attributes. Information of this type is relevant for describing the collection and forming a core collection.

EXPERIMENTAL DATA AND STATISTICAL PROCEDURE

The Tuxpeño collection in the Maize Germplasm Bank at the Centro International de Mejoramiento de Maíz y Trigo (CIMMYT) includes 848 accessions that have been evaluated in replicated field experiments. Of these, 175 have been selected based on lodging and adaptation for further evaluation

(Taba et al., 1992). These accessions were collected in two ecogeographical regions: 58 accessions from dry ecologies (< 1000 m above sea level) and 48 from wet ecologies (< 1000 m above sea level). The remaining 69 mixed-ecology accessions include materials from both ecogeographical regions, as well as some mid-altitude accessions. The 175 accessions were planted using a generalised lattice design with two replicates at two locations (Poza Rica and Tlaltizapan) in Mexico in 1991. The morphological and agronomic traits (attributes) recorded were silking and anthesis dates, plant and ear height, senescence (number of days to ear leaf senescence after silking), grain moisture (%), shelling (%), grain yield, ear length, ear diameter, kernel length, and root and stalk lodging. The last two traits were not included in the analysis.

Statistical analyses involved two stages: a classification to determine whether the accessions could be regarded as consisting of a number of partially dissociated groups (cluster analysis), and an ordination study of those accessions to examine their spatial relationships (PCA). Cluster analysis was performed using ISS and complete, single and average linkage strategies. One important objective was to determine the optimum number of groups obtained from the cluster analysis. In this study, the cutting point was based on the usefulness of the subgroups resulting from it. In general, we were conservative and used small subgroups from the dendrogram such that accessions within subgroups were homogeneous and non-diverse, phenotypically speaking. Thus, at least one or two accessions were randomly selected from each subgroup.

The results of the classification analysis were tested (and confirmed) with those obtained from PCA in a stepwise manner. Several accessions selected on the basis of dendrograms from cluster analysis were included in the following step for further analysis. At each step, about half of the accessions, or fewer, were advanced to the next step. The process ended when about 27% of the original collections (175) were retained in the core collection.

RESULTS

The aim was to determine, among the 175 accessions, those with good agronomic type that will represent, as much as possible, the existing morpho-agronomic diversity. This was done in steps to ensure that accessions were carefully selected on the basis of results obtained from classification and ordination at each step.

Combined analysis of variance across locations showed that genotype x environment interaction was significant ($P < 0.005$) for some attributes. Classification and ordination of the 175 Tuxpeño accessions were conducted for each location and across locations. The resulting subgroups were similar in both cases. Therefore, further analyses were conducted only on simple (or adjusted) means across locations. Clusters obtained from ISS and complete, single and average linkage methods were similar; therefore, only clusters of accessions based on ISS are presented. The data were standardised because attributes were measured on different scales.

Step 1

Cluster analysis of the 175 accessions using the ISS strategy placed most dry-ecology and wet-ecology accessions in two major subgroups (dendrogram not shown). The 69 mixed-ecology accessions were more or less evenly distributed between these two subgroups. This result indicates that dry-ecology and wet-ecology Tuxpeño accessions had distinct morphological and agronomic traits. On the other hand, the set of mixed-ecology accessions did not seem to constitute a separate set.

As mentioned before, two reasons for using ordination techniques are to study diversity of traits among accessions and to complement and confirm the grouping obtained through cluster analysis. In PCA, these objectives are achieved by spatial representation (diagram) of the accessions in the first two or three principal components axes.

Plotting of the first two principal components indicated that dry-ecology and wet-ecology Tuxpeño accessions had distinct agronomic and morphological attributes, although they overlapped to some degree (*see* Figure 1). Most dry-ecology accessions had negative first principal component scores. Accessions in the wet-ecology group seemed to be more diverse than their dry-ecology counterparts. Tuxpeño accessions in the mixed-ecology category were spread out among the other two groups and were also diverse. These results confirmed the two major subgroups obtained by cluster analysis. The first three principal components accounted for 60%, 14% and 11% of total variability, respectively.

Figure 1 Principal components analysis ordination of 175 accessions in a maize Tuxpeño race collection

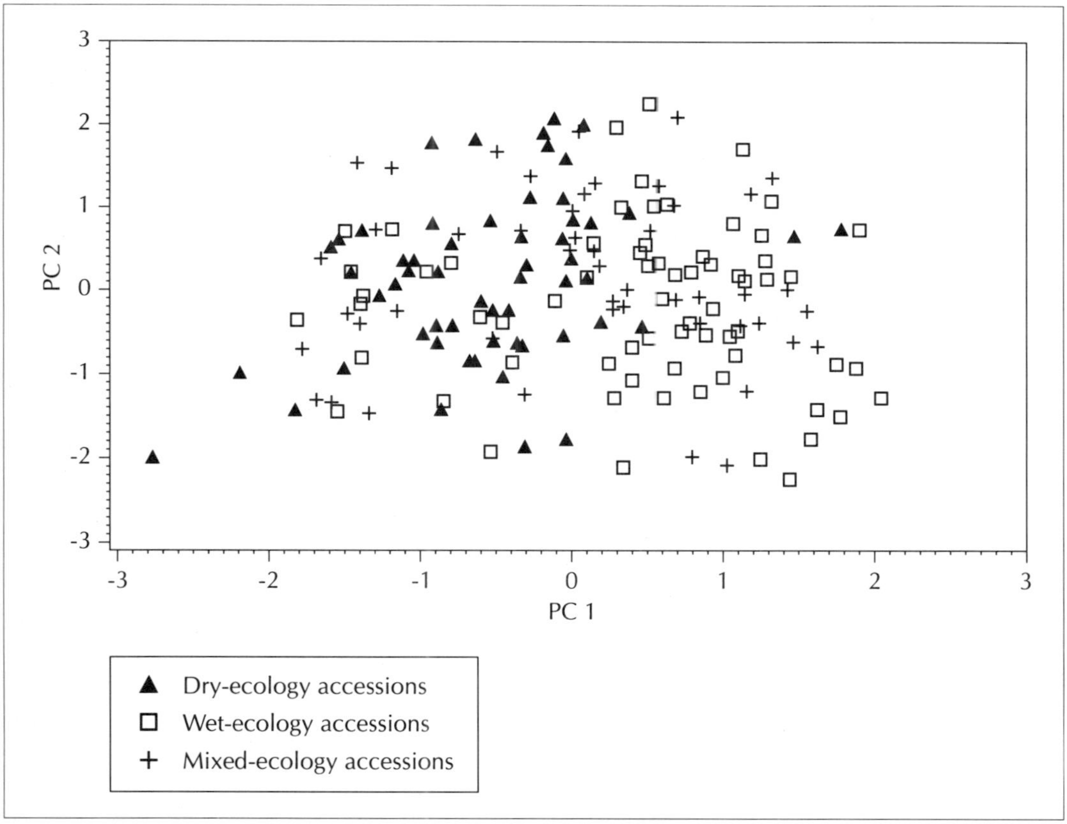

The magnitude of the eigenvectors indicates the importance of the kth attribute to one particular principal components axis. For example, traits separating accessions along the first principal components axis are (loads in parenthesis): silking (0.41), anthesis (0.42), plant height (0.40), ear height (0.40), grain moisture (0.36) and ear length (0.32). Traits that separate accessions along the second component are grain yield (0.74), senescence (0.50) and shelling (%) (0.30).

A total of 80 accessions was selected and advanced to the next step, where further reduction of the number of accessions will be performed based on recurrent results of classification and ordination. Of the 80 accessions, 26 are dry-ecology, 30 are wet-ecology and 24 are mixed-ecology accessions.

Step 2

The PCA of 80 accessions showed that the dry-ecology and wet-ecology Tuxpeño accessions had distinct agronomic and morphological traits (*see* Figure 2). Mixed-ecology accessions were spread out between the dry-ecology and wet-ecology subgroups.

Figure 2 **Principal components analysis ordination of 80 accessions in a maize Tuxpeño race collection**

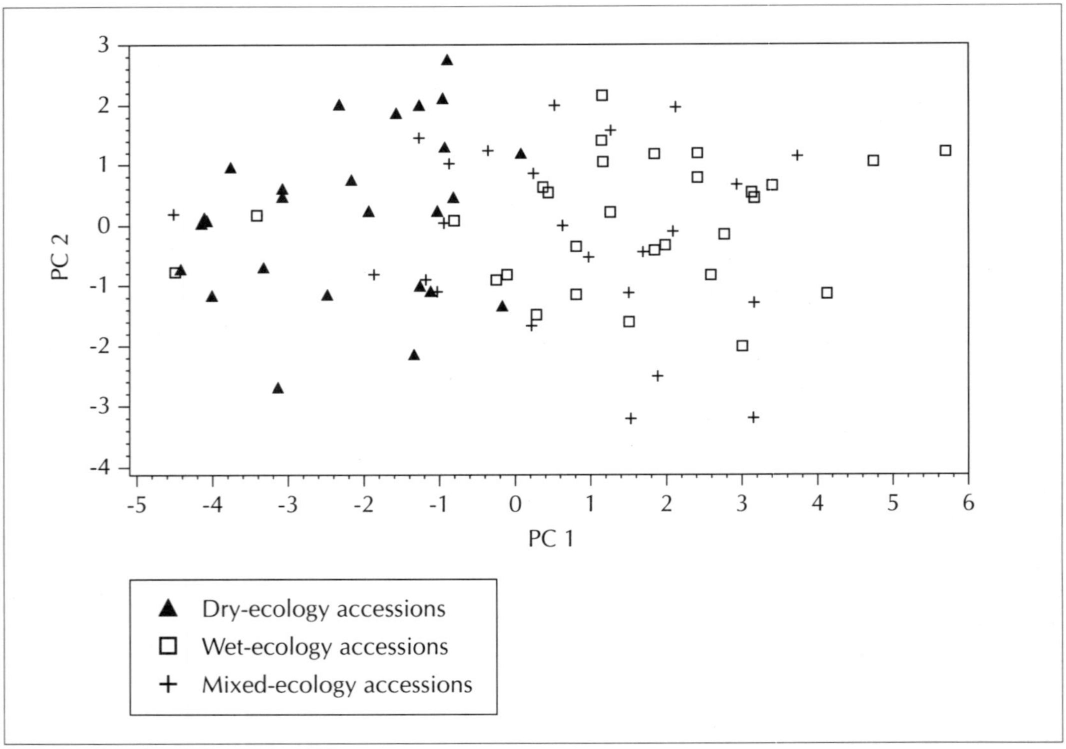

The first three principal components axes accounted for 53%, 15% and 11% of total variability, respectively. In general, PCA results confirmed those obtained by the ISS cluster analysis. Traits affecting accessions along the first principal components axis were the same as those in step 1: silking (0.40), anthesis (0.39), plant height (0.38), ear height (0.39), grain moisture (0.36) and ear length

(0.30). Traits that separated accessions along the second principal components axis were grain yield (0.41), senescence (0.22) and shelling (%) (-0.47).

The dendrogram illustrating the ISS cluster analysis of 80 accessions is shown in Figure 3. The last two branches of the dendrogram represent two major subgroups of accessions. One subgroup (cluster 1) contains 22 accessions: 14 wet-ecology and eight mixed-ecology accessions. The other subgroup includes all dry-ecology accessions as well as the remaining wet-ecology and mixed-ecology accessions. However, the final cutting point was set further down into the branches of the dendrogram where more homogeneous and less diverse clusters of accessions were found. The distribution of accessions within different race categories shows that most of the 22 accessions in cluster 1 are Tuxpeño, Tuxpeño 8, Tuxpeño-Tepecintle and Tepecintle-Tuxpeño (*see* Table 1). Most of the 14 accessions in cluster 2 are Tuxpeño, Tuxpeño-Vandeno and other types of Tuxpeño. The majority of the accessions in clusters 3 and 4 are from Tuxpeño-Vandeno and Tuxpeño-Olotillo, respectively. A total of 48 accessions was selected to form the core.

Table 1 **Distribution of 80 maize Tuxpeño race accessions characterised by race classification in each cluster**

Race classification		Cluster			
Primary	**Secondary**	**1**	**2**	**3**	**4**
Tuxpeño		6	4	1	4
Tuxpeño 8		5	0	2	6
Tuxpeño	Tepecintle	4	0	1	1
Tepecintle	Tuxpeño	4	0	1	0
Tuxpeño	Vandeno	1	4	11	0
Celaya	Tuxpeño	0	0	3	0
Tuxpeño	Olotillo	1	1	0	10
Tuxpeño	Conico Norteno	0	2	0	0
Tuxpeño	Other	1	3	1	3

Source: Taba et al. (1992)

Step 3

The core collection of the Tuxpeño racial complex comprises the 48 accessions (19 dry-ecology, 13 wet-ecology, and 16 mixed-ecology) selected in step 2. This represents 27% of the 175 accessions.

The PCA of the 48 accessions (*see* Figure 4) showed the distinct morphological and agronomic attributes of wet-ecology and dry-ecology accessions included in the core. As expected, mixed-ecology accessions overlapped both subgroups.

A discriminant analysis was conducted using the two ecogeographical subgroups (wet-ecology and dry-ecology) and the mixed-ecology accessions as classification criteria. Posterior probability of membership based on morpho-agronomic traits indicated that, of the original classification into the three groups, all dry-ecology accessions were correctly classified, 75% of the wet-ecology accessions were correctly classified and 25% were misclassified as mixed-ecology accessions, and 7% of the mixed-ecology accessions were classified as dry-ecology and 29% as wet-ecology.

Figure 3 **Dendrogram from the classification of 80 maize Tuxpeño race accessions using ISS or Ward's strategy**

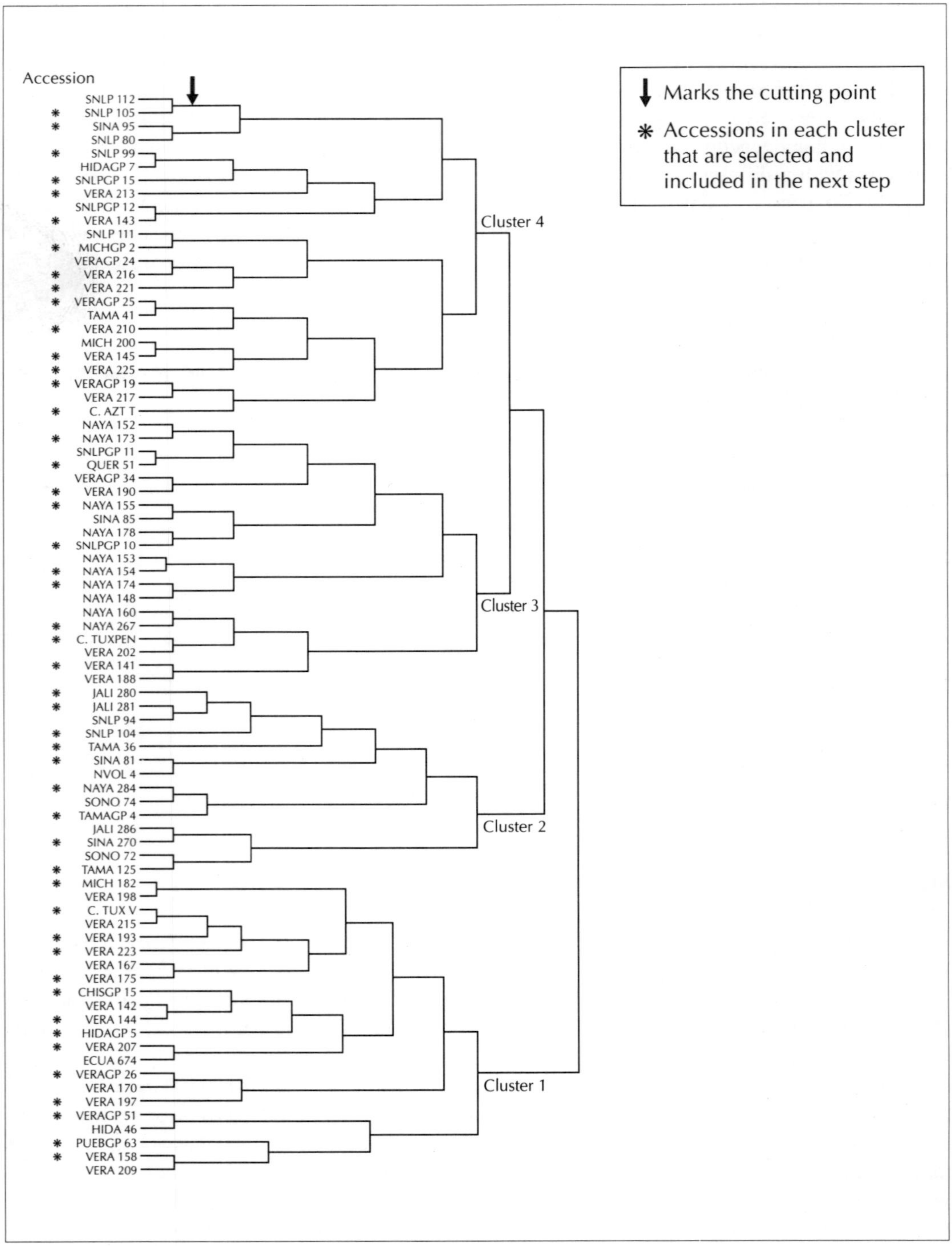

Figure 4 **Principal components analysis ordination of 48 accessions of a maize Tuxpeño race collection**

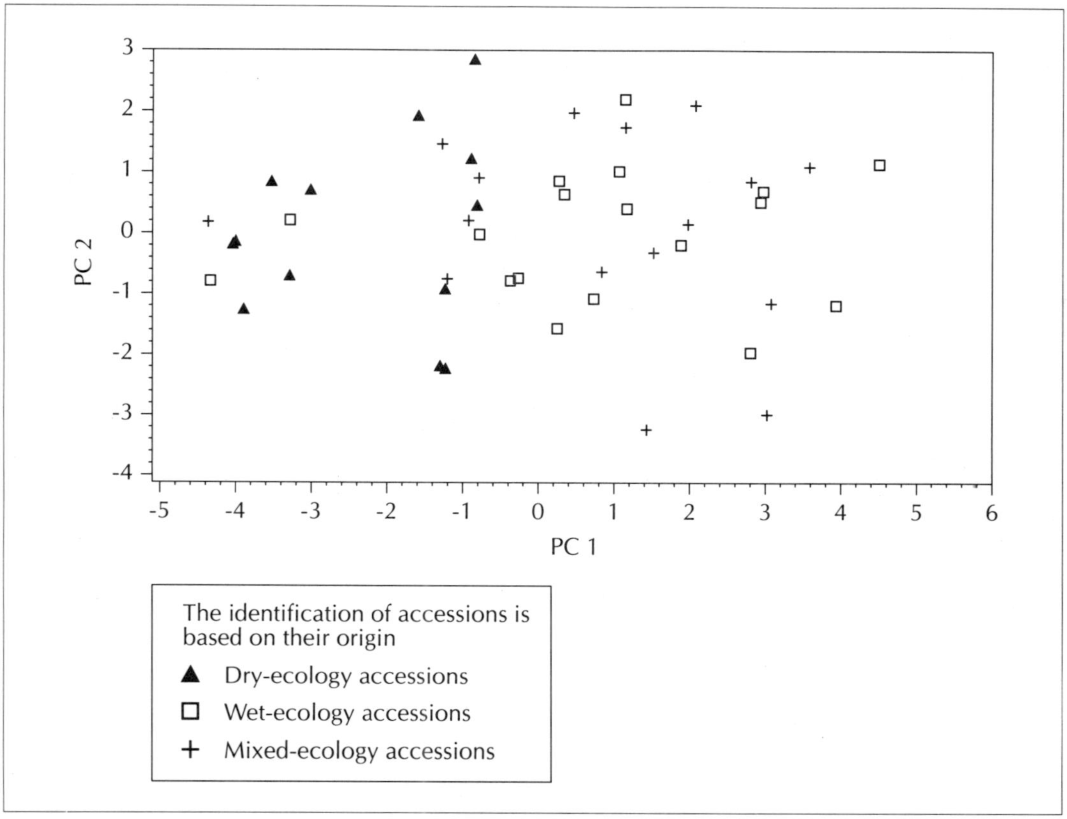

The canonical discriminant analysis provides a graphic output in fewer dimensions, illustrating the existence of morpho-agronomic subgroups. Good discrimination between 19 dry-ecology accessions and 13 wet-ecology accessions was obtained by plotting the first two canonical variables (*see* Figure 5).

DISCUSSION

Classification methods used in conjunction with ordination techniques were applied to a multivariate data set consisting of 11 morpho-agronomic attributes measured on 175 Tuxpeño accessions. Two different morpho-agronomic groups of accessions (dry-ecology and wet-ecology) were confirmed. Classification and ordination analyses were very accurate in discriminating by origin of accession. This agrees with results obtained by Peeters and Martinelli (1989), who suggested that origin is a simple and effective way to partition variation in germplasm collections. Mixed-ecology accessions form another subgroup that overlaps the others. The sequence of complementary classification-ordination multivariate analyses has proved useful in selecting accessions for the core collection.

Figure 5 **Canonical discriminant analysis of 48 accessions in a maize Tuxpeño race collection**

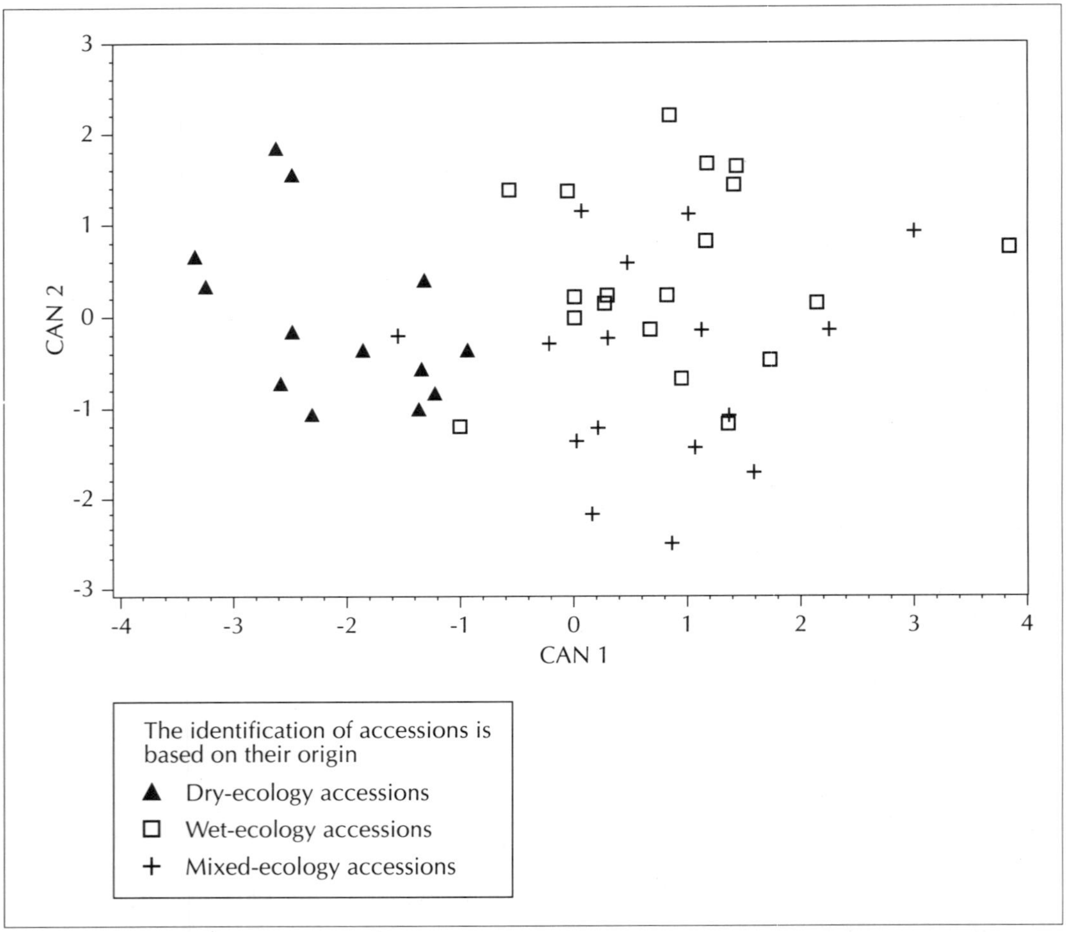

A system for forming core collections should also include identifying accessions based on molecular marker data. Clearly, any system for forming core collections of a germplasm collection should be confirmed by both biochemical and genetic methods (such as allozyme analysis) for estimating changes in allelic frequencies caused by the process of core collection formation. These methods should be integrated for estimating the loss of genetic diversity when forming the core collection and how well the core collection represents the range of diversity of the whole germplasm collection. Drastic changes in allelic frequencies would indicate that a different approach for selecting accessions for the core should be investigated.

A significant problem in the process of forming core collections arises when more than one classification has been proposed for the same set of accessions. This may be because classification is based on: different sets of attributes, the same set of attributes but groups are formed using different classificatory techniques, or different sets of attributes and different classificatory techniques. This is also a problem in biological systematics and gave rise to the concept of taxonomic congruence (Mickevich, 1978), which is the degree to which classifications of the same items produce the same groupings. One implication of this concept is related to the problem of finding attributes that show superiority over the others as indicators of taxonomic relationships among items. Crisci (1984) concluded that classification reflects the evolution that has occurred in certain sets of attributes; therefore, it is expected that when additional attributes are included in the study, the group originally formed will be altered. Peeters and Martinelli (1989) concluded that, in classification, the choice of attributes, as well as the algorithm used, affects the final outcome.

Simultaneous analyses including classification and ordination techniques have been proposed for dealing with multi-attribute genotype x environment data (Basford et al., 1991). Generalisations of cluster analysis (mixture maximum likelihood) and PCA for three-way data (three-mode PCA) have been described by Basford et al. (1990). We have investigated the usefulness of these techniques for analysing accessions x attributes x environments data in forming core collections and the results will be published elsewhere.

One of the main objectives of this study was to reduce the number of existing accessions in the Tuxpeño race collection by forming a core collection biased toward better agronomic types. This is a departure from the original proposal of a core collection based on a gene bank. However, our study has extended that original idea (Mackay, *Chapter 4.4, this volume*) by assembling a specific core collection that offers maize breeders in particular and national programmes in general a valuable opportunity to utilise it directly as a source of breeding material.

References

Basford, K.E., Kroonenberg, P.M., DeLacy, I.H. and Lawrence, P.K. 1990. Multi-attribute evaluation of regional cotton variety trials. *Theoretical and Applied Genetics* 79: 225-34.

Basford, K.E., Kroonenberg, P.M. and DeLacy, I.H. 1991. Three-way methods for multi-attribute genotype x environment data: An illustrated survey. *Field Crop Research.* 27: 131-57

Bird, R.Mck. and Goodman, M.M. 1978. The races of maize. V. Grouping maize races on the basis of ear morphology. *Economic Botany* 31: 471-81.

Brown, A.H.D. 1989. The case for core collections. In Brown, A.H.D., Frankel, O.H., Marshall, D.R. and Williams, J.T. (eds) *The Use of Plant Genetics Resources.* Cambridge, UK: Cambridge University Press.

Byth, D.E., Eisemann, R.L. and DeLacy, I.H. 1976. Two-way pattern analysis of a large data set to evaluate genotypic adaptation. *Heredity* 37: 215-30.

Crossa, J., Taba, S., Eberhart, S. and Bretting, P. 1994. Practical considerations for maintaining germplasm in maize. *Theoretical and Applied Genetics* (in press).

Crisci, J.V. 1984. Taxonomic congruence. *Taxon* 33(2): 233-39.

DeLacy, I.H. and Cooper, M. 1990. Pattern analysis of the regional variety trials. In Kang, M.S. (ed) *Genotype-by-Environmental Interaction and Plant Breeding.* Baton Rouge, Louisiana, UK: Louisana State University Agricultural Center.

DeLacy, I.H., Cooper, M. and Lawrence, P.K. 1990. Pattern analysis of the regional variety trials: Relationship among sites. In Kang, M.S. (ed) *Genotype-by-Environmental Interaction and Plant Breeding.* Baton Rouge, Louisiana, UK: Louisana State University Agricultural Center.

Digby, P.G.N. and Kempton, R.A. 1987. *Multivariate Analysis of Ecological Communities.* London, UK: Chapman and Hall.

Fisher, R.A. 1936. The use of multiple measurements in taxonomic problems. *Annals of Eugenics* 7: 179-88.

Frankel, O.H. 1984. Genetic perspective of germplasm conservation. In Arber, W.K., Llimensee, K., Peacock, W.J. and Starlinger, P. (eds) *Genetic Manipulation: Impact on Man and Society*. Cambridge, UK: Cambridge University Press.

Frankel, O.H. and Brown, A.H.D. 1984. Current plant genetic resources — a critical appraisal. In *Genetics: New Frontiers*. (vol. 4). New Delhi, India: Oxford and IBH Publishing Co.

Goodman, M.M. and Bird, R.Mck. 1977. The races of maize. IV. Tentative grouping of 219 Latin American races. *Economic Botany* 31: 204-21.

Gower, J.C. 1966. Some distance properties of latent root and vector methods used in multivariate analysis. *Biometrika* 53: 325-38.

Lance, G.N. and Williams, W.T. 1967. A general theory of classificatory sorting strategies. I. Hierarchical systems. *Comput. J.* 9: 373-80.

Mickevich, M.F. 1978. Taxonomic congruence. *Systematic Zoolology* 27: 143-50

Peeters, J.P. and Martinelli, J.A. 1989. Hierarchical cluster analysis as a tool to manage variation in germplasm collections. *Theoretical and Applied Genetics* 78: 42-48.

Perry, M.C. and McIntosh, M.S. 1991. Geographical patterns of variation in the USDA soybean germplasm collection. I. Morphological traits. *Crop Science* 31: 1350-55.

Perry, M.C., McIntosh, M.S. and Stoner, A.K. 1991. Geographical patterns of variation in the USDA soybean germplasm collection. I. Allozyme frequencies. *Crop Science* 31: 1356-60.

Singh, S.P., Gutierrez, J.A., Molina, A., Urreas, C. and Gepts, P. 1991. Genetic diversity in cultivated common bean. II. Marker-based analysis of morphological and agronomic traits. *Crop Science* 31: 23-29.

Sneath, P.H.A. and Sokal, R.R. 1973. *Numerical Taxonomy: The Principles and Practice of Numerical Classifications*. San Francisco, California, USA: W.H. Freeman and Co.

Taba., S., Pineda, F. and Crossa, J. 1992. Forming core subsets from Tuxpeno race complex. In *Abstracts of the First International Crop Science Congress*. Ames, Iowa, USA: Crop Science Society of America/ Iowa State University.

Williams W.T. (ed). 1976. *Pattern Analysis in Agricultural Science*. Amsterdam, Netherlands: Elsevier.

Part 3

USE OF DIFFERENT KINDS OF DATA IN DEVELOPING CORE COLLECTIONS

Core Collections of Plant Genetic Resources
Edited by T. Hodgkin, A.H.D. Brown, Th.J.L. van Hintum and E.A.V. Morales

A Wiley-Sayce Publication

3.1

The combined use of agroecological and characterisation data to establish the CIAT *Phaseolus vulgaris* core collection

J. TOHME, P. JONES, S. BEEBE and M. IWANAGA

Abstract

A core collection was formed from the 24 000 accessions which are available from the world common bean (*Phaseolus vulgaris* L.) collection held in CIAT, Cali, Colombia. A baseline of 10% of the reserve collection was set for representation by countries in the primary centres, but this was adjusted up or down according to specific situations such as duplication of accessions. Subsequently, a three-step process was followed. First, regions were prioritised, giving greater weight to traditional bean growing areas. Second, germplasm was classified according to agroecological origin. Four environmental parameters were identified as critical: length of growing season, photoperiod, soil type and moisture regime, with 3, 2, 3 and 3 levels, respectively. All possible combinations of these parameters yielded 54 possible environments, of which 49 were actually represented in the germplasm. Another minor class was created to represent cold environments of very long season. By using map coordinates of the germplasm collection sites, accessions were matched to their respective environmental class. The third criterion used was based on morpho-physiological data of growth habit and grain colour and size. Primitive types were weighted more heavily than modern commercial types. Having weighted the representation as such, a random selection was practised within the agroecological classes. A total of about 1000 accessions were identified from primary centres, and an additional 300 were chosen from secondary centres, in addition to 40 cultivars, 40 bred lines and 40 genetic stocks.

International germplasm collections have played a major role in securing genetic diversity and promoting its use. However, curators and users of major international germplasm banks are already facing the burden of properly conserving, characterising and evaluating a large number of accessions. The need for a reappraisal of plant genetic resources has been recognised (Frankel and Brown, 1984). Lack of easy access to germplasm and the sheer numbers of accessions of rice, wheat, maize and bean in their respective collections make it an almost impossible task to evaluate the germplasm properly. Passport data are lacking for at least 65% of all accessions in the world's germplasm banks, and some 80-95% of them lack characterisation or evaluation data (Plucknett et al., 1987).

The formation of a core collection, a recently proposed concept (Frankel and Brown, 1984; Brown, 1989a, b), is aimed at promoting the use of the germplasm and facilitating the study of its genetic diversity. Researchers working on problems that normally require the screening of a large number of accessions will be interested in looking first at a small number of entries but with maximum variability. Thus, a core collection would be the 'first look' at the germplasm collection for a researcher to identify quickly the desired trait. It will eventually serve as a guide to other sources held in the reserve collection. It is expected that the core collection will be requested by major national germplasm banks interested in widening their genetic variability without the burden of handling huge numbers.

The common bean (*Phaseolus vulgaris* L.) collection of some 24 000 accessions stored at Palmira, Colombia by the Centro Internacional de Agricultura Tropical (CIAT) is by far the largest collection of any food legume in the world. A common bean core collection, once established, will facilitate the exchange of germplasm with national gene banks and the characterisation of traits such as abiotic stress tolerance. The importance of properly maintaining and characterising the whole collection is well recognised and will still be the main responsibility of the Genetic Resources Unit at CIAT.

Keeping in mind that the germplasm bank does not and will not have access to the original variability (that is, in its original context), it is thus of paramount importance to sample the existing populations to try to preserve as much of the real genetic diversity as possible throughout the whole process of germplasm conservation. It is also considered important to promote the use of the germplasm in various breeding programmes and to facilitate the study of the genetic diversity of the crop. This prompted us to use a systematic method of establishing a core collection, with major considerations on how to sample genetic diversity from the large collection.

This chapter presents the methodology used to develop a bean core collection. The proposed model uses both evolutionary and agroecological approaches. The CIAT bean core collection will consist of a representative sample of the whole genetic variability of the germplasm of *P. vulgaris*. Only the work on the cultivated common bean will be discussed here, although a core collection for the whole *Phaseolus* genus, domesticated as well as wild, is being considered.

MODEL SCHEME FOR WELL-DOCUMENTED COLLECTIONS

A synthesis between genetic and ecogeographical approaches was followed, taking into account the structure of the germplasm collection and the agroecological areas of origin. The selection model takes six main points into account:

- genetic diversity is not randomly distributed
- crop domestication is important
- ecological adaptation is important
- local selection and preference play a role
- there are patterns of bean dissemination from the primary centres
- the collection has intrinsic characteristics

Based on archaeological, botanical and molecular evidence, two main independent domestication centres of *P. vulgaris* in the Mesoamerican and Andean regions have been proposed (Gepts et al., 1986;

Debouck and Tohme, 1989; Gepts and Debouck, 1991). Exactly when and where the domestications took place and how they proceeded over time is still to be determined. The present data suggest, however, that, like other crops, the genetic diversity is not randomly distributed. In the case of the common bean the multiple domestication centres have resulted in two main gene pools, a Mesoamerican one and an Andean one (Gepts, 1988; Sprecher, 1988; Khairallah et al., 1990; Gepts and Debouck, 1991; Singh et al., 1991a, b). Besides major differences at the molecular level, some evidence suggests that these two major gene pools may have different responses to ecological environments (Ghaderi et al., 1982). This suggests that adaptation to specific environmental niches played an important evolutionary role. For example, initial screening for the capacity to yield under conditions of low soil phosphorus availability indicates that wild *P. vulgaris* material is sensitive to phosphorus deficiency, in contrast to landraces from the same region. This observation suggests that the tolerance of cultivated *P. vulgaris* to low phosphorus might be the result of post-domestication selection (Beebe et al., 1992).

Modern bean landraces, although diverse and rich in genetic variability as attested by the variability in morphology, seed size and seed colour, have been derived from a small portion of the original variability in wild material (Debouck and Tohme, 1989). Such a reduction in genetic variability is known as the 'founder effect' (Ladizinsky, 1985).

The number of examples of interesting traits that are present only in original wild populations is increasing: resistance to bruchids (Schoonhoven et al., 1983; Osborn et al., 1986; Romero Andreas et al., 1986); and restriction of nodulation by some groups of *Rhizobium* (Kipe-Nolt et al., 1992). Such data indicated the importance of including wild materials in a core collection. However, since many breeders and agronomists with a very practical interest in a core collection may not wish to work with wild beans precisely because of their non-domesticated characteristics, a separate core was created specifically for wild materials.

The other consideration to be taken into account is related to seed dissemination. After the Spanish conquest, the common bean moved quite rapidly to Europe, Africa and Asia; these areas, together with the Caribbean, the USA and non-Andean South America, will be considered secondary centres of diversity. Routes of introductions have been determined using historical data and molecular markers (Martin and Adams, 1987a, b; Gepts, 1988; Gepts and Bliss, 1988; Gepts et al., 1988).

The germplasm from the secondary centres is related to the two main gene pools. However, the effect of new selection pressures and possible inter gene pool recombination on genetic variability is still to be determined. Preliminary data suggest that germplasm from the secondary centres is highly duplicative of that from primary centres. It was decided that germplasm from countries outside the primary centres would be sampled at a lower weight.

The definition of primary and secondary centres of diversity is, to some extent, arbitrary. It is clear that domestication sites, wherever they may be, constitute primary centres. Likewise it is clear that the regions of introduction in the post-Columbian age, such as Africa, are secondary centres. Regions of pre-Columbian dispersion, such as the American south-west and non-Andean South America, are more difficult to classify. Such areas may be the focus of special studies in the future, but for present purposes they are not included as primary centres. In the context of this study, Mexico, Central America and the Andean countries will be considered to constitute the primary centres.

STATUS OF THE BEAN GERMPLASM COLLECTION AT CIAT

Faced with the urgency to acquire a bean germplasm collection, especially with respect to landraces under threat of genetic erosion, germplasm acquisitions and collections were conducted in the past in a rather unorganised and opportunistic fashion. Germplasm from non-primary centres was over-represented, while germplasm from important areas in terms of genetic variability remained uncollected. More recently, a more systematic strategy of acquisition and collection was developed. Since 1987, CIAT and the International Board for Plant Genetic Resources (IBPGR) have jointly carried out collections to fill some major gaps (Debouck, 1987, 1988). Areas with rich bean genetic diversity, such as Peru, were targeted to enhance representation in the collection. Thus, the present CIAT collection is a truly global collection in comparison with all other major bean collections, in which germplasm from the Andean regions is under-represented. The present collection at CIAT can still be considered as over-representative of certain regions, having many duplicates. Half the accessions at CIAT come from non-primary centres. Germplasm from countries of recent plant introduction, such as Turkey and Iran, account for almost the same number of accessions as Peru. Moreover, half the accessions from these two countries belong to the same bean class in terms of seed colour and size. For this reason a sampling strategy using a random or proportional model based on origin would not adequately sample the genetic diversity.

The world collection of *P. vulgaris* at CIAT has already been screened for traits such as growth habit, seed size, seed colour and resistance to bean common mosaic virus (BCMV), *Empoasca* species, seed storage insects and bean common bacterial blight (BCBB). An attempt was made to screen the germplasm for photoperiod response (White and Laing, 1989). The data available indicated that these resistances were too scarce or too poorly distributed in the collection to be of use in the stratification scheme. Only growth habit, seed size and seed colour were used in the selection scheme for the core collection.

As mentioned earlier, it was decided to place special emphasis on germplasm originating from the primary centres. A major effort was thus devoted to compiling the available passport data of wild and cultivated bean germplasm, although such data are extremely variable with respect to precision. Priority for selection of the core collection was given to accessions with more complete passport data. So far, the accessions from Argentina, Bolivia, Colombia, Costa Rica, Ecuador, El Salvador, Guatemala, Honduras, Mexico, Nicaragua and Peru have been documented for the first time and to the degree that available data permitted (*see* Table1). They account for 11 000 accessions, almost half the present CIAT collection. The same exercise was carried out for wild *P. vulgaris* (Toro et al., 1990).

The accessions with proper coordinates of the collection site were plotted on an ecological map at CIAT's Agroecological Unit to classify the germplasm based on ecological regions rather than political boundaries (*see* Figure 1). The germplasm was then sampled based on ecological regions, on knowledge of the richness of genetic variability inherent in a region and on available morphological and molecular data. The accessions from the secondary centres were handled separately.

AGROECOLOGICAL CLASSIFICATION

For classification of the environments in a region, the data must be uniformly comparable across the region. While for some areas great detail is available in the climate and soil data sets, this is not universally true. These analyses are therefore based on the extensive database of climate data of long-term monthly means compiled by CIAT and on simple characteristics of soils as defined by the Food and Agriculture Organisation (FAO) soil map of the world, with a scale of 1:5 000 000 (UNEP/GRID, 1988).

The CIAT climate database now contains data from over 17 000 stations throughout the tropics, over 8000 of these being in South America. In general, the distribution of climate stations follows that

Table 1 **Number of germplasm accessions from countries in the primary centres of diversity, their respective contribution to the core collection, and selected passport data**

Country	Accessions in CIAT's world collection	Accessions in core collection	Origin by province	Origin by county	Origin by coordinates	Bred lines	Collected in market
Argentina	170	15	65	65	65		
Bolivia	115	25	95	84	66		
Colombia	828	75	689	593	369	95	120
Costa Rica	270	10	114	121	72		
Ecuador	761	60	665	503	458		65
El Salvador	214	5	61	82	8		
Guatemala	2181	70	781	686	406	50	
Honduras	486	6	245	287	67		
Mexico	4027	400	3728	3502	3182	87	620
Nicaragua	284	6	179	179	94		
Peru	2081	350	1795	1577	1451	222	62
Total	11417	1022	8417	7679	6238	454	867

of population. In practical terms this means that the areas with least information are of little interest with regard to bean germplasm. These include the Orinoco and Amazon lowlands and the eastern plains of Colombia, areas in which native bean germplasm is scarce or non-existent.

Nevertheless, the distribution of climate stations does not match that of bean accessions and interpolated climate values were needed for spaces between the stations. All stations in the database have measured rainfall records, about half have measured mean temperatures and rather less have measured diurnal temperature range. As the altitude of a collection site strongly affects temperature, the altitudes were taken from a digital altitude model (NOAA, 1988). This dataset provides the most common altitude (modal) inside a set of grid cells (pixels) dimensioned 10 minutes of latitude by 10 minutes longitude; these are approximately 18 x 18 km at the equator.

The method of interpolating the data to provide a representative value for each grid square was to search the database for the five stations closest to the centre of the grid square which possessed measured data for the variable. Hence the precision of the rainfall interpolation is greater than that for temperature, which in turn is greater than that for diurnal temperature range. In some cases all five stations fell within the grid square, while in others the data were wider spaced. Usually, the five stations were distributed around the estimation point, but in some cases (such as on a coastline) all would be to one side. The selection and interpolation procedure was automatic. The average of the five stations was calculated using the inverse of the square of the distance to the pixel centre as a weighting factor. This ensures that the interpolated surface passes smoothly through the exact values at the site of a station (Jones et al., 1991).

The effects of elevation on mean temperature were eliminated by correcting all the station temperature data to sea level value using a lapse rate model developed at CIAT and based on data from Riehl (1979). The interpolated temperature data were then corrected using the same model to the modal altitude from the altitude model.

Figure 1 **Bean germplasm distribution in Jalisco and Nayarit, Mexico, superimposed over the ecological classification**

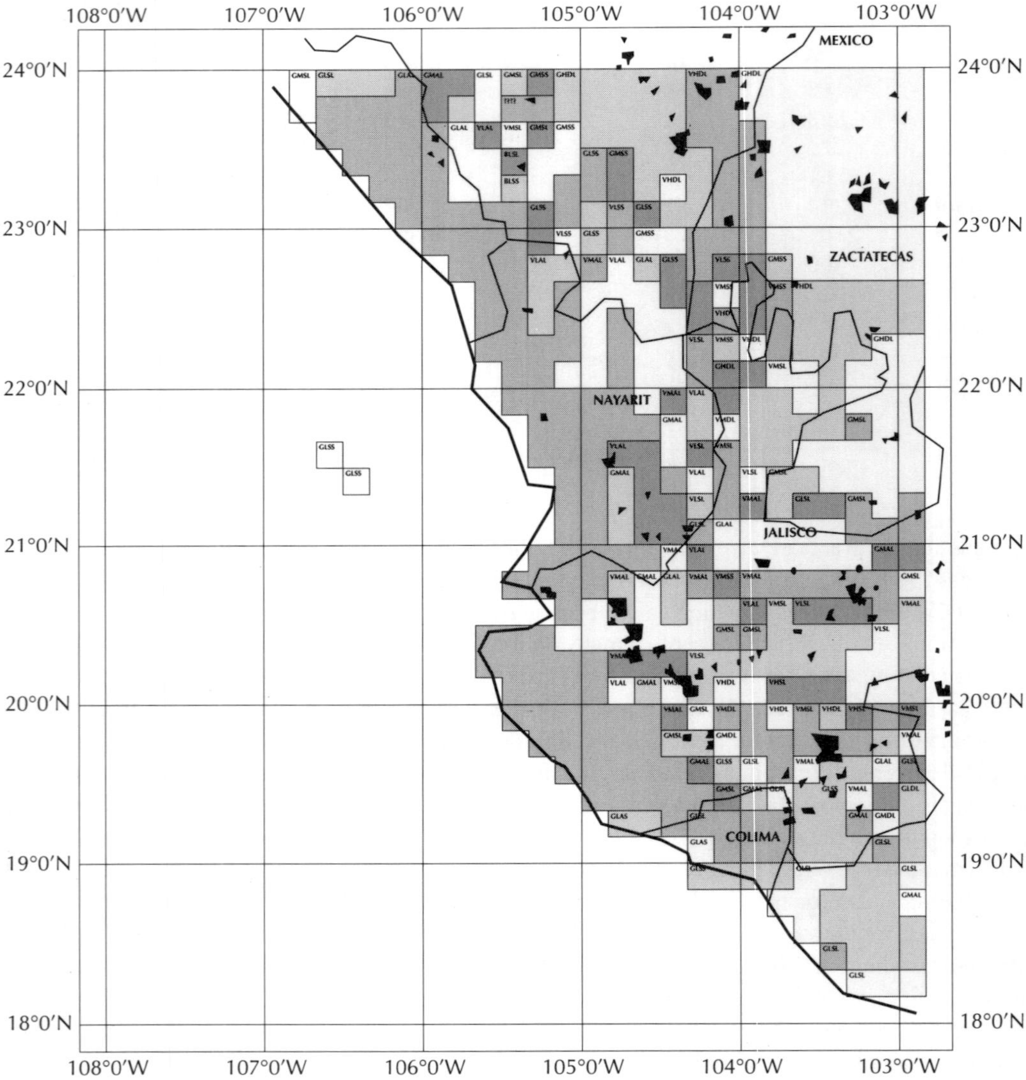

Black dots or clusters = bean collection sites
Shaded areas = different ecological classification zones

The interpolated climate and soil files were used to construct a classification file based on the 10-minute grid. The classification was developed in collaboration with members of the CIAT Bean Team and the Genetic Resources Unit. Four factors were chosen with a limited number of levels in order to keep the classification simple enough to be used: soil characteristics; expected phenology during the growing season; water stress; and photoperiod at flowering (*see* Table 2). All these factors require the use of the modal or most probable data from each pixel. Thus the classification gives an idea of the type of area and cannot be used to identify precisely the characteristics of a collection point for an individual accession.

Table 2 Factors used for the agroecological classification of continental South America

Factor	Description
Soil	Good mineral soil Poor mineral soil Volcanic soil
Phenology	75-85 days: lowland type 85-110 days: mid-altitude Over 110 days: highland
Water stress	Less than or equal to 10 stress days More than 10 stress days No season found (arid)
Photoperiod	Less than or equal to 13 hours of daylight More than 13 hours of daylight

The soil factor was created by reclassifying the 133 soil mapping unit codes from the digitised version of the FAO soil map. This version has a much higher precision than the 10-minute grid and therefore the centre point of the grid was taken as a point quadrant estimate of the most likely soil. The reclassification produced four soil classes (the last class was eliminated from the analysis):

- good mineral soils (good structure and drainage, and high base status)
- poor mineral soils (low base status and nutrient deficiencies)
- volcanic soils (Andosols)
- impossible soils (rock, salt, solonetz and solodic soils)

To calculate the other factors (phenology during growth, water stress and photoperiod at flowering), the long-term monthly mean data for rainfall and maximum and minimum temperature were interpolated to daily data using the Fourier algorithm described by Jones (1987). Daily evapotranspiration was estimated using the method described by Linacre (1977); the water balance was calculated using the algorithm WATBAL, modified from the method used by Reddy (1979) as reported in Jones (1987). Soil moisture-holding capacity, used for the water balance calculation, was estimated for representative profiles for each of the FAO soil map units.

The growing season was deemed to start when soil water content had exceeded 50% of potential for 5 days and to have finished when available soil water was less than 50% of potential for at least 8 consecutive days. The required growing period and time-to-flowering were calculated as a function of temperature based on data from Laing et al. (1984), ignoring photoperiod effects. The number of stress days was calculated as the number of days during the required growing period when actual soil moisture was less than 80% of soil water capacity. The photoperiod at the estimated time of flowering was calculated and recorded.

The possible combinations of all factor levels resulted in a potential of 54 distinct environmental classes (3 soils x 3 growing seasons x 3 moisture regimes x 2 photoperiods). A further class was added for a highland environment where the required phenology exceeded 365 days. This environment is extremely marginal and is classified as too cold. Of 4884 accessions, only 27 fall into this category.

The classification itself can be mapped accurately at many scales, not only for South America but also for Africa and Asia. This has important consequences for the development of core collections. Much experience has been gained in the use of the agroecological databases, and software now exists for assigning a germplasm accession to the relevant agroecological class if sufficient passport data points are available. Our approach points to the need for high quality, professionally recorded collection data on future accessions. To this end, the increasingly inexpensive Global Positioning Systems (GPS) could be an invaluable tool for germplasm explorations.

SELECTION SCHEME: A 'HYBRID' STRATIFIED-RANDOM MODEL

The model used takes into account the constitution of the collection, the agroecological classification and evolutionary knowledge. The selection was made in two phases. First, a baseline of 10% was used as the projected representation of each country in the primary centres. Depending upon specific factors or situations, the percentage was adjusted up or down according to a subjective weighted stratified model. Care was taken to ensure proper representation of the two major gene pools. Guatemala provides an example of a major adjustment of the percentage of representation in the core collection. The collection from Guatemala is believed to have been introduced into the CIAT gene bank several times and is thus highly repetitive, as confirmed by morphological observation. Thus the representation of Guatemalan accessions is strongly reduced in the core compared with the reserve collection.

Once the representation per country was determined, a three-step process was implemented to prioritise and classify accessions. First, an arbitrary weight was assigned to geographical areas delineated by similar adaptation zones within areas of the two major gene pools. Germplasm from areas with a long history of bean production were assigned a higher weight than germplasm from areas of recent introduction. For example, the dry region of central highland Mexico represents a sizable proportion of the national production and perhaps for this reason is well represented in the gene bank. However, extensive bean cultivation in this area is a phenomenon of only the past 50 years, and therefore materials from this area were given a lower weight.

The second step considered the classification of bean growing environments based on detailed agroecological data as described earlier. A complete map of the areas of interest in the continent was produced. Several 1:1 000 000 charts covering Mexico, Central America and the Andes were plotted, with the classification superimposed. The distribution of cultivated and wild accessions for which coordinates were available was plotted at exactly the same scale and projection (*see* Figure 1) and overlayed on the map. Although this step was not necessary to arrive at an agroecological classification *per se* of the germplasm, it was important in order to gain a visual appreciation of the distribution of collection points in relation to the agroecologically classified areas. In this way, we could readily see the relationship of collection points to physical and man-made features. It was observed that many points were clustered on major cities and towns. These accessions were assumed to be collections made in markets (*see* Table 1) and in this case the coordinates will not be representative of the agroecological conditions. Therefore these accessions were eliminated in the selection of the core.

Given the limited number of accessions with reliable coordinates, the germplasm database was subjected to a thorough investigation to identify geographic coordinates from other information. For example, if a passport collector had made a field collection in the neighbourhood of Arandas in Mexico and this city name was the only geographic indicator in the passport data, then the coordinates of the centre of Arandas would be entered in the database.

The map of the classified environments was then plotted on a 10-minute raster in order to relate it to the subset of germplasm accessions with coordinates of their site of origin. A FORTRAN programme was written to identify the corresponding pixel in the raster classification file for each accession. This provided us with the agroecological classification for each accession.

Of the 11 000 accessions, only 4800 had proper data. However, good coverage was obtained for Argentina, Bolivia, Colombia, Ecuador, Mexico and Peru, which represent the greater part of the primary centres. Table 3 presents a breakdown of the 2435 identifiable Mexican accessions by environmental classes.

The third level of stratification relied on morphophysiological data. Based on growth habit, seed size and seed colour data, typical landraces were given a higher weight than more 'modern' or

Table 3 Distribution over agroecological classes of the total Mexican germplasm and the accessions included in the core collection

			Soils					
			Good		Poor		Volcanic	
Phenology	Stress	Daylength	Total	Core	Total	Core	Total	Core
Lowland	Free	Short	13				41	10
		Long	95	19	18	6	12	4
	Stress	Short	67	7			15	3
		Long	104	15	2	1		
	Dry	Short	1	1				
		Long					2	1
Mid-altitude	Free	Short					74	4
		Long	54	6	9	2	175	33
	Stress	Short	94	16			8	2
		Long	80	9	4	3	143	17
	Dry	Short	24	2			3	
		Long	12	4			4	1
Highland	Free	Short					2	1
		Long	34	4			37	6
	Stress	Short	9	2			27	11
		Long	59	11	1		112	9
	Dry	Short	33	9				
		Long	931	86	1		135	19
Total			1610	191	35	12	790	121

commercial genotypes. The weights for these traits took into consideration the characteristic of each of the Mesoamerican and Andean gene pools. The final sampling was based on the different weights at each stratification level.

A preliminary comparison between isozymes in a sample of accessions from Peruvian provinces and those in the Peruvian accessions included in the core collection suggests that the sampling was valid (A.Valderrama and M. Iwanaga, unpubl.). The core collection now includes some 1000 accessions from countries in the primary centres and covers 49 out of the 55 possible ecological classes (*see* Table 4). Furthermore, 300 accessions from secondary centres have been selected. Since passport data are non-existent for most germplasm from secondary centres, a totally different approach was followed in its selection. As mentioned earlier, this will not be detailed here. Briefly, the broadest representation possible was sought within broad regions based on grain characteristics and plant habit. Additionally, 40 standard bred lines and 40 key landraces were included for purposes of comparison, as well as 40 genetic stocks based on isozymes, phaseolin, mtDNA or other molecular markers. Although this basic scheme was used to select the core collection, adaptations were made, depending upon the data available.We have noted some of these adaptations, and while these will be specific to the *Phaseolus* collections, we stress that researchers working with other species will encounter their own particular problems, according to the structure of their collections and the data at their disposal.

Table 4 **Number of ecological classes per country for the whole collection and the core collection from the countries in primary centres of diversity**

	Number of ecological classes	
Country	**Whole collection**	**Core collection**
Argentina	3	3
Bolivia	1	1
Colombia	20	13
Costa Rica	6	2
Ecuador	16	9
El Salvador	2	—
Guatemala	23	11
Honduras	7	1
Mexico	36	32
Nicaragua	7	1
Peru	21	13
Total	49	49

BEYOND THE CORE COLLECTION: THE NEED FOR AN INTEGRATED AND SYNTHETIC APPROACH TO BEAN BIODIVERSITY

The essential purpose of the core collection is to improve our understanding of the structure and distribution of genetic diversity so that researchers know where best to look for useful genes and, on finding multiple sources of a trait, which of the genes involved have the highest probability of being distinct and thus amenable to recombination. In serving this essential purpose, studies of the core collection will be undertaken which at the outset may appear academic but will contribute in the long term to the practical utilisation of germplasm.

The core collection will be characterised at morpho-physiological and molecular levels in order to study the genetic structure of *P. vulgaris* germplasm. Work is now being conducted on classifying the core collection for seed proteins such as lectin and phaseolin and on the use of restriction fragment length polymorphism (RFLP), random amplified polymorhpic DNA (RAPD) and hypervariable probes for fingerprinting the collection. Such characterisation will also serve to identify allele frequencies, detect duplicate groups and make adjustments in the constitution of the core. Germplasm from regions such as Central America and south-western USA needs to be further investigated to determine its importance.

The core collection will serve as a means of more efficiently identifying new traits and new sources of traits already studied. It will also serve to identify new sources of resistances, or known sources but in different gene combinations. Since we are dealing with two different gene pools we can expect that associations of traits will be different in each gene pool. Evidence for the usefulness of the core concept in identifying sources of traits is currently being documented at CIAT. Preliminary screenings for phosphorus efficiency has identified accessions from specific regions with desirable response. Screening of additional germplasm from these regions permitted the selection of a larger number of desirable genotypes.

It is expected that the core collection will be requested by major gene banks and breeding programmes in South America and Africa interested in broadening their genetic variability without the burden of handling huge numbers. However, a severe lack of adaptation between agroecological regions has been observed among genotypes from diverse regions for many years. The core collection will consist of a set of accessions representing agroecological regions cutting across the range of adaptation of *P. vulgaris*. For direct introduction to a new area it is therefore likely that transfer of the entire core collection will be inefficient since many of the materials may not even set seed in the area of introduction. A judicious pre-selection of the collection could avoid this waste of effort by sending only those materials that are likely to be adapted to the region.

Some type of pre-breeding may be important to achieve effective germplasm transfer. An example of this might be the transfer of phosphorus efficiency. Table 3 shows a large number of accessions coming from phosphorus-fixing volcanic soils in the mid-altitudes and highlands of Mexico. These materials show very poor adaptation at low altitudes where there are many phosphorus-poor soils. Prior breeding at higher altitudes in the appropriate agroecological situation and a gradual transfer to the target zone with breeding for adaptation to the lower elevations would result in an efficient transfer of useful traits to phosphorus-poor lowland regions, such as parts of Brazil. A network approach is recommended whereby data on different traits or data on the same traits under different environments could be pooled, providing a powerful tool for selecting parental materials for a pre-breeding effort.

To the extent that the bean core collection is a microcosm of both wild and cultivated *P. vulgaris*, it lends itself to further study of the evolution of the species as a crop. This would entail identifying putative wild populations which would have served as the raw material of domestication. This would help to understand better what limitations the initial founder effect had upon *P. vulgaris* as a crop. It would also give an appreciation of which traits have evolved since domestication and which traits are truly ancestral. One might expect little variability in the cultivated bean for an ancestral trait derived from a common ancestor. On the other hand, traits which have evolved since domestication may have evolved diversely in different environments.

It would be also very useful to know more about the variability of diseases and pests at the sites of origin, and to correlate this to the diversity in the bean material. Where there is evidence of co-evolution and local concentration of interesting traits, breeders and germplasm specialists should focus more on very primitive landraces and wild beans, since their distribution will best reflect the regional distribution of the biotic and abiotic constraints. The use of the core and reserve collections will require an integrated approach to build truly useful databases which should combine passport data with ecogeographical classification, ethnobotanical information and molecular data.

Acknowledgements

We wish to thank O. Toro for his contribution and dedication in compiling and processing the passport data, Dr D. Debouck from IBPGR for his valuable discussions on improving the weighting of the ecological regions, and Drs J. Lynch and J. Fairbairn for their help in the soil classifications. We also thank Drs C. Cardona and S. Singh for their suggestions during the development of this work, and R. Hidalgo for his assistance in the selection of accessions from the secondary centres.

References

Beebe, S., Lynch, J., Tohme, J. and Ochoa, I. 1992. Genetic diversity for phosphorus efficiency in landraces of *Phaseolus vulgaris* L. In *Abstracts of the First International Crop Science Congress.* Ames, Iowa, USA: Crop Science Society of America/Iowa State University.

Brown, A.H.D. 1989a. The case of core collections. In Brown, A.H.D., Frankel, O.H., Marshall. D.R. and Williams J.T. (eds) *The Use of Plant Genetic Resources.* Cambridge, UK: Cambridge University Press.

Brown, A.H.D. 1989b. Core collections: A practical approach to genetic resources management. *Genome* 31: 818-24.

Debouck, D.G. 1987. *Final Report: Collaborative Project IBPGR-CIAT on* Phaseolus *Germplasm Collection.* Rome, Italy: IBPGR.

Debouck, D.G. 1988. *Phaseolus* germplasm exploration. In *Genetic Resources of* Phaseolus *Beans.* Dordrecht, Netherlands: Kluwer.

Debouck, D.G. and Tohme, J. 1989. Implications for bean breeders of studies on the origins of common beans, *Phaseolus vulgaris* L. In Beebe, S. (ed) *Current Topics in Breeding of Common Bean.* Cali, Colombia: CIAT.

Frankel, O.H. and Brown, A.H.D. 1984. Plant genetic resources today: A critical appraisal. In *Crop Genetic Resources: Conservation and Evaluation.* London, UK: Allen and Unwin.

Ghaderi, A., Adams, M.W. and Saettler, A.W. 1982. Environmental response patterns in commercial classes of common bean (*Phaseolus vulgaris* L.). *Theoretical and Applied Genetics* 63: 17-22.

Gepts, P. 1988. A middle American and an Andean common bean gene pool. In *Genetic Resources of* Phaseolus *Beans.* Dordrecht, Netherlands: Kluwer.

Gepts, P. and Bliss, F.A. 1988. Dissemination pathways of common bean (*Phaseolus vulgaris*, Fabaceae) deduced from phaseolin variability. II. Europe and Africa. *Economic Botany* 42: 86-104.

Gepts, P. and Debouck, D.G. 1991. *Origin, Domestication, and Evolution of the Common Bean (*Phaseolus vulgaris *L.).* Wallingford, UK: CAB International.

Gepts, P., Osborn, T.C., Rashka, K and Bliss, F.A. 1986. Phaseolin protein variability in wild forms and landraces of the common bean (*Phaseolus vulgaris* L.): Evidence for multiple centers of domestication. *Economic Botany* 40: 451-68.

Gepts, P., Kmiecik, K., Pereira, P. and Bliss, F.A. 1988. Dissemination pathways of common bean (*Phaseolus vulgaris*, Fabaceae) deduced from phaseolin electrophoretic variability. 1. The Americas. *Economic Botany* 42: 73-85.

Jones, P.G. 1987. Current availability and deficiencies in data relevant to agroecological studies in the geographic area covered by the IARCs. In Bunting, A.H. (ed) *Agricultural Environments*. Wallingford, UK: CAB International.

Jones, P.G., Robinson, D.M. and Carter, S.E. 1991. A GIS approach to identifying research problems and opportunities in Natural Resource Management. In *CIAT in the 1990s and Beyond: A Strategic Plan*. Supplement. Cali, Colombia: CIAT

Khairallah, M.M., Adams, M.W. and Sears, B.B. 1990. Mitochondrial DNA polymorphisms of Malawian bean lines: Further evidence for two major gene pools. *Theoretical and Applied Genetics* 80: 753-61.

Kipe-Nolt, J.A., Montealegre, C.M. and Tohme, J. 1992. Restriction of nodulation by the broad host range *Rhizobium tropici* strain CIAT899 in wild accessions of *Phaseolus vulgaris* L. *New Phytologist* 120: 489-94.

Ladizinsky, G. 1985. Founder effect in crop-plant evolution. *Economic Botany* 39: 191-99.

Laing, D.R., Jones, P.G. and Davis J.H.C. 1984. Common bean (*Phaseolus vulgaris* L.). In Goldsworthy, P.R. and Fisher, N.M. (eds) *The Physiology of Tropical Field Crops*. New York, USA: John Wiley.

Linacre, E.J. 1977. A simple formula for estimating evaporation rates in various climates using temperature alone. *Agricultural Meteorology* 18: 409-24.

Martin, G.B. and Adams, M.W. 1987a. Landraces of *Phaseolus vulgaris* (Fabaceae) in northern Malawi. 1. Regional variation. *Economic Botany* 41: 190-203.

Martin, G.B. and Adams, M.W. 1987b. Landraces of *Phaseolus vulgaris* (Fabaceae) in northern Malawi. 2. Generation and maintenance of variability. *Economic Botany* 41: 204-15.

NOAA. 1988. TGP-OO6D digital topographic data on 10-minute grid. Boulder, Colorado, USA: National Oceanic and Atmospheric Administration.

Osborn, T.C., Blake, T., Gepts, P. and Bliss, F.A. 1986. Bean arcelin. 2. Genetic variation, inheritance and linkage relationships of a novel seed protein of *Phaseolus vulgaris* L. *Theoretical and Applied Genetics* 71: 847-55.

Plucknett, D.L., Smith, N.J.H., Williams, J.T. and Anishetty, N.M. 1987. *Gene Banks and the World's Food*. Princeton, New Jersey, USA: Princeton University Press.

Reddy, S.J. 1979. *Users Manual for the Water Balance Models*. Patancheru, India: ICRISAT.

Riehl, M. 1979. *Climate and Weather in the Tropics*. London, UK: Academic Press.

Romero Andreas, J., Yandell, B.S. and Bliss, F.A. 1986. Bean arcelin. 1. Inheritance of a novel seed protein of *Phaseolus vulgaris* L. and its effect on seed composition. *Theoretical and Applied Genetics* 72: 123-28.

Schoonhoven, A.V., Cardona, C. and Valor, J. 1983. Resistance to the bean weevil and the Mexican bean weevil (Coleoptera: Bruchidae) in non-cultivated common bean accessions. *J. Economic Entomology* 76: 1255-59.

Singh, S.P., Gepts, P.L. and Debouck, D.G. 1991a. Races of common bean (*Phaseolus vulgaris*, Fabaceae). *Economic Botany* 45: 379-96.

Singh, S.P., Nodari, R. and Gepts, P. 1991b. Genetic diversity in cultivated common bean: I. Allozymes. *Crop Science* 31: 19-23.

Sprecher, S.L. 1988. *Allozyme Differentiation between Gene Pools in Common Bean (*Phaseolus vulgaris *L.), with Special Reference to Malawian Germplasm*. East Lansing, Michigan, USA: Michigan State University.

Toro, O., Tohme, J. and Debouck, D.G. 1990. *Wild Bean (*Phaseolus vulgaris *L.): Description and Distribution*. Cali, Colombia/Rome, Italy: CIAT/IBPGR.

UNEP/GRID. 1988. *FAO Soils Map of the World Digitised at 30 Seconds Resolution*. Nairobi, Kenya: Global Resource Information Database.

White, J.W. and Laing, D.R. 1989. Photoperiod response of flowering in diverse genotypes of common bean (*Phaseolus vulgaris* L.). *Field Crops Research* 22: 113-28.

Core Collections of Plant Genetic Resources
Edited by T. Hodgkin, A.H.D. Brown, Th.J.L. van Hintum and E.A.V. Morales

A Wiley-Sayce Publication

3.2

The use of characterisation data in developing a core collection of sorghum

K.E. Prasada Rao and V. Ramanatha Rao

Abstract

The range of genetic variability in cultivated sorghum is very extensive. The International Crops Research Institute for the Semi-Arid Tropics (ICRISAT) conserves over 33 100 accessions of sorghum from 86 countries, an appropriate size for a world sorghum collection in view of the plant's geographic and taxonomic diversity. Although several subsets of the total world collection (such as a 'working collection' and a 'basic collection') were developed for utilisation by sorghum scientists, it became clear that these subsets were location specific and did not give a fair representation of the world collection. This necessitated the development of another 'small collection' which would effectively represent the genetic diversity in the world collection. The concept of a core collection seemed to serve this purpose; such a collection will provide plant breeders with a gateway to the world collection. The use of a small collection, similar to a core collection, was tested in Ethiopia and resulted in the release of two improved cultivars.

A core collection was established at ICRISAT by stratifying the total world collection geographically and taxonomically into subgroups. Accessions in each subgroup were then clustered into closely related groups based on characterisation data, using principal components analysis. Representative accessions from each cluster were drawn in proportion to the total number of accessions present in that subgroup. Thus, a sorghum core collection of 3475 accessions (approximately 10% of the total world collection) was formed. The core collection offers scientists a way of making a relatively rapid assessment of the diversity present so that a greater number of related germplasm accessions can be tested subsequently to identify promising accessions for utilisation in breeding programmes or for direct release as improved cultivars. The core collection will not affect the conservation of the world collection of sorghum germplasm at ICRISAT. Such conservation, both at ICRISAT and at other centres where duplicate sets are conserved, will continue.

Sorghum (*Sorghum bicolor* [L.] Moench) is fifth in importance among the world's cereals. It is a species of tropical origin, but in recent history it has been adapted, through selection, to temperate regions. It remains the staple food of many countries in Africa and Asia and is now a major feed grain crop in Argentina, Australia, Mexico, South Africa and the USA. It was probably domesticated in north-eastern Africa, in an area extending from the Ethiopian-Sudanese border westwards to Chad (Doggett, 1970; de Wet et al., 1976). From this area it spread to India, China, the Middle East and Europe soon after its domestication (Doggett, 1965).

Sorghum is an immensely variable genus and is subdivided into the sections Chaetosorghum, Heterosorghum, Parasorghum, Stiposorghum and Sorghum. The Sorghum section includes cultivated grain sorghum, a complex of closely related annual taxa from Africa and a complex of perennial taxa from southern Europe and Asia (de Wet, 1978). The range of genetic variability available in cultivated sorghum races and their wild relatives is very extensive. Extreme types are so different as to appear to be separate species (Prasada Rao and Mengesha, 1988). The collection and conservation of the available diversity attracted the attention of breeders and botanists about three decades ago. Since then, with the disappearance of many landraces and the rapid, large-scale destruction of natural habitats of wild and weedy relatives of sorghum through urbanisation and industrialisation, the conservation of genetic resources of sorghum has become increasingly important.

Establishing a system for conserving sorghum germplasm for present use and for future generations is the global responsibility of the International Crops Research Institute for the Semi-Arid Tropics (ICRISAT). The Genetic Resources Unit at ICRISAT currently conserves over 33 100 accessions of sorghum, collected from 86 countries. It is expected that this number may reach 45 000 by the end of this decade.

SIZE OF GERMPLASM COLLECTIONS

The sizes of germplasm collections have grown markedly. Many collections are so large that extensive evaluation is impossible for all but a few characters which are readily and rapidly discernible on single plants (for example, some morphological traits, chemical differences detectable by spot tests, and major genes for disease resistance). Greater use of germplasm collections could be made, particularly for a wider range of characters, if a small number of accessions were to be given priority in evaluation and utilisation. There are limits to the number of samples that can be handled effectively in programmes for the evaluation and utilisation of genetic resources. These limits are imposed primarily by the financial and manpower resources available.

The size of the present sorghum world collection at ICRISAT is appropriate in view the geographic and taxonomic diversity of the plant. However, this size is imposing constraints on evaluation, maintenance and use of the collection.

Basic collection

To facilitate the use of the world sorghum collection at ICRISAT, a basic collection of 1400 lines were carefully chosen and stratified by race, subrace, geographical distribution, and ecological adaptation (Harlan, 1972). The selection was made at ICRISAT Center (near Patancheru, in Andhra Pradesh, India, at 18°N 78°E). At this location, many tropical landraces did not flower or flowered very late, giving a poor expression of agronomic potential. Ultimately, it became clear that the basic collection was location specific and did not give a fair representation of photoperiod-sensitive germplasm from such countries as Cameroon, Ethiopia, Nigeria, Sudan and Yemen, where most landraces occur. This highlighted the need to develop another 'small collection' which would effectively represent the genetic diversity present in the world collection.

Core collection

Given the need for reduced size of germplasm collections, Frankel (1984) argued that a collection could be pruned to a 'core collection' that would represent the genetic diversity of a crop species and its wild relatives. The accessions not included in the core would not be discarded but retained as the 'reserve collection' (Brown, 1989). The main purpose of the core is to provide efficient access to the whole collection. It is important that the major kinds of diversity present in the whole collection are represented in the core collection, thus providing the breeder with the means of appraising relatively rapidly the diversity available. This would then guide the breeder to other related sources held in the world collection. For example, while working in Ethiopia with a small portion of ICRISAT's collection (selected sorghum accessions for the grain mould nursery), breeders found that the sorghum race from South Africa ('kafirs') did well under the high rainfall and intermediate altitude conditions of Ethiopia; this led to the testing of a complete set of kafirs from South Africa (278 accessions) and ultimately two accessions, IS 9302 and IS 9323, were released for large-scale production in the Bako, Birr Valley and Jimma areas of Ethiopia, where mechanised farming is possible (Menkir and Kebede, 1984).

Experienced sorghum breeders who are familiar with both the sorghum collection at ICRISAT and the local environmental conditions are able to specify their requirements in such a way as to make it relatively simple for ICRISAT's gene bank personnel to conduct a computerised search and provide a list of appropriate accessions. For scientists in national agricultural research systems, however, the process is less straightforward; many of them do not make use of the facility at ICRISAT because of their lack of familiarity with the collection and/or with detailed data on the agroclimatic conditions in the areas for which they need to breed improved varieties and hybrids. A core collection would make it easier to provide such scientists with an indication of the genetic diversity available, enabling them to select desirable accessions for their location. Once a core collection starts producing anticipated results, a larger quantity of seed from it can be regenerated for distribution, thus reducing the costs of multiplication and distribution; this would also reduce the frequency of regeneration and thus help to minimise genetic drift.

Setting up a core collection at ICRISAT will not alter the conservation strategy for the world collection. Conservation of the world collection will continue at ICRISAT and at other centres where duplicate sets are maintained.

ESTABLISHING THE SORGHUM CORE COLLECTION

Having recognised the need to establish a core collection, the next step was to select the core entries. The first major issue to be addressed was the number of entries. In sorghum, taking into account the present size of the world collection and the genetic diversity present in the crop species, 3500 seemed to be an appropriate number (roughly 10% of the world collection).

The decision was then taken to exclude wild and weedy races of sorghum because of problems in handling them. Some are difficult to grow, while others are serious weeds, especially those with rhizomes. These wild races may have value as sources of resistance or tolerance to pests and diseases, but they can pose serious threats to agriculture and should be handled with great care. Moreover, they need different types of descriptors and descriptor states for characterisation.

The next step was to analyse the world collection to assess the genetic diversity. Characterisation data were used to divide the collection into related groups from which representative samples could be taken to form the core collection. To identify these groups, three types of data were used:

- country of origin (representing geographical diversity)
- taxonomic group (representing taxonomic diversity)
- agronomic traits (using the numerical data for cluster analysis)

Geographical data

The origin, domestication and early distribution of sorghum is particularly important for plant breeders. The geographical origins of the plant have played an important role in the evolution of the diverse races and subraces of cultivated sorghum. Distinct races occur in different agroclimatic regions and differ from each other in morphology and some aspects of physiology. Natural and human selection have acted together to produce cultivars that are physiologically suited to the region where selection occurred, and culturally suited to the needs and expectations of people in areas where sorghum has been grown for centuries. An important consequence of this selection is that each race of sorghum has become ecologically and culturally specialised. Moreover, there appears to have been very little exchange of sorghum varieties of different races among groups of people (Stemler et al., 1975). Because of this geographical diversification, the place of origin of each accession played a key role in the establishment of the core collection.

Taxonomic data

Cultivated species of the genus *Sorghum* are frequently more variable morphologically than related wild species. Snowden (1936) recognised 28 cultivated species, which he further divided into 156 varieties and numerous forms. These species were combined into the following basic and intermediate races by Harlan and de Wet (1972):

Basic races		*Intermediate races*	
Race 1	*bicolor*	Race 6	*guinea-bicolor*
Race 2	*guinea*	Race 7	*caudatum-bicolor*
Race 3	*caudatum*	Race 8	*kafir-bicolor*
Race 4	*kafir*	Race 9	*durra-bicolor*
Race 5	*durra*	Race 10	*guinea-caudatum*
		Race 11	*guinea-kafir*
		Race 12	*guinea-durra*
		Race 13	*kafir-caudatum*
		Race 14	*durra-caudatum*
		Race 15	*kafir-durra*

Intermediate races bear a fairly close resemblance to the basic race but they also incorporate characteristics of the associated basic race in their spikelet morphology. At ICRISAT, the entire collection was classified into these five basic and 10 intermediate races, using characters such as spikelet morphology, panicle shape and compactness. Based on this classification, representative samples were taken for the core collection.

Agronomic data

The sorghum germplasm accessions were sown at ICRISAT (on Vertisols) during the rainy season in the second fortnight of June and harvested in November-December. Observations on days-to-flowering, plant height, and basal and nodal tillering were recorded only in the rainy season as most of the tropical germplasm is photoperiod-sensitive during this season because of longer daylength.

The same accessions were sown again in the post-rainy season in the second fortnight of September. With the shorter daylength in this season, all accessions, including those from tropical African countries, flowered comparatively early. During this season, plant, panicle and grain characters were recorded. Observations on midrib colour were recorded at flag leaf stage, and other characters between flowering and maturity. Grain characters were recorded after harvest in the laboratory. The data were documented and computerised along with passport information, using a VAX 11/780. These data are maintained under the ICRISAT Data Management Retrieval System (IDMRS).

All the sorghum germplasm accessions available at ICRISAT were characterised using 29 descriptors (passport, qualitative, quantitative and classification). The total world collection was stratified geographically and taxonomically into subgroups. The following descriptors were used for clustering the accessions within each subgroup, using principal components analysis (PCA):

- *Days-to-flowering:* This is number of days from mean emergence date to the date when 50% plants have started flowering. This character is very important as it indicates the cessation of the vegetative phase. The photoperiod influence on flowering behaviour during the rainy (long day) season is very high and some of the tropical accessions of West African origin did not flower or flowered very late.

- *Plant height:* This is the length of the main stalk (in cm) at 50% flowering. Ten selected plants were measured and the mean height was computed. This is another important character influenced by genetic and environmental variation. Photoperiod influences this character in the rainy (long day) season.

- *Inflorescence exsertion:* In sorghum inflorescence, exsertion is measured as the amount of exposed peduncle from the flag leaf to the base of panicle. The greater the peduncle exsertion, the more desirable the accession for plant breeding purposes.

- *Inflorescence length and width:* These are the two basic characters that affect yield. Openness of the panicle is linked with the taxonomic race. For example, relatively high yields will be obtained from an accession belonging to the *caudatum* race with a compact elliptic panicle having maximum length and width.

- *Grain covering:* This character indicates the amount of grain covered by glumes at maturity. This is one of the distinguishing characters used in the racial classification of cultivated sorghum. The variation ranges between completely open grain and completely closed grain (scored on a 1-5 scale).

- *Grain weight:* This is the weight in grams of 100 grains at a moisture content of 12%.

The variation among accessions for all these descriptors is given in Table 1.

Table 1 **Range and standard deviation of the quantitative characters used in principal components analysis for developing a core collection of sorghum germplasm**

Character	Range	Standard deviation
Days-to-flowering	33.00-199.00	92.00
Plant height	34.00-655.00	322.34
Inflorescence exsertion	00.00-71.00	16.27
Inflorescence length	2.50-86.00	22.31
Inflorescence width	1.00-80.00	9.26
Grain covering	1.00-5.00	2.73
Grain weight	0.72-8.92	3.05

Grouping the germplasm accessions

In the characterisation of germplasm, each accession is represented by a point in a seven-dimensional space, the coordinates of which provide an indication of the scores for each character. Thus, accessions with similar responses among these characters can be grouped in close proximity in the seven-dimensional space. In such a situation, PCA allows a reduction in dimensionality. In most cases, the first two principal components cover most of the information, and allow the data to be presented as different points in a two-dimensional space. The groups can be formed with nearby genotypes, visually (SAS, 1982). For further selection, individual accessions can be drawn from each group proportionately to represent genetic diversity. For example, in the subgroup Sudan-*durra* 224 accessions were divided into 11 groups and 22 accessions were randomly selected from these groups to represent the genetic diversity. Similarly, in the subgroup Gambia-*guinea* 46 accessions were divided into five groups and five accessions were selected from these groups. Using this procedure, a sorghum core collection consisting of 3475 accessions, approximately 10% of the total world collection, has been established at ICRISAT.

Acknowledgements

The authors wish to acknowledge the guidance and assistance of Dr K.V.S. Rao (Senior Statistician, ICRISAT), P. Venkateswarlu (Research Associate in ICRISAT's Statistics Unit) and V. Gopal Reddy (Senior Research Associate in ICRISAT's Genetic Resources Unit) in the PCA work with sorghum germplasm characterisation data and in the selection of core accessions.

References

Brown, A.H.D. 1989. The case for core collections. In Brown, A.H.D., Frankel, O.H., Marshall, D.R. and Williams, J.T. (eds) *The Use of Plant Genetic Resources*. Cambridge, UK: Cambridge University Press.
de Wet, J.M.J. 1978. Systematics and evolution of *Sorghum* Sect. Sorghum (Gramineae). *American J. Botany* 65:477-84.
de Wet, J.M.J., Harlan, J.R. and Price, E.G. 1976. Variability in *Sorghum bicolor*. In Harlan, J.R., de Wet, J.M.J. and Stemler, A.B.L. (eds) *Origins of African Plant Domestication*. The Hague, Netherlands: Mountain Press.
Doggett, H. 1965. The development of cultivated sorghums. In Hutchinson, J. (ed) *Crop Plant Evolution*. Cambridge, UK: Cambridge University Press.
Doggett, H. 1970. *Sorghum*. London, UK: Longmans.
Frankel, O.H. 1984. Genetic perspectives of germplasm conservation. In Arber, W., Llimensee, K., Peacock, W.J. and Starlinger, P. (eds) *Genetic Manipulation: Impact on Man and Society*. Cambridge, UK: Cambridge University Press.
Harlan, J.R. 1972. Genetic resources in sorghum. In Rao, N.G.P. and House, L.R. (eds) *Sorghum in the Seventies*. New Delhi, India: Oxford and IBH Publishing Co.
Harlan, J.R. and de Wet, J.M.J. 1972. A simplified classification of cultivated sorghum. *Crop Science* 12: 172-76.
Menkir, A. and Kebede, Y. 1984. High rainfall, intermediate altitude sorghum (HRIAS). *Sorghum Newsletter*. 27: 3-4.
Prasada Rao, K.E. and Mengesha, M.H. 1988. Sorghum genetic resources: Synthesis of available diversity and its utilisation. In Paroda, R.S., Arora, R.K. and Chandel, K.P.S. (eds) *Plant Genetic Resources — Indian Perspective*. New Delhi, India: National Bureau of Plant Genetic Resources.
SAS Institute. 1982. *SAS User's Guide: Statistics*. Cary, North Carolina, USA: SAS Institute Inc.
Snowden, J.D. 1936. *The Cultivated Races of Sorghum*. London, UK: Adlard.
Stemler, A.B., Harlan, J.R. and de Wet, J.M.J. 1975. *Caudatum* sorghums and speakers of chari-Nile languages in Africa. *J. African History* 16: 161-83.

Core Collections of Plant Genetic Resources
Edited by T. Hodgkin, A.H.D. Brown, Th.J.L. van Hintum and E.A.V. Morales

A Wiley-Sayce Publication

3.3

Developing a coffee core collection using the principal components score strategy with quantitative data

S. HAMON, M. NOIROT and F. ANTHONY

Abstract

The genetic organisation of the coffee gene pool was examined at three levels: biogeography, genetic resources and available data. This investigation indicated that a core collection for coffee should consist of 88 genetic diversity groups of three types, according to their genetic history and the available genetic knowledge: a *Coffea arabica* category, a category containing well-studied species such as *C. liberica* and *C. canephora*, and a category with a large number of neglected species. Different strategies were applied to the three categories. After defining the diversity groups, tests were conducted, using data obtained for *C. liberica*, on a new method (principal components score strategy) for developing a core collection using quantitative data. The results showed that about half the inertia was obtained when 10% of the 338 genotypes were selected, and 90% of the total inertia was obtained with a sample of 50% of these genotypes.

Setting up a core collection implies the selection of a limited number of accessions which are genetically representative of the cultivated species and its wild relatives (Frankel and Brown, 1984). The main objectives of establishing a core collection are to facilitate the management of germplasm, reduce the cost of conservation, and promote the diffusion and the use of genetic resources. Brown (1989a, b) suggested, on the basis of allele frequencies, two types of strategy: the first, fully random sampling, can be used when the genetic organisation is unknown; the second, stratified sampling with three sampling possibilities (constant, proportional or logarithmic) is appropriate when information is available on the target species.

Creation of a core collection for a given crop is difficult when the gene pool is large and there are different evolutionary histories, ploidy levels and breeding behaviours, and when the extent of knowledge of genetic diversity is not the same for each species, several being well studied but others still undescribed botanically. This is not helped by the level of information available in world collections. It has been estimated that some 65% of accessions in world collections have no passport data, 80% are not characterised and only 1% have been extensively evaluated (Peeters and Williams, 1984). So, despite the simple formulation of the core collection concept, the strategy to be used is difficult to define precisely.

Coffee has been widely collected (Berthaud and Charrier, 1988; Anthony, 1992). In terms of storage possibilities coffee seeds are considered as intermediate — they currently cannot be conserved for longer than 1 year — and so genetic resources are conserved in field collections. The cultivated species, *Coffea arabica*, is well known, but other species are still undescribed. The maintenance of coffee genetic resources in the field is expensive, time consuming and subject to the risk of loss. In this chapter we will first define the diversity groups of coffee and then describe the application of the principal components score strategy (PCSS), designed to maximise the selected inertia, to a well-studied category of the diversity groups.

DEFINITION OF THE DIVERSITY GROUPS OF COFFEE

Main biogeographic groups

In Africa and Madagascar coffee trees grow under the canopy of tropical forest. Species are distributed according to three biogeographic units delimited by the Mozambican canal and the dorsal of Kiwu located in the eastern part of Zaire. About 100 taxonomic units have been described by many botanists (for example, Chevalier, 1947; Bridson and Verdcourt, 1988).

The most economically important species, *C. arabica*, is native to and geographically isolated in southern Ethiopia, northern Kenya and southern Sudan. In West Africa, coffee species, particularly in the area with a guineo-congolese climate, have a tree-like shape and long, thin leaves. The period between flowering and ripening ranges from 7 to 15 months and the coffee beans have a relatively high level of caffeine content (0.5-4% dry weight). In East Africa, coffee species have characteristics of xerophyllus adaptation. Plants are bushy with small thick leaves. Beans have a lower content of caffeine (0-2% dry weight). Apart from this subcontinental differentiation, well-diversified groups are also found at lower geographic levels (Berthaud, 1986; Anthony, 1992).

Collected genetic resources

Apart from *C. arabica*, which is allotetraploid (2n = 44) and self-fertile, all *Coffea* species are diploid (2n = 22) and self-sterile. The diploid species all have the same genome. Because *C. arabica* is very different both at the genetic and the economic level, coffee genetic resources can be divided in two categories: *C.arabica* and the other species.

For *C. arabica*, a collecting mission was conducted in 1964 in Ethiopia, the centre of origin and domestication. The mission to collect cultivated forms was organised by the Food and Agriculture Organisation (FAO) and an association of coffee producers (FAO, 1968). A collecting mission, focusing particularly on 'native types', was undertaken in 1966 by the Office de la Recherche Scientifique et Technique Outre Mer (ORSTOM). Some 196 samples were collected by FAO and 70 by ORSTOM (IFCC, 1978). With regard to the actual field collections, the base collection is located in Jimma, Ethiopia, and a large number of other collections, partially duplicates, are located in Cameroon, Costa Rica, Côte d'Ivoire and Kenya. Isozyme studies have shown that the amphiploid, self-fertile *C. arabica* is characterised by a low level of genetic diversity, with a polymorphism index of 0.003 compared with 0.05 for the diploid species (Berthou and Trouslot, 1977).

All diploid species are native to Africa and Madagascar. Collecting missions were conducted in eight countries by ORSTOM and the Institut de Recherche sur le Café et le Cacao (IRCC); for a review, the reader is referred to Berthaud and Charrier (1988). In West Africa, collections were made in Côte d'Ivoire and Guinea; in Central Africa they were made in Cameroon, the Central African Republic and the Congo; and in East Africa they were made in Kenya and Tanzania. About 17 500 genotypes were collected, representing 75 species. The only field collections are in Côte d'Ivoire for all African accessions and in Madagascar for all Madagascar species.

Definition of diversity groups

To build up the core collection we used a database (BASECAFE) with passport and characterisation data (Anthony, 1992). Passport data are systematic and include taxonomic identification, origin, site and type of collected material. Characterisation data are also available, as are data on the location of the plant in the field, the year of planting and the nature of root system. The list of descriptors is presented in Table 1. The list includes both morphological and biochemical data. As for other crops, the availability of characterisation data depends upon the economic importance of the species and the priorities of the genetic resources centre.

For the identification of the coffee diversity groups we summarised collecting and characterisation reports, and synthesised the following genetic evaluation data (Charrier, 1978; Hamon et al., 1984;

Table 1 Coffee descriptors used for characterisation, recorded in the BASECAFE database

Characters	Descriptors	Coding
Morphology	Leaf dimensions	mm
	Fruit stalk length	mm
	Fruit dimensions	mm
	Bean dimensions	mm
Isozymes[a]	Esterases alpha, beta cathodic (EST)	1/0
	Isocitrate dehydrogenase (ICD)	1/0
	Malate dehydrogenase (MDH)	1/0
	Phospho-gluco-isomerase (PGI)	1/0
	Phospho-gluco-mutase (PGM)	1/0
Flowering	Intensity (estimation of flower number)	5 classes
Production	Mature berry weight by passage	kg
Technology	100 beans weight at 12% moisture content	g
	Commercial coffee yield	%
Biochemistry	Caffeine content in green bean	% dm
	Caracoli bean rate	%
Fertility	Filling ovule rate	%
	Berry filling	%

Note: a Presence of an allele (1); lack of an allele (0)

Table 2 Main diversity groups for the core collection of *Coffea*

Africa (38 groups)

C. arabica (3 groups)(1[a])

C. arabica	(cultivated, wild[a], Mt Marsabit)

Simple groups, botanically described (5 groups) (4[a])

C. fadenii	*C. racemosa*[a]
C. humilis[a]	*C. salvatrix*[a]
C. pseudozanguebariae[a]	

Simple groups, botanically undescribed (8 groups) (7[a])

C. sp. F[a]	*C.* sp. 'Ngongo 2'[a]
C. sp. 'Bakossi'[a]	*C.* sp. 'Ngongo 3'[a]
C. sp. 'Congo'[a]	*C.* sp. 'Nkoumbala'[a]
C. sp. 'Mayombe'[a]	*C.* sp. 'Song-Mbong'

Complex groups (22 groups) (19[a])

C. brevipes	(Mt Cameroun[a], Kumba-Loum[a], var. *heterocalyx*[a])
C. canephora	(guinean[a], congolian[a], cameroonese[a], Nana[a])
C. congensis	(centrafrican[a], cameroonese[a], congolian[a])
C. eugenioides	(kenyan, var. *kivuensis*)
C. kapakata	(tanzanian, *'brésilian'*)
C. liberica	(guinean[a], congolian[a], Koto[a])
C. sessiliflora	(Shimba[a], Kitulangalo[a])
C. stenophylla	(Assabli[a], Ira[a])
C. sp. 'Moloundou'	(Moloundou[a], Souanké[a])

Madagascar (50 groups)

C. andrambovatensis	*C. mauritiana* (Réunion Island)
C. arenesiana	*C. millotii*
C. augagneuri	*C. mogeneti*
C. bertrandi	*C. perrieri*
C. boiviniana	*C. pervilleana*
C. buxifolia	*C. resinosa*
C. dolichophylla	*C. richardii*
C. dubardi	*C. sahafariensis*
C. farafanganensis	*C. sakarahae*
C. heimii	*C. tetragona*
C. humblotiana (Comores Islands)	*C. tsirananae*
C. homollei	*C. vatovavyensis*
C. jumellei	*C. vaughanii*
C. kianjavatensis	*C. vianneyi*
C. lancifolia	*C. ind.* (20 undescribed taxons)
C. mangoroensis	

Note: a Diversity groups represented in the current *Coffea* core collection

Berthaud, 1986; Berthaud and Charrier, 1988; Anthony, 1992, Rakotomalala, 1993). This resulted in 88 groups for the core collection (*see* Table 2). There are 64 simple groups that correspond to 64 taxa. A further 24 groups are complex; these groups correspond to nine species, structured mainly according to ecological regions or to morphotypes.

THE PRINCIPAL COMPONENTS SCORE STRATEGY

Statistical procedure

To avoid ambiguity, we will first define the difference between diversity, variability and inertia and describe the assumptions on which the strategy is based.

Diversity relates to all that is potentially diverse and is a function of the genetic material and the tools used for its estimation. In this context, variability, in terms of the statistical parameter 'variance', is a part of the diversity restricted to quantitative data. Inertia is the generalised sum of squares of standardised and independent variables. In a given diversity group we assume that there is no reproductive barrier, apart from self-incompatibility alleles, and that the principle of general additivity for quantitative data could be applied. We consider that a cross between two extreme genotypes is possible and that all intermediate phenotypic forms can be obtained by recombination.

Using quantitative descriptors, the variability of a given set of genotypes depends upon the differences recorded between individuals. Here, the distance used to describe differences between individuals has the following characteristics: it is metric, gives the same weight to the descriptors, avoids the co-linearity between descriptors, and removes the residual variability. PCSS consists of three steps: the application of principal components analysis (PCA) to the quantitative data; the computerisation of the weighted Euclidean distance between individuals; and the selection of genotypes that maximise subset inertia.

The choice of the weighted Euclidean distance allows the two first conditions to be fulfilled. The distance d_{ik} between two individuals (i and k) is:

$$d_{ik} = \sqrt{\sum_{j=1}^{J} [(X_{ij} - X_{kj}) \cdot \sigma_j^{-1}]^2}$$

where:

J = number of descriptors

s_j = standard deviation of the jth descriptor

The co-linearity between variables and the residual variability are removed by the use of PCA. This step applies a PCA to the quantitative data table in order to obtain new uncorrelated variables (the eigenvectors) and new coordinates of each individual. The residual variability is removed only by considering an axis with an eigenvalue greater than 1 (by definition the number is L).

The second step computes the weighted Euclidean distance between individuals, using new coordinates on the L first axis. Weights are given by the square root of eigenvalues. Thus, the weighted Euclidean distance between two genotypes is:

$$d_{ik} = \sqrt{\sum_{j=1}^{L} [(X_{ij} - X_{kj}) \cdot \sqrt{\lambda_j^{-1}}]^2}$$

where:

l_j = the eigenvalue value of the jth eigenvector

The third step applies the following selection. Let us consider the total inertia of an N*L table, where N is the number of genotypes and L is the number of uncorrelated and reduced variables. It is equal to the product N x L (Lebart et al., 1977). The inertia of one genotype (P_i) is the sum of squares of their factorial coordinates for the L factors:

$$P_i = \sum_{j=1}^{L} X_{ij}^2$$

The relative contribution of each genotype (RC_i) is therefore:

$$RC_i = P_i / (N \cdot L)$$

The selection procedure involves searching for genotypes which have the higher RC_i. Then, step by step, we add the genotype that gives the highest RC score. This procedure can be curtailed at any moment, either at a pre-selected level of inertia or when a certain percentage of genotypes is selected. Theoretically, the percentage of the selected inertia using PCSS, with random sampling, a large number of descriptors and N infinite, increases linearly (*see* Figure 1, curve c).

Testing the PCSS in the *Coffea liberica* guinean group

We applied the selection procedure to a set of 338 genotypes of *C.liberica* which originated in the Central African Republic. They were characterised for 11 quantitative descriptors including morphology, technology, fertility, biochemistry and agronomy (*see* Table 3).

Apart from caffeine content, most descriptors are not too far from normal random variables. The coefficient of variation is generally 10-20% except for some descriptors such as coffee production and

Table 3 Quantitative descriptors used for characterising the *Coffea liberica* guinean group

Characters	Descriptors	Code	Mean	CV (%)	Skewness	Kurtosis
Morphology	Leaf length (cm)	LOFE	21.8	10.2	-0.3	1.2
	Leaf width (cm)	LAFE	10.4	13.1	0.3	0.4
	Stem diameter (cm) 6 years after planting	CCOL	30.4	16.5	-0.2	0.6
	Height or first persist branch (cm)	HPLA	49.5	21.9		0.6
	Maximum tree diameter at HPLA (15 classes)	JUPE	9.6	20.0	-0.7	1.7
	Plant total height (18 classes)	HAUT	10.5	28.1	-0.4	
Production	Mature berry weight by passage (kg)	PROD	31.4	53.5	0.4	0.6
Technology	100 bean weight at 12% moisture content (g)	P100	14.4	20.3	0.4	0.2
	Commercial coffee yield (%)	TRDM	15.7	15.5	-0.2	0.6
Biochemistry	Caffeine content in green bean (%)	CAFE	1.2	21.7	1.4	3.3
Fertility	Caracoli bean rate (%)	TCAR	28.7	50.3	0.9	0.8
	Berry filling (%)	TREM	71.8	13.1	-0.4	

Table 4 **Correlations between descriptors and partial contributions of the descriptors to the principal component axes**

Variable	LOFE	LAFE	DCOL	CCOL	HAUT	JUPE	HPLA	P100	TRDM	TREM	TCAR	CAFE	PROD
LOFE													
LAFE	714												
DCOL	78	102											
CCOL	135	174	884										
HAUT	210	259	692	711									
JUPE	83	120	516	549	411								
HPLA	61	15	302	278	252	123							
P100	9	127	211	274	285	226	-19						
TRDM	53	125	155	173	276	154	54	555					
TREM	132	103	159	134	251	91	105	167	549				
TCAR	-50	-40	-58	-45	-155	18	83	1	-351	-785			
CAFE	128	139	4	-10	42	69	25	104	71	75	18		
PROD	136	98	420	480	426	324	165	203	232	251	-230	-104	
Axis/descriptor													
1 (29.9%)	90	117	641	696	662	364	114	203	251	230	110	4	394
2 (15.2%)	10	10	153	157	30	99	21	25	320	580	550	20	
3 (12.8%)	707	703	23	8		2	13	3	17	20	37	103	19
Total	807	830	817	861	692	465	148	231	588	830	697	127	413

caracoli bean rate, where it reaches 50%. Correlations between the descriptors and the PCA axes indicate that the first three main axes have the following features: axis 1 (30%) plant shape and vigour; axis 2 (15%) plant fertility; and axis 3 (13%) essentially the leaf polymorphism (*see* Table 4). The distribution of the genotypes along these axes did not reveal visually defined subgroups. The group test, made by clustering methods followed by a discriminant analysis, confirmed the lack of a robust subgroup and was thus in accord with the hypothesis of a single diversity group.

The effect of selection on the inertia retained is presented in Figure 1. Curve (b) shows that the selection of the first 10% of the most variable genotypes corresponds to 30% of the total inertia. The level of 50% inertia is obtained with 25% of the genotypes. When we repeated the analysis with only the three most informative descriptors (that is, those with the maximum contribution to axis 1 — stem diameter, berry filling and leave width), we observed (curve a) that 10% of the most variable genotypes corresponded to 45% of the total inertia and 50% of the genotypes corresponded to 90% of the total inertia. In both cases, the selected inertia using PCSS was greater than that with the strictly random sampling (curve c), which is theoretically linear when N trends to infinite.

DISCUSSION

Global definition of genetic diversity groups

A good core collection needs to define genetic diversity groups precisely and thus it is important to have standardised descriptors. The available data (passport and characterisation) are very different in

Figure 1 **Principal components score selection within the *Coffea liberica* guinean group**

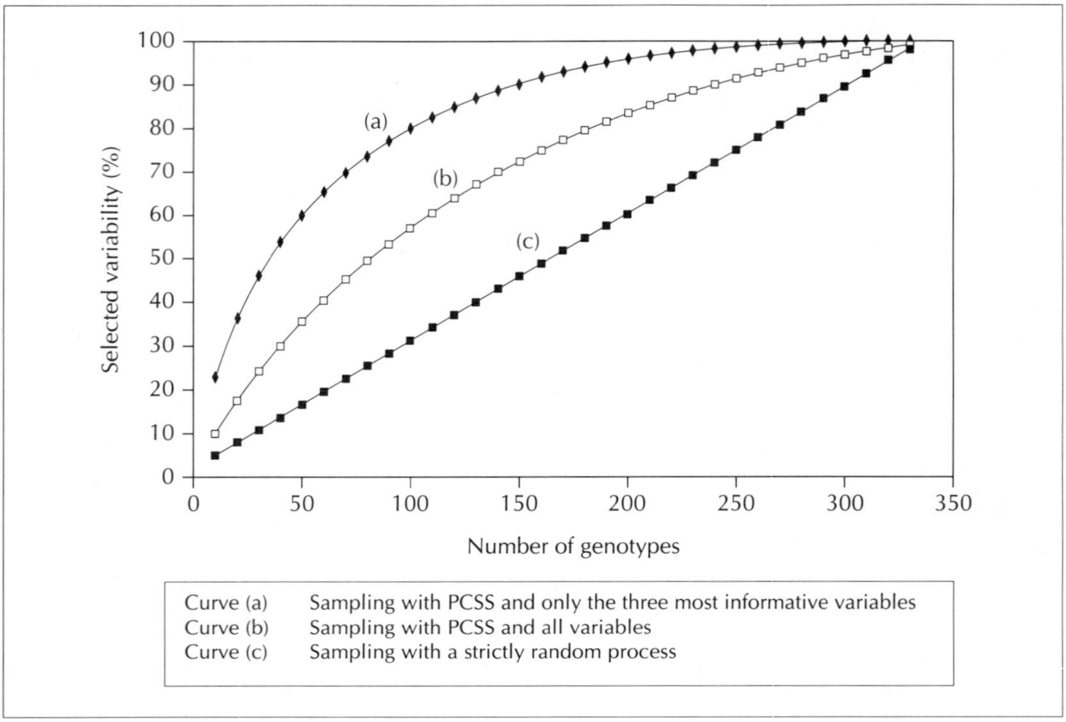

nature, quantity and reliability. The starting point is certainly the biological species. When available, botanical data must be confirmed by inter-crossing tests in order to specify the level and nature of putative reproductive barriers. This almost corresponds to the former definition of the gene pools and compartments (Harlan, 1975; Pernès, 1984). The definition of the diversity group must be more precise. Biogeographic data on the original sampling site could be used to emphasise ecological subgroups. Climatic and pedological data on the sites of origin are of great importance, but unfortunately they are often missing from or inadequately covered in collection reports (Peeters and Williams, 1984).

The genetic structure of plant populations is influenced mainly by biological and ecological factors (Loveless and Hamrick, 1984). Crawford (1985) reports that rapid and recent speciation is correlated with little or no allozyme divergence. Davis and Gilmartin (1985) stress that morphological differentiation could be associated with negligible isozymic difference. Despite this, these factors are often useful for design strategies and the choice of samples for in-depth molecular analysis (Brown, 1990). Taking *Coffea* as an example, let us consider the differences between theory and practice.

Establishing a coffee core collection: One strategy or more?

As mentioned earlier in this chapter, we propose that, for coffee species, 88 diversity groups divided into three categories (based on knowledge of their genetic diversity) need to be represented in the core collection. For each of these three categories, different strategies need to be employed in forming the core collection:

- *C. arabica* category: The economically important species, *C. arabica* (three diversity groups), has to be studied separately. This species is amphiploid, autogamous and has a recent domestication history. In addition, its level of isozymic polymorphism (IE = 0.005) is particularly low (Berthou and Trouslot, 1977). The genetic basis is narrow and most accessions in collections are genetically closely related. For this species, workers select cultivated original mutants and variants. To remove redundancy in the collection, detailed identification methods have to be used, such as fingerprinting (Weising et al., 1992) and random amplified polymorphic DNA (RAPD) (Hadrys et al., 1992). Currently, collections are working or breeders' collections, described mainly by quantitative characters, for which PCSS would be appropriate.

- A category containing little-studied wild species: For a large number of wild species, few studies have been conducted and some species remain botanically undescribed (84 diversity groups). These species, diploid and self-sterile, are highly heterozygous. Brown's (1989b) logarithmic strategy is certainly the most suitable strategy for this category. Unfortunately, in practice it is difficult to conserve such large diversity. The genetic resources of Madagascar species (50 diversity groups) are not available outside the country. So, we considered only the African species (35 diversity groups), for which we chose, as a first step, to conduct a random selection of a constant number of genotypes (40) for each diversity group (groups marked with an 'a' in Table 2).

- A category containing well-studied species, characterised by quantitative data: For this category, we suggest that the sampling strategy is improved by using PCSS, a procedure which is based on a simple concept and is easy to use with a standard computer. It allows the choice of the most variable genotypes. As for other methods, however, we need to be sure that we are actually in a diversity group. Consequently, the main limitation for the global core collection strategy is the definition and the validity test of a diversity group. Chloroplastic DNA is useful to test phylogenic hypothesis between genera of the Rubiaceae family (Bremer, 1991). Perhaps the systematic study of chloroplastic DNA within a particular genus (that is, *Coffea*) could be useful for defining these diversity groups. Despite this, the PCSS procedure, which maximises the selected inertia, introduces *de facto* a statistical sampling bias. If we assume that there is no correlation between morphological and molecular markers and that the postulate of general additivity can be applied within a diversity group, the selection of 10% of the most extreme genotypes, using PCSS, theoretically does not produce less neutral diversity than a completely random selection method (Brown's strategy). Thus, for the *C. liberica* guinean group, 15% of 338 studied genotypes (not far from the Brown's 10%) could capture almost 60% of the total inertia using PCSS.

CONCLUSION

Several different strategies, rather than a single one, should be used to establish a core collection, depending not only upon the organisation of the gene pool but also upon the level of knowledge. A reliable definition of genetic diversity groups is then the first priority. When the groups have been defined, different sampling approaches can be applied, according to the particular level of knowledge and type of available data.When a group has been well studied in terms of quantitative characters, PCSS could be helpful in improving the selected inertia.

References

Anthony, F. 1992. *Les Ressources Génétiques des Caféiers: Collecte, Gestion d'un Conservatoire et Evaluation de la Diversité Génétique*. Série TDM 81. Paris, France: ORSTOM.

Berthaud, J. 1986. *Les Ressources Génétiques pour l'Amélioration des Caféiers Africains Diploïdes: Evaluation de la Richesse Génétique des Populations Sylvestres et de ses Mécanismes Organisateurs. Conséquences pour l'Application*. Série TDM 188. Paris, France: ORSTOM

Berthaud, J. and Charrier A. 1988. Genetic resources of *Coffea*. In Clarke, R.J.and Macrae, R. (eds) *Coffee. 4. Agronomy*. London, UK: Elsevier.

Berthou, F. and Trouslot, P. 1977. L'analyse du polymorphisme enzymatique dans le genre *Coffea*: Adaptation d'une méthode d'électrophorèse en série, premiers résultats. In *Actes du 8ème Colloque de l'ASIC, Abidjan, Côte d'Ivoire*. Paris, France: ASIC.

Bridson, D. and Verdcourt, B. 1988. *Coffea*. In Polhill, R.M. and Balkema, A.A. (eds) *Flora of Tropical East Africa. Rubiaceae. Part 2.* Rotterdam, Netherlands: A.A. Balkema.

Bremer, B. 1991. Restriction data from chloroplast DNA for phylogenetic reconstruction: Is there only one way of scoring? *Plant Systematics and Evolution* 175: 39-54.

Brown, A.H.D. 1989a. The case for core collections. In Brown, A.H.D., Frankel, O.H., Marshall, D.R. and Williams, J.T. (eds) *The Use of Plant Genetic Resources*. Cambridge, UK: Cambridge University Press.

Brown, A.H.D. 1989b. Core collection: A practical approach to genetic resources management. *Genome* 30: 818-24.

Brown, A.H.D. 1990. The role of isozyme studies in molecular systematics. *Australian Systematic Botany* 3: 39-46.

Charrier, A. 1978. *La Structure Génétique des Caféiers Spontanés de la Région Malgache (Mascarocoffea). Leurs Relations avec les Caféiers d'Origine Africaine (Eucoffea)*. Série Mémoires 87. France: ORSTOM.

Chevalier, A. 1947. Les cafeiers du globe. Systématique des caféiers et faux caféiers. Maladies et insectes nuisibles. In *Encycl. Biol. XXVIII, Fas. III*. Paris, France: Lechevalier.

Crawford, D.J. 1985. Electrophoretic data and plant speciation. *Systematic Botany* 10: 405-16.

Davis, J.I. and Gilmartin, A.J. 1985. Morphological variation and speciation. *Systematic Botany*. 10: 417-25.

FAO. 1968. *FAO Coffee Mission to Ethiopia 1964-1965*. Roma, Italy: FAO.

Frankel, O.H.and Brown, A.H.D. 1984. Plant genetic resources today: A critical appraisal. In Holden, J.H.W. and Williams, J.T. (eds) *Crop Genetic Resources: Conservation and Evaluation*. London, UK: Allen and Unwin.

Hadrys, H., Balick, M. and Schierwater, B. 1992. Applications of random amplified polymorphic DNA (RAPD) in molecular ecology. *Molecular Ecology* 1: 55-63.

Hamon, S., Anthony, F. and Le Pierrès, D. 1984. La variabilité génétique des caféiers de la section Mozambicoffea A. Chev. Précisions sur 2 espèces affines: *C. pseudozanguebariae* Bridson et *C.* sp. Bridson. *Bull. Mus. Hist. Nat., Andansonia* 4 (série 6): 207-23.

Harlan, J.R. 1975. *Crops and Man*. Madison, Wisconsin, USA: American Society of Agronomy/Crop Science Society of America.

IFCC. 1978. Etude de la structure et de la variabilité génétique des caféiers: Résultats des études et des expérimentations réalisées au Cameroun, en Côte d'Ivoire et à Madagascar sur l'espèce *Coffea arabica* L. collectée en Ethiopie par une mission ORSTOM en 1966. *Bulletin IFCC* 14.

Lebart, L., Morineau, A. and Tabart, N. 1977. *Techniques de la Description Statistique: Méthodes et Logiciels pour l'Analyse des Grands Tableaux*. Paris, France: Dunod.

Loveless, M.D. and Hamrick, J.L. 1984. Ecological determinants of genetic structure in plant populations. *Annual Review of Ecology and Systematics* 15: 65-95.

Peeters, J.P. and Williams, T. 1984. Towards better use of gene banks with reference to information. *IBPGR Plant Genetic Resources Newsletter* 60: 22-32.

Pernès, J. 1984. *Gestion des Ressources Génétiques des Plantes. 2. Organisation des Complexes d'Espèces*. Paris, France: ACCT.

Rakotomalala, J.J. 1993. *Analyse de la Diversité Biochimique des Caféiers Malgaches*. Série TDM. Paris, France: ORSTOM.

Weising, K., Kaemmer, D., Weigand, F., Epplen, J.T. and Kahl, G. 1992. Oligonucleotide fingerprinting reveals probe-dependent levels of informativeness in chickpea (*Cicer arietinum*). *Genome* 35: 436-42.

Core Collections of Plant Genetic Resources
Edited by T. Hodgkin, A.H.D. Brown, Th.J.L. van Hintum and E.A.V. Morales

A Wiley-Sayce Publication

3.4

Genetic markers and core collections

P. Gepts

Abstract

Molecular and biochemical markers such as isozymes, RFLPs and seed proteins have been used to characterise genetic diversity in germplasm collections. In turn, this information can help in the selection of a core collection that is more representative of the main collection. Genetic markers reveal patterns and levels of genetic diversity that reflect the evolutionary relationships of individual accessions and can thus assist in identifying groups of accessions that are related by common ancestry. The various categories of markers differ for several attributes: level of polymorphism, degree of environmental stability, number of loci, molecular basis of the polymorphism, and practicality. One or other of these markers can be chosen, depending upon the objective of the study. Using these markers it has been possible to identify or confirm centres of domestication. In some cases, markers have been able to identify more specifically than other traits: which area was the most likely centre of domestication; multiple areas of domestication; and, in polyploid crops, the female or male parent. Markers have been used to subdivide the cultivated gene pool and thus determine in some cases whether phenotypic similarity was attributable to common ancestry or other causes such as hybridisation or polyphyly. Genetic marker studies generally show a reduction of diversity during domestication. The principal advantage of molecular and biochemical markers is that they are genotypic markers and, hence, will reflect the actual genetic distance between accessions and their common ancestry more accurately than phenotypic markers. Their main disadvantage is that they are cumbersome, and ways to circumvent this disadvantage are discussed.

According to the definition proposed by Frankel (1984), a core collection is a subset of the general (or main) collection that represents the genetic diversity of a crop species and its relatives with a minimum of repetitiveness. There are therefore two essential steps in the establishment of a core collection: the first involves characterising the genetic diversity of a species and its wild relatives; the second involves the choice of accessions to be included in the core collection based on genetic diversity data. For both these steps a sampling strategy is needed. In particular, sampling in the first step involves choosing the criteria by which genetic diversity is characterised. Plant genomes may contain some 100 000 genes (Kamalay and Goldberg, 1980). Therefore, when one chooses a particular class of traits or markers to characterise diversity, one performs, in effect, a sampling of the plant genome.

Various criteria have been used to analyse genetic diversity in crop plants, including morphological, agronomic, ecogeographical and biochemical or molecular traits or markers. Each of these criteria have their advantages and disadvantages as markers of genetic diversity. Molecular markers include any markers reflecting direct changes at the DNA sequence level, principally restriction fragment length polymorphisms (RFLPs), but also other markers such as random amplified polymorphic DNA (RAPD) and minisatellite markers. Because crop evolution studies require adequate sampling of the diversity contained in the species, direct DNA sequencing has rarely been used, if at all, because of its inherent cumbersomeness.

Biochemical markers have included principally isozymes and seed proteins. The major advantage of molecular and biochemical markers is their presumed selective neutrality, although cases of non-neutrality have been reported, such as *Adh* in *Drosophila* (Anderson and McDonald, 1983). This general neutrality allows us to distinguish those similarities that are attributable to common ancestry from similarities caused by convergence. In the past few years, an increasing amount of data on crop evolution has accumulated based on molecular and biochemical markers (reviewed in Doebley, 1989, 1992; Clegg, 1990; Gepts, 1990, 1993)

In this chapter, molecular and biochemical markers will be discussed as indicators of patterns and levels of diversity. After a discussion of the attributes of various types of markers, some of their contributions to our knowledge of the organisation of genetic diversity in crop plants will be presented. This type of knowledge is an essential element in the establishment of core collections.

ATTRIBUTES OF MOLECULAR AND BIOCHEMICAL MARKERS AS GENETIC DIVERSITY INDICATORS

Molecular and biochemical markers have been used in a substantial number of studies of genetic diversity in crop plants. A survey of the literature shows that isozymes, RFLPs and seed proteins have been used most often in these studies (reviewed in Doebley, 1989, 1992; Clegg, 1990; Gepts, 1990). More recently, RAPDs and sequences hybridising to minisatellite markers have also been used.

There are several attributes by which one can assess the potential usefulness of a particular category of markers. These include level of polymorphism, environmental stability, the number of loci, molecular basis of the polymorphism and the ease and cost of analysis (*see* Table 1). Depending upon the objective of the study, a certain level of polymorphism is required: at higher taxonomic levels (species or above) more conserved markers are needed, whereas at the population level more variable markers are desirable. Electrophoretic patterns should be free of environmental influence to confirm that observed differences are genotypic differences. The number of loci should be as high as possible and preferably should be distributed at random in the genome to ensure adequate genome coverage. The molecular basis of the polymorphism (for example, nucleotide substitutions, insertions and deletions, and co- and post-translational modifications) should be known so that genetic distance and diversity parameters can be determined. Markers for which the polymorphism results from simple changes at the molecular level, such as nucleotide substitutions (for example, isozymes, RFLPs), are therefore more amenable to such quantitative analysis than markers for which the polymorphism involves a more complex series of events (for example, seed proteins). On the other hand, the probability of homoplasy is higher for the former class of markers than for the latter (homoplasy can be defined as similarity in character states attributable to causes other than common ancestry, such as reversal or convergence.) The methodology employed should be as inexpensive and simple as possible to permit the analysis of samples of adequate size.

Table 1 Comparison of molecular electrophoretic markers in evolutionary, genetic and breeding studies

	Allozymes	RFLPs	Seed proteins	Minisatellite sequences	RAPDs
Polymorphism	low	low-high	high	very high	low-high
Environmental stability	moderate	high	high	high	high
Number of loci	moderate (< 50 loci)	high	low (< 10 loci)	moderate	high
Molecular basis of polymorphism	simple	intermediate	complex	complex	complex
Practicality	quick, cheap	slow, expensive	quick, cheap	intermediate	quick expensive

Source: Gepts (1993)

In essence, each class of molecular or biochemical markers possesses advantages and disadvantages. Depending upon the goals of the study, one or the other, or a combination of markers, can be used. In general, however, isozymes and RFLPs are preferred because they allow us to recognise homologies and determine genetic distance and diversity parameters.

Isozymes

The main attributes of isozymes include the simplicity and low cost of the analysis, the simple molecular basis of their polymorphism and a reasonable genome coverage (10-50 loci per species). In addition, standardised experimental conditions allow us to detect genotypic differences although isozymes are phenotypic markers. A disadvantage is the general but not universal lower level of polymorphism (Doebley, 1989; Weeden and Wendel, 1989; Wendel and Weeden, 1989; Hamrick and Godt, 1990).

RFLP markers

RFLP markers can display a wide range of levels of polymorphism, depending upon the species (Nodari et al., 1992), the genome (such as cytoplasmic vs. nuclear; Curtis and Clegg, 1984; Palmer, 1987; Wolfe et al., 1987; Zurawski and Clegg, 1987) or the particular sequence. In general, however, RFLPs are more polymorphic than isozymes. For example, in a direct comparison between the two classes of markers in the same set of maize genotypes, Messmer et al. (1991) found that RFLPs were more polymorphic than isozymes both in terms of the number of polymorphic loci (94 vs. 68 %) and the average number of variants per polymorphic locus (3.4 vs. 2.5). Additional advantages of RFLPs include their better genome coverage and environmental stability. The molecular basis of RFLPs can be as simple as a single nucleotide change which can lead to a restriction site loss or, less likely, a site gain. The likelihood of a repeat mutation involving a nucleotide change may be higher than that involving a rearrangement characterised by its size as well as location. Through restriction mapping, it is possible to identify the molecular basis of the polymorphism, although this limits the number of sequences that can be analysed (Gepts and Clegg, 1989). RFLP technology is cumbersome and costly, which effectively limits the sample size.

Seed proteins

Attributes of seed proteins as electrophoretic markers include their high level of polymorphism, their high level of environmental stability, although a few exceptions have been reported (for example, Gayler and Sykes, 1985), and the complex molecular basis of the electrophoretic patterns that includes nucleotide substitutions, insertions and deletions, and co- and post-translational modifications. This complexity at the molecular level, however, makes it difficult to relate phenotypic changes in the electrophoretic banding pattern to changes at the molecular level. Hence, one is usually limited to phenetic analyses with this category of markers. A disadvantage is the low number of loci involved (usually less than 10) (reviewed in Gepts, 1990).

Minisatellite markers

Minisatellite markers, mainly those revealed through cross-hybridisation with human minisatellites or M13-derived sequences (for example, Dallas, 1988; Stockton et al., 1992), often reveal very high levels of polymorphism. The fingerprinting pattern does not appear to be influenced by environmental conditions (G. Sonnante et al., unpubl.). Little is known about the actual molecular basis of hypervariable sequences in plants. The complexity of the fingerprinting pattern and actual mapping data (T. Stockton et al., unpubl.) suggest, however, that a number of loci are involved, although the genome coverage may not be as extensive as that of RFLPs or RAPDs. The technology is similar to RFLP technology, with the exception that several loci can be sampled at once.

RAPD markers

The advantages of RAPD markers are that they can be more polymorphic than RFLPs (Williams et al., 1990), they offer genome coverage equivalent to that of RFLPs, and their methodology can be quite simple provided that a rigorously standardised methodology with the necessary controls is adhered to. Disadvantages include limited information about the environmental stability of the polymorphism and the molecular basis of RAPDs.

INFORMATION ON GENETIC DIVERSITY PROVIDED BY MOLECULAR AND BIOCHEMICAL MARKERS

Studies using molecular and biochemical markers have to address the following issues affecting organisation and distribution of genetic diversity: identification of the wild ancestor and the pattern of domestication, divergence within the cultivated gene pool, gene flow between wild ancestor and cultivated descendant, and fate of genetic diversity during domestication.

Identification of the wild ancestor and the pattern of domestication

Identification of the wild ancestor is important in crop genetic resources conservation because it tells us which wild taxon contributed genetic material to the cultivated gene pool. Conversely, it also tells us which wild taxa did not contribute genetic material to the cultivated gene pool. Both these aspects are important when identifying accessions for a core collection that is supposed to reflect the spectrum of genetic diversity for the improvement of the crop.

Traditionally, the wild ancestor has been identified on the basis of the overall morphological similarity between the crop and a wild taxon. More recently, other arguments have been used, including cytogenetic, crossability and chemotaxonomic data (for example, Simmonds, 1976). It is clear from these studies that the wild ancestor, although sometimes very different morphologically from its cultivated descendant, is conspecific with it and belongs to its primary gene pool (Harlan and de Wet, 1972).

Molecular and biochemical markers have usually been used to confirm the identity of the ancestral taxon and in some cases to identify more specifically in which geographic area domestication may have occurred (*see* Tables 2 and 3).

Three cases — *Zea mays*, *Phaseolus vulgaris* and *Brassica* spp. — provide examples of the power of molecular markers.

Case 1 Molecular and biochemical markers have confirmed teosinte as the actual and immediate progenitor of maize. Multivariate analyses of allozyme frequencies in the genus *Zea* by Doebley et al. (1984) showed that the wild taxon most similar isoenzymatically to the cultivars is *Z. mays* ssp. *parviglumis* (a short-spikelet annual teosinte adapted to mesic intermediate altitudes). Other teosintes, such as *Z. mays* ssp. *mexicana* (a large-spikelet annual teosinte adapted to arid high altitudes), *Z. luxurians* (an annual teosinte from south-eastern Guatemala) and *Z. perennis* and *Z. diploperennis* (perennial teosintes), were more distantly related to the cultivars.

Chloroplast DNA restriction site analyses also confirmed that annual teosintes were closely related to maize; however, these analyses could not distinguish between the subspecies *parviglumis* and *mexicana* as the most likely ancestral form (Doebley et al., 1987). Further confirmation that teosinte is the ancestor of maize was provided by rDNA restriction site analysis (Zimmer et al., 1988).

Isozyme analyses by Doebley (1990) further suggested that the *Z. mays* ssp. *parviglumis* populations of the Balsas and Jalisco regions in Mexico were the most probable progenitors of cultivated maize. Isozyme data did not agree with morphological data in that they showed *Z. mays* ssp. *parviglumis* to be the most closely related to maize, whereas morphologically *Z. mays* ssp. *mexicana* is most similar (maizoid) to maize (Doebley, 1990).

Case 2 Wild *P. vulgaris* has a very large distribution, extending from northern Mexico to northern Argentina (Brücher, 1988; Delgado Salinas et al., 1988). In addition, the archaeological record included finds of approximately equal ages in Mexico and Peru (Kaplan and Kaplan, 1988). A survey of morphological diversity among cultivars suggested the existence of two groups, one distributed in Mexico and Central America and the other in the Andes (Evans, 1976). These data, however, were inconclusive with regard to the domestication pattern of common bean because they did not tell us whether these two groups arose from separate domestications or through divergent selection after a single domestication. Molecular and

Table 2 **Examples of closest wild relative of diploid species identified or confirmed using molecular and biochemical markers**

Crop	Closest wild relative	Markers used	Source
Avocado (*Persea americana* var. *guatemalensis*: one of three cultivated avocado races)	*P. steyermarkii* (female) x *P. nubigena* (male)	cpDNA, rDNA cellulase	Furnier et al. (1990)
Barley (*Hordeum vulgare* ssp. *vulgare*)	*H. vulgare* ssp. *spontaneum*	Isozymes, cpDNA, mtDNA	Jørgensen (1976); Nevo et al. (1979, 1986); Clegg et al. (1984); Neale et al. (1986); Holwerda et al. (1986)
Common bean (*Phaseolus vulgaris* var. *vulgaris*	*P. vulgaris* var. *aborigineus* and *mexicanus*	Isozymes, seed proteins, mtDNA	Gepts et al. (1986)
		nuclear single-copy RFLPs	Koenig and Gepts (1989); Khairallah et al. (1990); Singh et al. (1991a)
Lentil (*Lens culinaris* ssp. *culinaris*)	*L. culinaris* ssp. *orientalis*	Isozymes, nuclear single-copy RFLPs	Pinkas et al. (1985); Hoffman et al. (1986); Havey and Muehlbauer (1989)
Lettuce (*Lactuca sativa*)	*L. serriola* and other unidentified wild taxa	Nuclear single-copy RFLPs	Kesseli et al. (1991)
Maize (*Zea mays*)	*Z. mays* ssp. *parviglumis* (Balsas and Jalisco)	Isozymes	Doebley et al. (1984)
Pea (*Pisum sativum*)	*P. humile*	cpDNA	Palmer et al. (1985)

Table 2 (contd.)

Crop	Closest wild relative	Markers used	Source
Peanut (*Arachis hypogea*)	*A. monticola*	Nuclear single-copy RFLPs, RAPDs	Halward et al. (1991); Kochert et al. (1991)
Pearl millet (*Pennisetum glaucum*)	*P. glaucum* ssp. *monodii*	cpDNA	Gepts and Clegg (1989)
Peppers		Isozymes	McLeod et al. (1982, 1983)
Capsicum baccatum var. *pendulum*	*C. baccatum* var. *baccatum*		
C. pubescens	*C. eximium* (presumably not the wild ancestor but a closely related species)		
C. annuum, *C. frutescens*, *C. chinense*	*C. annuum* var. *aviculare*		
Rice		Isozymes, cpDNA	Second et al. (1982); Cordesse et al. (1990); Dally and Second (1990)
Oryza sativa	*O. rufipogon*		
O. glaberrima	*O. breviligulata*		
Sorghum (*Sorghum bicolor*)	*S. bicolor* ssp. *arundinaceum*	cpDNA	Duvall and Doebley (1990)
Soybean (*Glycine max*)	*G. soja*	rDNA	Doyle and Beachy (1985)
Tomato (*Lycopersicon esculentum*)	*L. esculentum* var. *cerasiforme*	Isozymes, mtDNA, nuclear single-copy RFLPs	Rick and Fobes (1975); McClean and Hanson (1986); Miller and Tanksley (1990)

Table 3 Origin of some polyploid crops as determined by molecular and biochemical markers

Crop	Parental taxa	Markers used	Source
Brassica spp.			
Brassica carinata (n = 17)	*B. nigra* (n = 8) x *B. oleracea* (n = 9)	cpDNA, nuclear single-copy RFLPs, rDNA	Erickson et al. (1983); Palmer et al. (1983); Song et al. (1988); Delseny et al. (1990);Hosaka et al. (1990)
B. juncea (n = 18)	*B. campestris* (n = 10) x *B. nigra* (n = 8)		
B. napus (n = 19)	*B. oleracea* (n = 9) x *B. campestris* (n = 10)		
Cotton			
Gossypium barbadense, *G. hirsutum*	A genome species (Old World; female) x D genome species (New World; male)	cpDNA	Wendel (1992)
Potato (tetraploid):		cpDNA	
Solanum tuberosum var. *andigena*	Cultivated diploids (*S. goniocalyx* *S. phureja*, *S. stenotomum* and a wild diploid (*Solanum* spp.)		Hosaka and Hanneman (1988a, b)
Wheat			
Diploid: *Triticum monococcum* (AA)	*T. monococcum* ssp. *aegilopoides*	Isozymes	Asins and Carbonell (1986); Jaaska (1980, 1981)
Tetraploid: *T. turgidum* var. *turgidum*	*T. turgidum* var. *diccoides*	cpDNA, mtDNA	Ogihara and Tsunewaki (1988); Graur et al. (1989)
A genome donor B genome donor	*T. urartu* *T. speltoides* or related species	Nuclear repetitive RFLPs and RSAPs; rDNA	Dvorák and Appels (1982); Dvorák et al. (1988); Dvorák and Zhang (1990)
Hexaploid: *T. aestivum*			
AB genome donor	*T. turgidum* var. *turgidum*	cpDNA, mtDNA	Ogihara and Tsunewaki (1988); Graur et al. (1989)
D genome donor	*T. tauschii* ssp. *strangulata*	Isozymes	Asins and Carbonell (1986); Jaaska (1980, 1981)

biochemical markers, on the other hand, have revealed that the conspecific wild ancestor has diverged into two major groups, the Middle American and Andean groups, and that separate domestications in the Middle American and Andean areas have led to two groups of distinct cultivars (reviewed in Gepts, 1990a, 1993).

Case 3 The relationships among the nuclear genomes of cultivated *Brassica* species are summarised by the triangle proposed by U (1935). In this triangle, the diploid species — *B. rapa* (syn. *campestris*), n = 10; *B. nigra*, n = 8; and *B. oleracea*, n = 9 — occupy the apices, and the amphidiploid species — *B. carinata*, n = 17; *B. juncea*, n = 18; and *B. napus*, n = 19 — occupy the sides between their respective progenitors.

Because of maternal inheritance of cpDNA, it was possible to identify the maternal parent of the amphidiploids (Erickson et al., 1983; Palmer et al., 1983). *B. carinata* and *B. juncea* derived their cytoplasm from *B. nigra* and *B. campestris*, respectively. Part of *B. napus* derived its cytoplasm from *B. oleracea*, whereas the other part may have derived its cytoplasm through introgressive hybridisation from another *Brassica* species (Palmer et al., 1983). The identity of the parents of the amphidiploids was confirmed by Song et al. (1988a) and Hosaka et al. (1990) on the basis of nuclear genome-specific RFLP markers and by Delseny et al. (1990) on the basis of nuclear rRNA gene polymorphism. The molecular data also confirmed previous results obtained by various approaches such as cytogenetics, isozymes and artificial resynthesis (*see* Hosaka et al., 1990, for references).

The following lessons can be drawn from these studies:

- Markers can confirm or identify the wild ancestral taxon of crop plants.

- This capability allows us to identify more specifically than with most other traits the wild populations that are genetically most similar to the crop. The ancestors of such populations were presumably involved in the domestication process. For polyploid crops with uniparental maternal cytoplasmic inheritance, molecular and biochemical markers allow us to identify the female parent.

- Results from molecular or biochemical marker and morphological trait studies are not always correlated. Discrepancies may be attributed to possible selective effects more likely to be associated with morphological traits than with molecular markers. This underscores the fact that more than one category of markers or traits should be used to characterise genetic diversity.

Wild ancestral populations or their immediate progeny populations are generally conspecific with their cultivated descendants. Because of the lack of reduced viability of fertility in crosses with cultivated materials, they belong to the primary gene pool of crops and could be more readily used as a source of genetic variation than other wild relatives belonging to different species. Molecular analyses can tell us which wild populations are most closely related to the cultivars and, perhaps more importantly, which populations are more distantly related and could therefore make a more novel contribution to the cultivated gene pool. It is suggested here that wild ancestral populations should be included in a core collection because of their conspecificity with the cultivated materials and the reduction in diversity observed during domestication (*see overleaf*). Genetic data from molecular marker analyses may help select specific wild populations.

Table 4 **Examples of studies involving molecular and biochemical markers on divergence within cultivated gene pools**

Crop	Divergence within the cultivated gene pools	Markers used	Source
Brassica spp. *B. rapa*	2 groups: Indo-European (turnip, turnip rape, spring broccoli raab, sarson) vs. East Asian (pak choi, Chinese cabbage)	Nuclear single-copy RFLPs	Song et al. (1988b, 1990)
Common bean (*Phaseolus vulgaris* L.)	3 Middle American (Durango, Jalisco, and Peru) and Andean (Nueva Granada, Peru, and Chile) races differentiated by morphological, agronomic and ecological characteristics	Isozymes, phaseolin seed protein	Singh et al. (1991a, b, c)
Cotton (*Gossypium hirsutum*)	Geographic differentiation that does not support a previous subdivision in 7 races on morphological grounds	Isozymes	Hutchinson (1951); Wendel et al. (1992)
Maize (*Zea mays*)	3 complexes of morphological and ecologically similar races: 1) high elevation Mexican pyramidal complex: (e.g., Palomero Toluqueño, Chalqueño); 2) northern complex (e.g., Azul, Apachito); 3) remainder: southern and western lowland dents and flours: (e.g., Tuxpeño, Tabloncillo)	Isozymes	Doebley et al. (1985)
Sorghum (*Sorghum bicolor*)	Geographic differentiation that does not support a previous classification based on morphological grounds	Isozymes	Harlan and de Wet (1972); Morden et al. (1990)

Divergence within the cultivated gene pool

Divergence within the cultivated gene pool according to morphological, agronomic and ecological factors will provide additional sampling criteria for the establishment of a core collection. In addition to the information mentioned earlier for several crops, some more detailed studies have been performed on a number of crops (*see* Table 4).

Case 1 Wellhausen et al. (1952) established a racial classification of Mexican maize on morphological grounds. Doebley et al. (1985) provided evidence that these maize races can be grouped into three complexes of morphologically and ecologically similar races: the high elevation, Mexican pyramidal complex, including races Palomero Toluqueño, Chalqueño and Cónico; the northern complex, which includes races Azul and Apachito; and the remaining bulk of races, including the southern and western lowland dents and flours such as Tuxpeño and Tabloncillo. Frequencies of 22 isozyme alleles were correlated with altitude, which, in Mexico, is associated with precipitation and temperature (that is, the lowlands are hotter and moister whereas the highlands tend to be cooler and drier).

Case 2 In common bean, when phaseolin or isozymes are used as *a priori* classification criteria, multivariate analyses such as canonical discriminant analysis reveal correlations with phenotypic traits such as growth habit, internode length, leaf and seed size, phenology, disease resistance and general ecological adaptations (*see* Figure 1). These data have led to a proposal for six races or groups of related cultivars, three in each major gene pool (Singh et al., 1991c). In the Middle American gene pool, race Jalisco represents the predominantly climbing cultivars of the southern, humid highlands of Mexico and Central America. Race Durango includes cultivars with a prostrate growth habit from the northern, arid highlands of Mexico, and race Mesoamerica includes the indeterminate bush cultivars from the humid, hotter lowlands of Mexico, Central America and South America.

In the Andean gene pool, race Nueva Granada includes cultivars with determinate bush or indeterminate climbing growth habits adapted to moderate altitudes. Race Peru consists of cultivars with a climbing growth habit adapted to higher altitudes, whereas race Chile includes cultivars with a prostrate growth habit. Interestingly, races Durango and Chile display a similar phenotype, which includes medium-sized seeds and light pigmentation, in addition to the prostrate growth habit. These two races diverge, however, at the molecular and biochemical levels (Gepts et al., 1986; Singh et al., 1991b; O.M. Paredes et al., unpubl.). Their similar phenotype may have resulted from convergence caused by selection for adaptation to arid environments prevalent in northern Mexico and Chile.

Figure 1 Canonical discriminant analysis of morpho-agronomic diversity in the Midde American gene pool of *Phaseolus vulgaris*

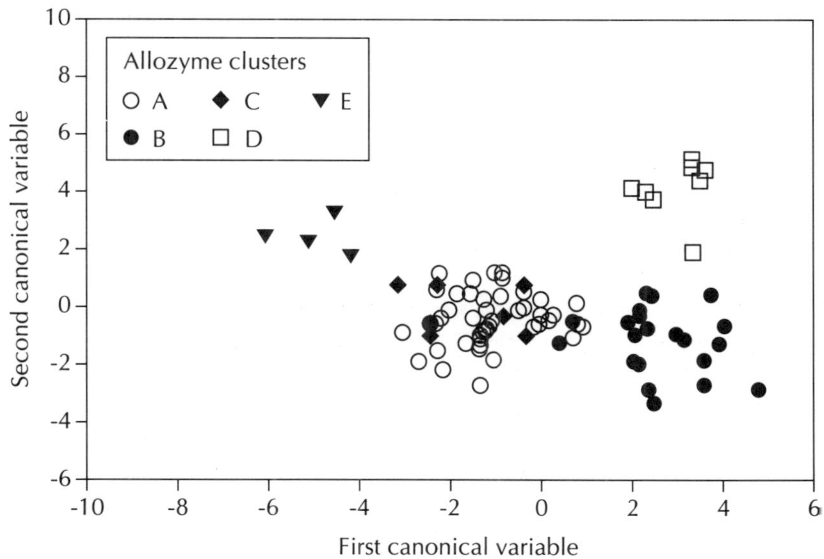

Source: Singh et al. (1991a, b)

In summary, it is possible to identify examples that show or do not show concordance between patterns of diversity identified either by molecular and biochemical markers or morpho-agronomic traits. Discrepancies can be attributed to outcrosses between wild ancestors and cultivated descendants or within the cultivated gene pools. Alternatively, polyphyletic origins for the morpho-agronomic traits could also account for dissimilar results. When establishing a core collection, representatives of the various evolutionary lineages identifed with molecular markers should be represented in order to help ensure that the core is broadly based.

Gene flow between wild ancestor and cultivated descendant

Gene flow between wild ancestor and cultivated descendant will introduce additional genetic diversity into the cultivated gene pool. Methodologically, however, the identification of these introgressants is somewhat difficult (Doebley, 1989; Gepts, 1993).

Case 1 As mentioned earlier, the progenitor of cultivated maize is an annual teosinte, *Z. mays* ssp. *parviglumis*. A distinct annual teosinte, *Z. mays* ssp. *mexicana*, is not the direct progenitor of maize but may have contributed genetic diversity to the cultivated gene pool; in the region where maize and the subspecies *mexicana* are sympatric, isozyme alleles characteristic of the wild taxon were observed in cultivars. These alleles were absent from the cultivars elsewhere (Doebley et al., 1984). Evidence for introgression in the other direction is provided by the perennial teosinte *Z. diploperennis* in which a plant showed isozyme alleles at two linked loci characteristic of the cultivated gene pool, suggesting that the chromosome segment marked by the two loci was introduced from maize (Doebley et al., 1984).

Case 2 Cultivated pearl millet (*Pennisetum glaucum*) is sympatric with its wild ancestor in the Sahel. Viable and fertile hybrids can be formed, although male sterility in these crosses has been described (Marchais and Pernès, 1985). Isozyme analyses revealed that pollen from wild forms was preferentially involved in fertilisation on wild pistils in pollination experiments with mixtures of pollen from wild and cultivated forms. Conversely, pollen from cultivated forms preferentially effected fertilisation on cultivated pistils after mixed pollinations with wild pollen (Robert et al., 1991). This type of reproductive isolation may explain, at least in part, why wild and cultivated pearl millet maintain their phenotypic integrity in spite of their sympatry and predominant allogamy.

Case 3 In *P. vulgaris*, most wild and cultivated genotypes possess the $Mdh\text{-}2^{100}$ allele. Exceptions to this are a limited number of wild-growing accessions and race Jalisco cultivars from the southern highlands of Mexico, which show the $Mdh\text{-}2^{102}$ allele. The rarity and the narrow geographic localisation of the $Mdh\text{-}2^{102}$ allele argue in favour of gene flow, although it is not possible to determine the direction of the gene flow (Singh et al., 1991b). It is interesting to note, however, that the phenotype of the cultivars involved does not display any hint of past hybridisation with wild beans (S. Singh, pers. comm.). Phaseolin data also provide evidence for occasional outcrosses between wild and cultivated beans. All cultivated accessions from Mexico analysed so far show the 'S' phaseolin type, with the exception of one accession with an 'M' phaseolin type characteristic of wild-growing accessions of the same region. The latter accession also displays morphological signs of hybridisation with wild beans, such as smaller seeds and the striping and spotting pattern characteristic of wild bean seeds (Koenig et al., 1990). Some wild accessions from Colombia, Ecuador and northern Peru also show signs of introgression from cultivars based on phaseolin data (Gepts and Bliss, 1986; Gepts et al., 1986; O.M. Paredes and P. Gepts, unpubl.).

Cultivated germplasm that may have resulted from gene flow with wild populations deserves to be included in a core collection because it may contain alleles that are not represented in the bulk of the cultivated gene pool. Molecular markers provide ways, in addition to morphological markers, of detecting introgressants from the wild gene pool.

Fate of genetic diversity during domestication

Doebley (1989) surveyed isozyme studies examining differences in diversity between wild ancestor and cultivated descendants. Total heterozygosity (H_t) decreased from 0.24 in wild forms to 0.19 in cultivated forms, the proportion of polymorphic loci per population from 0.32 to 0.25, the proportion of polymorphic loci per taxon from 0.58 to 0.49, and the number of alleles per locus per taxon from 2.47 to 2.15. Chloroplast DNA data surveyed by Doebley (1992) also showed a reduction in diversity (calculated as the average probability per variable site that two accessions will differ; Clegg et al., 1984) and the proportion of polymorphic restriction sites (*see* Table 5).

Table 5 Chloroplast DNA diversity in wild ancestors and cultivated descendants in crop species

Progenitor/Crop	N[a]	TS	P	D	Source
Glycine soja/	8	2	1.0	0.47	Close et al.
G. max	46	2	1.0	0.20	(1989)
Helianthus annuus wild/weedy/	11	4	1.0	0.25	Rieseberg and
H. annuus cultivated	23	4	0.0	0.00	Seiler (1990)
Hordeum vulgare ssp. *spontaneum/*	11	5	1.00	0.43	Clegg et al.
H. vulgare ssp. *vulgare*	9	5	0.20	0.08	(1984)
Hordeum vulgare ssp. *spontaneum/*	30	3	1.00	0.45	Neale et at.
H. vulgare ssp. *vulgare*	51	3	0.33	0.03	(1988)
Pisum humilis/	4	6	0.67	0.43	Palmer et al.
P. sativum	13	6	0.50	0.10	(1985)
Sorghum bicolor ssp. *arundinaceum/*	6	8	1.00	0.35	Duvall and
S. bicolor ssp. *bicolor*	3	8	0.13	0.06	Doebley (1990)
Zea mays ssp. *parviglumis* var. *parviglumis/*	31	3	1.00	0.38	Doebley
Z. mays ssp. *mays*	80	3	0.67	0.22	(1990)

Note: a N = number of accessions analysed; TS = total number of polymorphic sites; P = proportion of total sites that are polymorphic; D = diversity (that is, average probability that two cpDNAs will differ at a polymorphic site (Clegg et al., 1984)

Source: Doebley (1992)

Nuclear DNA markers such as genes for ribosomal RNA and single-copy RFLPs were also characterised by a reduction in diversity (Allard, 1988; Doyle, 1988; Gepts and Clegg 1989; Havey and Muehlbauer, 1989; Cordesse et al., 1990). A few apparent exceptions to this trend have been reported. In barley, isozymes and cpDNA showed equivalent levels of diversity in wild and cultivated accessions, whereas mtDNA diversity was higher in the cultivars than in the wild ancestor (Holwerda et al., 1986; Jana and Pietrzack 1988). These results contrast with those of Brown and Munday (1982) for isozymes and Clegg et al. (1984) and Neale et al. (1986) for cpDNA. It is not clear what the cause is of these discrepancies. One possibility is the sampling of the plant material, which differed among these studies for the total number of accessions, the relative number of wild and cultivated accessions and their respective geographic origins.

In pearl millet, RFLPs of genes for ribosomal RNA and the *Adh-1* locus analysed on the same set of wild and cultivated accessions showed contrasting trends in diversity: whereas the former showed a clear reduction in diversity between the wild ancestor and the cultivated descendant, the latter showed comparable levels of diversity (Gepts and Clegg, 1989). This discrepancy may be attributable to differential selection pressures operating on the *Adh-1* and rRNA loci or loci linked to them. For example, it is possible that the diversity at the *Adh-1* locus in the cultivars is maintained through introgression with the sympatric wild ancestor. For rRNA genes, this mechanism would not operate if these genes were tightly linked to an essential feature of the cultivated phenotype; hence, after outcrossing with a wild population, selection for the cultivated phenotype in subsequent generations would eliminate introgressed rRNA genes from wild populations. Information on the actual linkage relationships between these molecular markers and genes controlling the cultivated phenotype is needed to verify this hypothesis.

In common bean, a comparison of trends in genetic diversity revealed by isozymes, low-copy nuclear RFLPs and sequences hybridising to M13-derived sequences also showed a reduction in diversity in the cultivars compared with wild *P. vulgaris* (*see* Table 6). The RFLP data for the Andean gene pool are an exception; for these, a slight increase in diversity was observed. Possible explanations include a sampling effect and the lack of representation of wild populations from certain areas in the Andes.

In summary, available data suggest a reduction in diversity averaged over the whole genome in the cultivated gene pool compared with the wild ancestral gene pool. Some parts of the genome, however,

Table 6 Trends in genetic diversity in *Phaseolus vulgaris* based on isozyme, RFLP and M13 homologous sequences

			Average heterozygosity		
		N	Isozymes[a]	RFLPS[b]	M13[c]
Middle American	Wild	11	0.13	0.33	0.24
	Cultivated	36	0.09	0.27	0.20
Andean	Wild	11	0.07	0.25	0.20
	Cultivated	26	0.03	0.27	0.16

Note: a 19 loci
b 13 probes for nuclear sequences, 3 restriction enzymes
c 1 probe, 1 restriction enzyme
Source: Koenig and Gepts (1989); Singh et al. (1991b); V. Becerra et al. (unpubl.)

may show increased variation caused by selection under domestication. As mentioned earlier, wild forms represent a source of additional diversity and should be represented in some way in the core collection of a crop.

DISCUSSION

The major advantage of molecular and biochemical markers is their genotypic nature — that is, they are a direct reflection of changes at the DNA level. Morphological traits, on the other hand, are phenotypic traits. Because the same phenotype can be achieved by different genotypes, it is often difficult to equate phenotype and genotype. Accessions with a similar phenotype may sometimes be evolutionarily unrelated. For example, in common bean, races Durango and Chile show similar phenotypes yet biochemical markers indicate that they belong to different gene pools (Middle America and Andes, respectively) of common bean (Singh et al., 1991c). An additional advantage of markers is that they are generally, although not always, selectively neutral and thus will show patterns of diversity that reflect ancestral relationships among gene pools and are independent of selection. As a consequence, molecular markers will allow us to measure genetic diversity levels and, particularly, genetic distances between populations or lines. This information, in turn, can then be used to form groups of evolutionarily related materials, which should be represented in a core collection. Levels of diversity in the various groups thus revealed can also be used as a measure of the sampling intensity to be applied to each group when establishing a core collection.

A disadvantage of molecular and biochemical markers can be their cumbersomeness. It is obviously unrealistic to subject an entire collection or even a large fraction of it to molecular and biochemical analyses. Simplifications of the procedures such as the introduction of the polymerase chain reaction (PCR) may help alleviate this problem. Another possible solution is to identify associations between correlate molecular/biochemical markers and morphological markers. For example, in common bean, certain alleles of morphological markers are found at high frequency in certain gene pools or races. Races can be distinguished by their leaflet and bracteole shape and size, seed size and shape, pod beak insertion and pigmentation patterns in flower and seed. The *Mf* gene causes the appearance of purple veination at the bottom of the flower and is found at high frequency in race Mesoamerica. *Punc* (coloured dots on seeds), *Mt* (seed mottling), *Str* (seed striping) and *Cor* (seed corona) are Andean alleles. Andean materials generally show larger leaves, longer internodes and larger seeds (Gepts et al., 1986; Singh et al., 1991a).

Multivariate statistical methods could be used to assign individual accessions on the basis of phenotypic data to specific groups (such as gene pools and races) that have been recognised previously using a marker-based analysis of diversity. Posterior probabilities of membership in the Andean or Middle American group obtained by discriminant analysis of morpho-agronomic traits were above 95% (when the prior classification criterion was Middle American vs. Andean phaseolin type). Posterior probabilities of membership ranged between 65% and 85% when the prior classification was the phaseolin type itself (P. Gepts, unpubl.).

This approach would involve a dynamic analysis consisting of multiple rounds in each of which diversity is analysed at the molecular and morphological levels until a satisfactory grouping of genotypes is achieved such that most accessions can be assigned to evolutionarily distinct groups. Further experiments are needed to corroborate this approach.

Acknowledgements

The research on molecular markers in *Phaseolus* and *Vigna* is supported in part by the International Board for Plant Genetic Resources (IBPGR) and the United States Agency for International Development (USAID).

References

Allard, R.W. 1988. Genetic changes associated with the evolution of adaptedness in cultivated plants and their wild relatives. *J. Heredity* 79: 225-38.

Anderson, S.M. and McDonald, J.F. 1983. Biochemical and molecular analysis of naturally occurring *Adh* variants in *Drosophila melanogaster*. *Proc. National Academy of Science, USA* 80: 4798-802.

Asins, M.J. and Carbonell, E.A. 1986. A comparative study on variability and phylogeny of *Triticum* species. 2. Interspecific relationships. *Theoretical and Applied Genetics* 72: 551-58.

Brown, A.H.D. and Munday, J. 1982. Population genetic structure and optimal sampling of landraces of barley from Iran. *Genetica* 58: 85-96.

Brücher, H. 1988. The wild ancestor of *Phaseolus vulgaris* in South America. In Gepts, P. (ed) *Genetic Resources of* Phaseolus *beans*. Dordrecht, Netherlands: Kluwer.

Clegg, M.T. 1990. Molecular diversity in plant populations. In Brown, A.H.D., Clegg, M.T., Kahler, A.L. and Weir, B.S. (eds) *Plant Population Genetics, Breeding, and Genetic Resources*. Sunderland, Massachusetts, USA: Sinauer.

Clegg, M.T., Brown, A.H.D. and Whitfeld, P.R. 1984. Chloroplast DNA diversity in wild and cultivated barley: Implications for genetic conservation. *Genetics Research* 43: 339- 43.

Close, P.S., Shoemaker, R.C. and Keim, P. 1989. Distribution of restriction site polymorphism within the chloroplast genome of the genus *Glycine*, subgenus *Soja*. *Theoretical and Applied Genetics* 77: 768-76.

Cordesse, F., Second, G. and Delseny, M. 1990. Ribosomal gene spacer length variability in cultivated and wild rice species. *Theoretical and Applied Genetics* 79: 81-88.

Curtis, S.E. and Clegg, M.T. 1984. Molecular evolution of chloroplast DNA sequences. *Molecular Biology and Evolution* 1: 291-301.

Dallas, J.F. 1988. Detection of DNA 'fingerprints' of cultivated rice by hybridisation with a human minisatellite DNA probe. *Proc. National Academy of Science, USA* 85: 6831-35.

Dally, A.M. and Second, G. 1990. Chloroplast DNA diversity in wild and cultivated species of rice (genus *Oryza*, section Oryza). Cladistic-mutation and genetic-distance analysis. *Theoretical and Applied Genetics* 80: 209-22.

Delgado Salinas, A., Bonet, A. and Gepts, P. 1988. The wild relative of *Phaseolus vulgaris* in Middle America. In Gepts, P. (ed) *Genetic Resources of* Phaseolus *Beans*. Dordrecht, Netherlands: Kluwer.

Delseny, M., McGrath, J.M., This, P., Chevre, A.M. and Quiros, C.F. 1990. Ribosomal RNA genes in diploid and amphidiploid *Brassica* and related species: Organisation, polymorphism, and evolution. *Genome* 33: 733-44.

Doebley, J. 1989. Isozymic evidence and the evolution of crop plants. In Soltis, D.E. and Soltis, P.S. (eds) *Isozymes in Plant Biology*. Portland, Oregon, USA: Dioscorides.

Doebley, J. 1990. Molecular systematics of *Zea* (Gramineae). *Maydica* 35: 143-50.

Doebley, J. 1992. Molecular systematics and crop evolution. In Soltis, P.S., Soltis, D.E. and Doyle, J.J. (eds) *Molecular Systematics of Plants*. New York, USA: Chapman Hall.

Doebley, J.F., Goodman, M.M. and Stuber, C.W. 1984. Isoenzymatic variation in *Zea* (Gramineae). *Systematic Botany* 9: 203-18.

Doebley, J.F., Goodman, M.M. and Stuber, C.W. 1985. Isozyme variation in the races of maize from Mexico. *American J. Botany* 72: 629-39.

Doebley, J., Renfroe, W. and Blanton, A. 1987. Restriction site variation in the *Zea* chloroplast genome. *Genetics* 117: 139-47.

Doyle, J.J. 1988. 5S ribosomal gene variation in the soybean and its progenitor. *Theoretical and Applied Genetics* 75: 621-24.
Doyle, J.J. and Beachy, R.N. 1985. Ribosomal gene variation in soybean (*Glycine*) and its relatives. *Theoretical and Applied Genetics* 70: 369-76.
Duvall, M.R. and Doebley, J.F. 1990. Restriction site variation in the chloroplast genome of *Sorghum* (Poaceae). *Systematic Botany* 15: 472-80.
Dvorák, J. and Appels, R. 1982. Genome and nucleotide differentiation in genomes of polyploid *Triticum* species. *Theoretical and Applied Genetics* 63: 349-60.
Dvorák, J. and Zhang, H.-B. 1990. Variation in repeated nucleotide sequences sheds light on the phylogeny of the wheat B and G genomes. *Proc. National Academy of Science, USA* 87: 9640-44.
Dvorák, J., McGuire, P.E. and Cassidy, B. 1988. Apparent sources of the A genomes of wheat inferred from polymorphism in abundance and restriction fragment length of repeated nucleotide sequences. *Genome* 30: 680-89.
Erickson, L.R., Straus, N.A. and Beversdorf, W.D. 1983. Restriction patterns reveal origins of chloroplast genomes in *Brassica* amphiploids. *Theoretical and Applied Genetics* 65: 201-06.
Evans, A.M. 1976. Beans. In Simmonds, N.W. (ed) *Evolution of Crop Plants*. London, UK: Longman.
Frankel, O.H. 1984. Genetic perspectives of germplasm conservation. In Arber, W.K., Llimensee, K., Peacock, W.J. and Starlinger, P. (eds) *Genetic Manipulation: Impact on Man and Society*. Cambridge, UK: Cambridge University Press.
Furnier, G., Cummings, M.P. and Clegg, M.T. 1990. Evolution of the avocados as revealed by DNA restriction fragment variation. *J. Heredity* 81: 183-88.
Gayler, K.R. and Sykes, G.E. 1985. Effect of nutritional stress on the storage proteins of soybeans. *Plant Physiology* 78: 582-85.
Gepts, P. 1990. Genetic diversity of seed storage proteins in plants. In Brown, A.H.D., Clegg, M.T., Kahler, A.L. and Weir, B.S. (eds) *Plant Population Genetics, Breeding, and Genetic Resources*. Sunderland, Massachusetts, USA: Sinauer.
Gepts, P. 1993. The use of molecular and biochemical markers in crop evolution studies. *Evolutionary Biology* 27: 51-94.
Gepts, P. and Bliss, F.A. 1986. Phaseolin variability among wild and cultivated common beans (*Phaseolus vulgaris*) from Colombia. *Economic Botany* 40: 469-78.
Gepts, P. and Clegg, M.T. 1989. Genetic diversity in pearl millet (*Pennisetum glaucum* [L.] R.Br.) at the DNA sequence level. *J. Heredity* 80: 203-08.
Gepts, P., Osborn, T.C., Rashka, K. and Bliss, F.A. 1986. Phaseolin-protein variability in wild forms and landraces of the common bean (*Phaseolus vulgaris*): Evidence for multiple centers of domestication. *Economic Botany* 40: 451-68.
Graur, D., Bogher, M. and Brieman, A. 1989. Restriction endonuclease profiles of mitochondrial DNA and the origin of the B genome of wheat, *Triticum aestivum*. *Heredity* 62: 335-42.
Halward, T.M., Stalker, H.T., LaRue, E.A. and Kochert, G. 1991. Genetic variation detectable with molecular markers among unadapted germplasm resources of cultivated peanut and related wild species. *Genome* 34: 1013-20.
Hamrick, J.L. and Godt, M.J.W. 1990. Allozyme diversity in plant species. In Brown, A.H.D., Clegg, M.T., Kahler, A.L., and Weir, B.S. (eds) *Plant Population Genetics, Breeding, and Genetic Resources*. Sunderland, Massachusetts, USA: Sinauer.
Harlan, J.R. and de Wet, J.M.J. 1972. A simple classification of cultivated sorghum. *Crop Science* 12: 172-76.
Havey, M.J. and Muehlbauer, F.J. 1989. Variability for restriction fragment lengths and phylogenies in lentil. *Theoretical and Applied Genetics* 77: 839-43.
Hoffman, D.L., Soltis, D.E., Muehlbauer, F.J. and Ladizinsky, G. 1986. Isozyme polymorphism in *Lens* (Leguminosae). *Systematic Botany* 11: 392-402.
Holwerda, B.C., Jana, S. and Crosby, W.L. 1986. Chloroplast and mitochondrial DNA variation in *Hordeum vulgare* and *Hordeum spontaneum*. *Genetics* 114: 1271-91.

Hosaka, K. and Hanneman, R.E. 1988a. The origin of the cultivated tetraploid potato based on chloroplast DNA. *Theoretical and Applied Genetics* 76: 172-76.

Hosaka, K. and Hanneman, R.E. 1988b. Origin of chloroplast DNA diversity in the Andean potatoes. *Theoretical and Applied Genetics* 76: 333-40.

Hosaka, K., Kianian, S.F., McGrath, J.M., Quiros, C.F. 1990. Development and chromosomal localization of genome-specific DNA markers of *Brassica* and the evolution of amphidiploids and n = 9 diploid species. *Genome* 33: 131-42.

Hutchinson, J.B. 1951. Intra-specific differentiation in *Gossypium hirsutum*. *Heredity* 5: 161-93.

Jaaska, V. 1980. Electrophoretic survey of seedling esterases in wheats in relation to their phylogeny. *Theoretical and Applied Genetics* 56: 273-84.

Jaaska, V. 1981. Aspartate aminotransferase and alcohol dehydrogenase isoenzymes: Intraspecific differentiation in *Aegilops tauschii* and the origin of the D genome polyploids in the wheat group. *Plant Systematics and Evolution* 137: 259-73.

Jana, S. and Pietrzak, L. 1988. Comparative assessment of genetic diversity in wild and primitive cultivated barley in a center of diversity. *Genetics* 119: 981-90.

Jorgensen, R.B. 1976. Relationships in the barley genus (*Hordeum*): An electrophoretic examination of proteins. *Hereditas* 104: 273-91.

Kamalay, J.C. and Goldberg, R.B. 1980. Regulation of structural gene expression in tobacco. *Cell* 19: 935-46.

Kaplan, L. and Kaplan, L.N. 1988. *Phaseolus* in archaeology. In Gepts, P. (ed) *Genetic Resources of* Phaseolus *Beans*. Dordrecht, Netherlands: Kluwer.

Kesseli, R., Ochoa, O. and Michelmore, R.W. 1991. Variation at RFLP loci in *Lactuca* spp. and origin of cultivated lettuce (*L. sativa*). *Genome* 34: 430-36.

Khairallah, M.M., Adams, M.W. and Sears, B.B. 1990. Mitochondrial DNA polymorphisms of Malawian bean lines: Further evidence for two major gene pools. *Theoretical and Applied Genetics* 80: 753-61.

Kochert, G., Halward, T., Branch, W.D. and Simpson, C.E. 1991. RFLP variability in peanut (*Arachis hypogea* L.) cultivars and wild species. *Theoretical and Applied Genetics* 81: 565-70.

Koenig, R. and Gepts, P. 1989. Allozyme diversity in wild *Phaseolus vulgaris*: Further evidence for two major centers of diversity. *Theoretical and Applied Genetics* 78: 809-17.

Koenig, R., Singh, S.P. and Gepts, P. 1990. Novel phaseolin types in wild and cultivated common bean (*Phaseolus vulgaris*, Fabaceae). *Economic Botany* 44: 50-60.

Marchais, L. and Pernès, J. 1985. Genetic divergence between wild and cultivated pearl millets (*Pennisetum typhoides*). I. Male sterility. *Zeitschrift für Pflanzenzüchtung* 95: 103-12.

McClean, P.E. and Hanson, M.R. 1986. Mitochondrial DNA sequence divergence among *Lycopersicon* and related *Solanum* species. *Genetics* 112: 649-67.

McLeod, M.J., Guttman, S.I. and Eshbaugh, W.H. 1982. Early evolution of chili peppers (*Capsicum*). *Economic Botany* 36: 361-68.

McLeod, M.J., Guttman, S.I., Eshbaugh, W.H. and Rayle, R.E. 1983. An electrophoretic study of evolution in *Capsicum* (Solanaceae). *Evolution* 37: 562-74.

Messmer, M.M., Melchinger, A.E., Lee, M., Woodman, W.L., Lee, E.A. and Lamkey, K.R. 1991. Genetic diversity among progenitors and elite lines from the Iowa Stiff Stalk Synthetic (BSSS) maize population: Comparison of allozyme and RFLP data. *Theoretical and Applied Genetics* 83: 97-107.

Miller, J.C. and Tanksley, S.D. 1990. RFLP analysis of phylogenetic relationships and genetic variation in the genus *Lycopersicon*. *Theoretical and Applied Genetics* 80: 437-48.

Morden, C.W., Doebley, J. and Schertz, K.F. 1990. Allozyme variation among the spontaneous species of *Sorghum* section Sorghum (Poaceae). *Theoretical and Applied Genetics* 80: 296-304.

Neale, D.B., Saghai-Maroof, M.A., Allard, R.W., Zhang, Q. and Jorgensen, R.A. 1986. Chloroplast DNA diversity in populations of wild and cultivated barley. *Genetics* 120: 1105-10.

Nevo, E., Zohary, D., Brown, A.H.D. and Haber, M. 1979. Genetic diversity and environmental associations of wild barley, *Hordeum spontaneum*, in Israel. *Evolution* 33: 815-33.

Nevo, E., Beiles, A. and Zohary, D. 1986. Genetic resources of wild barley in the Near East: Structure, evolution and application in breeding. *Biology J. Linnaean Society* 27: 355-80.

Nodari, R.O., Koinange, E.M.K., Kelly, J.D. and Gepts, P. 1992. Towards an integrated linkage map of common bean. I. Development of genomic DNA probes and levels of restriction fragment length polymorphism. *Theoretical and Applied Genetics* 84: 186-92.

Ogihara, Y. and Tsunewaki, K. 1988. Diversity and evolution of chloroplast DNA in *Triticum* and *Aegilops* as revealed by restriction fragment analysis. *Theoretical and Applied Genetics* 76: 321-32.

Palmer, J.D. 1987. Chloroplast DNA evolution and biosystematic uses of chloroplast DNA variation. *American Naturalist* 130: S6-S29.

Palmer, J.D., Shields, C.R., Cohen, D.B. and Orton, T.J. 1983. Chloroplast DNA evolution and the origin of amphidiploid *Brassica* species. *Theoretical and Applied Genetics* 65: 181-89.

Palmer, J.D., Jorgensen, R.A. and Thompson, W.F. 1985. Chloroplast DNA variation and evolution in *Pisum*: Patterns of change and phylogenetic analysis. *Genetics* 109: 195-213.

Pinkas, R., Zamir, D. and Ladizinsky, G. 1985. Allozyme divergence and evolution in the genus *Lens*. *Plant Systematics and Evolution* 151: 131-40.

Rick, C.M. and Fobes, J.F. 1975. Allozyme variation in the cultivated tomato and closely related species. *Bull. Torrey Botanical Club* 102: 376-84.

Rieseberg, L.H. and Seiler, G. 1990. Molecular evidence and the origin and development of the domesticated sunflower (*Helianthus annuus* L.). *Economic Botany* 44 (3S): 79-91.

Robert, T., Lespinasse, R., Pernès, J. and Sarr, A. 1991. Gametophytic competition as influencing gene flow between wild and cultivated forms of pearl millet (*Pennisetum typhoides*). *Genome* 34: 195-200.

Second, G. 1982. Origin of the genetic diversity of cultivated rice (*Oryza* spp.): Study of the polymorphism scored at 40 enzyme loci. *Japanese J. Genetics* 57: 25-57.

Simmonds, N.W. 1976. *Evolution of Crop Plants*. London, UK: Longman.

Singh, S.P., Gepts, P. and Debouck, D.G. 1991a. Races of common bean (*Phaseolus vulgaris* L., Fabaceae). *Economic Botany* 45: 379-96.

Singh, S.P., Gutiérrez, J.A., Molina, A., Urrea, C. and Gepts, P. 1991b. Genetic diversity in cultivated common bean. II. Marker-based analysis of morphological and agronomic traits. *Crop Science* 31: 23-29.

Singh, S.P., Nodari, R. and Gepts, P. 1991c. Genetic diversity in cultivated common bean. I. Allozymes. *Crop Science* 31: 19-23.

Song, K.M., Osborn, T.C. and Williams, P.H. 1988a. *Brassica* taxonomy based on nuclear restriction fragment length polymorphisms (RFLPs). 1. Genome evolution of diploid and amphidiploid species. *Theoretical and Applied Genetics* 75: 784-94.

Song, K.M., Osborn, T.C. and Williams, P.H. 1988b. *Brassica* taxonomy based on nuclear restriction fragment length polymorphisms (RFLPs). 2. Preliminary analysis of subspecies within *B. rapa* (syn. *campestris*) and (*B. oleracea*). *Theoretical and Applied Genetics* 76: 593-600.

Song, K.M, Osborn, T.C. and Williams, P.H. 1990. *Brassica* taxonomy based on nuclear restriction fragment length polymorphisms (RFLPs). 3. Genome relationships in *Brassica* and related genera and the origin of *B. oleracea* and *B. rapa* (syn. *campestris*). *Theoretical and Applied Genetics* 79: 497-506.

Stockton, T., Sonnante, G. and Gepts, P. 1992. Detection of minisatellite sequences in *Phaseolus vulgaris*. *Plant Molecular Biology Reporter* 10: 47-59.

U, N. 1935. Genome analysis in *Brassica* with special reference to the experimental formation of *B. napus* and peculiar mode of fertilisation. *Japanese J. Botany* 7: 389-452.

Weeden, N.F. and Wendel, J.F. 1989. Genetics of plant isozymes. In Soltis, D.E. and Soltis, P.S. (eds) *Isozymes in Plant Biology*. Portland, Oregon, USA: Dioscorides.

Wellhausen, E.J., Roberts, L.M. and Hernández X.E. 1952. *Races of Maize in Mexico*. Cambridge, Massachusetts, USA: Harvard University.

Wendel, J.F. and Weeden, N.F. 1989. Visualisation and interpretation of plant isozymes. In Soltis, D.E. and Soltis, P.S. (eds) *Isozymes in Plant Biology*. Portland, Oregon, USA: Dioscorides.

Wendel, J.F., Brubaker, C.L. and Percival, A.E. 1992. Genetic diversity in *Gossypium hirsutum* and the origin of upland cotton. *American J. Botany* 79: 1291-310.

Williams, J.G.K., Kubelik, A.R., Livak, K.J., Rafalski, J.A. and Tingey, S.V. 1990. DNA polymorphisms amplified by arbitrary primers are useful as genetic markers. *Nucleic Acids Research* 18: 6531-35.

Wolfe, K.H., Li, W.-H. and Sharp, P.M. 1987. Rates of nucleotide substitution vary greatly among plant mitochondrial, chloroplast, and nuclear DNAs. *Proc. National Academy of Science, USA* 84: 9054-58.

Zimmer, E.A., Jupe, E.R. and Walbot, V. 1988. Ribosomal gene structure, variation, and inheritance in maize and its ancestors. *Genetics* 120: 1125-36.

Zurawski, G. and Clegg, M.T. 1987. Evolution of higher plant chloroplast DNA-encoded genes: Implications for structure-function and phylogenetic studies. *Annual Review of Plant Physiology* 38: 391-418.

Core Collections of Plant Genetic Resources
Edited by T. Hodgkin, A.H.D. Brown, Th.J.L. van Hintum and E.A.V. Morales

A Wiley-Sayce Publication

3.5

Integrating different types of information to develop core collections, with particular reference to *Brassica oleracea* and *Malus* x *domestica*

S. Kresovich, W.F. Lamboy, J.R. McFerson and P.L. Forsline

Abstract

The fundamental responsibilities of a gene bank curator are to acquire, maintain and make accessible useful and representative genetic diversity of the crop of interest. This requires an understanding of the genetic and agronomic/horticultural framework of the variation exhibited by the crop. Genetic measures of identity, structure, relationship and action/location are needed for sound conservation and use. A curator may implement various strategies and procedures suited to establishing these characteristics of accessions and taxa. Initially, he/she must be able to formulate the appropriate question to be answered, and then integrate different sources and types of phenotypic and genotypic data (including whole plant, biochemical and molecular) to address immediate and long-term needs.

The objective of this chapter is to highlight the dynamic nature of genetic resources conservation and use. Issues of particular relevance to *ex situ* conservation and characterisation are discussed, including: specific situations commonly encountered by curators, formulating questions to be addressed, identifying potentially relevant data sets for decision making and planning, devising approaches for data analysis, generating priority options for action, and examining the advantages and limitations of the proposed actions. The authors' experience with collections of *Brassica* and *Malus* species is used to illustrate various points. In the conclusion, they discuss the evolution of core collections in the light of theoretical and technical advances in molecular genetics and biology, biometry, database management and computing science, and plant breeding.

The genesis, evaluation and implementation of the core collection concept (Frankel, 1984; Brown, 1989a, b) allows curators and users of genetic resources to make more accessible the potential value of the broad array of entries maintained in gene banks. Recent advances in technologies and models of molecular and population genetics have encouraged the implementation of many approaches, methods and analyses to improve our understanding of the genetic basis and value of diversity. However, this abundance of potential opportunities to resolve and use genetic variation has created other types of problems for the management of *ex situ* collections.

This chapter focuses on situations commonly encountered within an *ex situ* gene bank, including questions which may arise, data available for problem resolution, analytical approaches, and subsequent actions. We describe problems encountered during out attempt to develop core collections and their application for improved management of our collections of *Brassica* and *Malus* species. We also discuss other examples drawn from collection maintenance activities at the Plant Genetic Resources Unit (PGRU) at Cornell University in New York, USA. In the concluding section of the chapter, we discuss the evolution of core collections based on progress in allied fields concerned with improved resolution and quantification of genetic variation.

ESTABLISHING CORE COLLECTIONS AND CHARACTERISING GENETIC VARIATION

Many biological and non-biological considerations affect the management of plant genetic resources (*see* Table 1). Together, these issues make the task of an *ex situ* gene bank curator a complex one (Kresovich, 1992; Kresovich and McFerson, 1992). Within the framework of these considerations,

Table 1 Considerations to be taken into account in the management of plant genetic resources

Biological	**Non-biological**
Systematic considerations: Evolutionary history of the taxon and related taxa Genome organisation and complexity	Operations: Location of the active collection Information availability Current level of ongoing research Other collections Number, quality and cost of personnel employed
Agroecological considerations: Geographic and niche range of the taxon and related taxa Populational diversity and accessibility	Infrastructure: Regeneration facilities Storage facilities
Genetic considerations: Life habit Time to, and criteria for, reproduction Pollination system Breeding system	Finances: Level of support Timing
Agronomic/horticultural considerations: Plant size Morphological variation End use Method of preservation Phytosanitary status Method of distribution	Tradition: History Previous paradigms for plant introduction Access to cooperators

examples are presented here on ways of resolving difficulties in determining genetic identity, relatedness and structure in the establishment of a core collection.

Genetic identity

The genetic identity of an accession is important in the establishment of any core subset or the main collection itself (Kresovich et al., 1992, 1993). It will be of value in such critical issues of global genetic resources conservation as ownership, intellectual property and the movement and exploitation of genes and genotypes. Investigations are being conducted at the PGRU to identify and establish the within-accession heterogeneity of similarly named entries in the *Brassica oleracea* L. collection. As isozymic analysis lacked the resolving power needed, we now use simple molecular techniques including random amplified polymorphic DNA (RAPD) (Kresovich et al., 1992) and molecular probing with short, interspersed repetitive DNA (SINE) with sampling strategies to calculate genetic identity of the entries recorded as 'Golden Acre'. This historically important selection is represented more than 30 times across collections held at the PGRU, at Wellesbourne in the UK and at Wageningen in the Netherlands. Because of the great expense involved in the maintenance and regeneration of small-seeded, allogamous biennials, the information obtained will help the curator handle the array of 'Golden Acre' entries.

Genetic relatedness

Two examples serve to illustrated the usefulness of a core collection in the maintenance and evaluation of *Malus* x *domestica* Borkh. and related species. In the first case, we have established and used a database supported by pedigree records, morphological and horticultural traits, and isozyme analysis (seven enzyme systems encoded by 15 loci) for over 2000 horticulturally important apple clones and related species (Dickson et al., 1991). Using multivariate techniques (hierarchical agglomerative clustering and principal components analysis) and based on which material had been sanitised previously, priorities have been set for virus eradication using chemotherapy and thermotherapy complemented by *in vitro* techniques and subsequent field planting. In addition, isozyme screening of the collection has yielded a marker at the *Pgm-2*/*Pgm-3* loci which has proved useful in discriminating wild from domesticated species of *Malus* (Dickson et al., 1991). Further analyses may identify other useful markers.

In the second case, the same database and approach has been employed to support a genetic resources evaluation network. Because of the great demand for phenotype in adaptation and pest screening studies in apple, material is propagated clonally and difficulties are often encountered with movement because of quarantine regulations stemming from concern about the spread of latent viruses. Therefore, a core of 200 clones representing an apple collection of 2500 entries was established for a replicated multi-site national network for disease screening in the USA.

Genetic structure

Genetic structure within an accession or within a collection serves as an important criterion for the development of the core collection. We have used genetic structure to help in three major activities at the PGRU. The first is an intensive programme to eradicate pea seed-borne mosaic virus (PSbMV) from the *Pisum sativum* L. collection. The second involves establishing some measure of genetic organisation in landraces of *B. oleracea* (particularly cabbage and kale) recently collected from the major cropping regions of Portugal. The third activity is designed to establish the genetic structure of a large test population of the crop types of *B. oleracea* as a whole.

Eradication of pea seed-borne mosaic virus

This is an issue of great concern globally as pressures increase to exchange genetic resources. As will be noted later, the use of a core collection for both eradication and screening for disease susceptibility may prove to have been a wise alternative approach.

The virus was first described in 1968 and since then has been found to occur not only in production fields but also in genetic resource collections across the world. It has been a problem in the collection which has been maintained by the PGRU for almost 25 years. In collaboration with three plant pathologists/breeders in Washington and Oregon, USA, an intensive programme was initiated to eradicate PSbMV from the collection and to increase the seed of the virus-free accessions. The PSbMV eradication phase included obtaining a seed stock source for each plant introduction with the minimum potential of PSbMV infection, growing 20-30 plants per accession under aphid-free, greenhouse conditions, and roguing individual plants showing symptoms visually or through an enzyme linked immunosorbent assay (ELISA). Seed harvested from the PSbMV-free plants was bulked to form the increase and serve as sanitised stock for future increase and distribution. We felt it would be instructive to conduct a detailed follow-up study in order to assess the degree of genetic shift which had taken place in a broad range of morphological and biochemical markers during an *actual* virus eradication programme.

Fifty accessions comprising heterogeneous landraces and cultivars which represented diverse geographic origins and had undergone the virus eradication programme were selected for study. Forty plants for each stock of each entry were grown in the greenhouse and analysed for 12 biochemical (isozyme) and 20 morphological markers. In addition, all plants were tested for the presence of PSbMV 21 days after germination, using ELISA.

Based on these 32 characters, F-similarity coefficients were calculated for the pea accessions under study. The results suggested that significant genetic shift had occurred between the source and sanitised stocks in all accessions. As might be expected, the genetic structures of heterogeneous landraces were more greatly affected than those of the more uniform cultivars. Quantitative characters exhibited the least amount of shift, while qualitative and isozymic markers were highly affected by the eradication and regeneration processes. Additional statistical analyses were used to determine the characters most affected. Nei's diversity index was applied to all isozyme data and pooled across all accessions. Two loci, *Aat-3* and *Pgd-1*, showed a high degree of diversity with no observable shifts occurring between source and sanitised stocks. The loci *Gpi-2*, *Idh-1* and *Lap-1* were monomorphic for both source and sanitised populations. However, five isozyme loci (*Aat-2*, *Dia-3*, *Pgd-2*, *Pgm-1* and *Pgm-2*) displayed significant genetic shifts. Furthermore, allelic diversity at two loci (*Lap-1* and *Mdh-2*) was completely eliminated by the eradication and regeneration processes. For 10 morphological characters, the Shannon-Weaver index was applied to each pair of stocks within an accession and averaged across all accessions. Among the characters studied, diversity and subsequent allelic frequency shifts were generally greatest among flower- and seed-related traits and lowest among pod-related traits.

These observations underline the need for caution in undertaking large-scale disease eradication and seed regeneration programmes, particularly when maintenance of broad genetic and phenotypic diversity is paramount. Changes in genetic diversity might be minimised by other types of eradication strategies, such as therapy, rather than roguing of infected materials. However, curators ultimately must consider the cost of disease eradication programmes from both the financial and scientific perspectives.

Measuring genetic organisation of Brassica *landraces*

The second case involves the establishment of a set of accessions that represent the genetic variation collected from a particular country. Unique vegetable types of *B. oleracea* have evolved throughout Portugal as a result of particular selection pressures, including marketing considerations, cultivation methods and agroecological factors. Landraces in these regions exhibit considerable phenotypic variation in morphology and disease resistance. Despite the great horticultural importance of these diverse groups, little documentation exists on their genetic relationships. Isozyme analysis (starch and cellulose acetate gel electrophoretic techniques) and multivariate analysis (hierarchical agglomerative clustering and principal components analysis) were used to assess levels of genetic variation within, and genetic distance between, landraces collected from various regions of the country (Silva-Dias et al., 1994).

As expected for populations of an allogamous species, the landraces showed considerable genetic variation within and between accessions. Nine enzyme systems encoded by 22 putative loci yielded 107 alleles in the investigation. Of the 22 loci, only one was found to be monomorphic for the 20 plants of each of the 52 entries (including 36 landraces) studied. Among the particularly informative loci were *Gpi-2*, *Idh-4*, *Lap-1*, *Pgm-3* and *Pgm-4*. Using various types of multivariate analysis, we were able to identify genetically unique entries among the landraces. The analyses revealed patterns of diversity among accessions and the findings were complemented by concurrent horticultural and restriction fragment length polymorphism (RFLP) studies of the same material. Independent of the method used, it was found that crop form (for example, cabbage or kale) was less important in partitioning genetic variation than source of the accession (agroecological niche, such as cropping system, or edaphic and meteorological factors). Therefore, if one wished to establish a core subset of the Portuguese entries, it would be more rational to establish it based on niche. This finding, although surprising, dictated the most prudent strategy to be pursued.

Establishing the genetic structure within Brassica oleracea

The third activity was undertaken to establish the genetic structure within *B. oleracea* using isozyme data obtained from starch gel electrophoresis (Lamboy et al., 1993). Fifty-six accessions were grown from seed stored at the PGRU. They were selected to provide a cross-section of the species' variation and included 14 cultivar groups (such as cabbage, broccoli, cauliflower, Jersey kale and kohlrabi) which had been determined based on morphology. Four enzyme systems were analysed (*Gpi*, *Lap*, *Pgd*, and *Pgm*). Allelic frequencies were analysed phenetically using Nei's genetic identity measures.

Six genetically defined loci were found in the four enzyme systems analysed: *Gpi-2*, *Lap*, *Pgd-1*, *Pgm-1*, *Pgm-2* and *Pgm-3*. All accessions were monomorphic at *Pgm-1*; the other loci were polymorphic. Four of the 56 accessions (three of cauliflower and one of cabbage) had low mean heterozygosities of less than 0.1. All four accessions were highly selected commercial lines. There were, however, other equally highly selected lines with greater diversity. Nine of the 56 accessions were particularly variable, with mean heterozygosities greater than 0.4. These accessions were either landraces or weakly selected lines of diverse commercial origin. In view of this finding, a curator or user wishing to maximise diversity based on a *finite number of accessions* may emphasise the need to include landraces in a core collection of *B. oleracea*. Conversely, these results also suggest that a core collection represented by a *finite number of plants* of *B. oleracea* may require the inclusion of a few seeds representing a diverse range of accessions of highly selected commercial lines, or the inclusion of a greater number of seeds derived from fewer heterogeneous landraces of particular origin (with possible exposure to the desired selection pressure).

Perhaps the most important result of the genetic analysis of the *B. oleracea* isozyme loci was found by partitioning the variation within and between crop types. Using Wright's F-statistics, it was found that, when averaged over all loci, more than 93% of the variation was attributable to differences between accessions (within crop types) and only about 7% of the diversity resulted from differences between varieties. Thus, if a core collection is to represent faithfully the diversity within *B. oleracea*, then *several* accessions must be selected from each crop type based on other factors affected by selection. This finding is in contrast to a more simple situation in which almost all the variation in a species is between crop types, in which case a core collection made up of just a *single* accession of each type might well be sufficient to represent the diversity in the species. As in the above-mentioned study of Portuguese entries, factors such as level of crop improvement and agroecological niche play a greater role in affecting variation than does crop type.

EVOLUTION OF THE CORE COLLECTION

The proposal to develop core collections has served to clarify the conceptual basis of genetic resources conservation and use, and has challenged curators to become more effective in establishing priorities. In essence, the formation of core collections has emphasised the value of *genetics* in effective genetic resources management.

As noted in many other chapters in this volume, core collections will continue to evolve to serve curators and users, whether they are involved with fundamental or applied plant sciences. Criticisms of the approach, such as reduced trait screening efficiency because of poor detection of rare alleles, and potential biases of core establishment, will have minimal impact on curatorial activities.

We suggest, apart from refinements of the core collection concept, other major advances will be made in related fields of research, particularly those concerned with intensive and extensive characterisation of entire collections. We also consider that the evolution of the discipline will occur more rapidly and be more accessible than currently believed (Kresovich et al., 1992, 1993).

The 15-year, US$3 billion effort known as the Human Genome Project will generate more data and technology than any single project to date in biology. Its impact on biology and related fields such as agriculture will be as significant as the impact of the space programme on the physical and computing sciences. Technologies which make use of DNA amplification and sequencing, particularly those automated for specific uses, may revolutionise approaches to genetic resources conservation and use. Methods employing random and semi-random priming for DNA amplification (Welsh and McClelland, 1990, 1991; Williams et al., 1990, 1993; Caetano-Anolles et al., 1991a, b; Weining and Langridge, 1991; Akopyanz et al., 1992) and repeat DNA polymorphisms (Litt and Luty, 1989; Edwards et al., 1991; Jacob et al., 1991; Jeffreys et al., 1991) can greatly assist in the development and maintenance of higher quality collections. In particular, these molecular techniques may be used to address additional curatorial needs such as determination of genetic identity and representation, partitioning of variation, monitoring of seed regeneration, and targeting of useful genes. This progress at the molecular level will enhance rather than reduce the need for more detailed agronomic and horticultural evaluations. Therefore, continued close linkage to advances in plant breeding is also essential.

In the longer term, researchers may screen collections directly for useful genes by using diagnostic methods such as the ligase chain reaction (LCR) (Nickerson et al., 1990; Barany, 1991). For this to become reality, we must be able to extract and store high-quality genomic and complementary DNA, understand gene and genome organisation and regulation, locate and sequence useful genes, and develop automated systems to support these efforts. All these avenues are being vigorously pursued by various research groups worldwide.

Researchers associated with genetic resources management must confront challenges and find new avenues to explore in order to devise new ways of making more effective use of the world's finite resources.

Acknowledgements

The authors gratefully acknowledge the valued efforts of the numerous PGRU cooperators (both nationally and globally) and personnel. The senior author also wishes to thank those researchers who served as hosts during his recent fellowship at the Division of Biology and Biomedical Sciences, Washington University, St Louis, USA.

References

Akopyanz, N., Bukanov, N.O., Westblom, U.T., Kresovich, S. and Berg, D. 1992. DNA diversity among clinical isolates of *Helicobacter pylori* detected by PCR-based RAPD fingerprinting. *Nucleic Acids Research* 20: 5137-42.

Barany, F. 1991. Genetic disease detection and DNA amplification using cloned thermostable ligase. *Proc. National Academy of Science, USA* 88: 189-93.

Brown, A.H.D. 1989a. The case for core collections. In Brown, A.H.D., Frankel, O.H., Marshall D.R. and Williams, J.T. (eds) *The Use of Plant Genetic Resources*. Cambridge, UK: Cambridge University Press.

Brown, A.H.D. 1989b. Core collections: A practical approach to genetic resources management. *Genome* 31: 818-24.

Caetano-Anolles, G., Bassam, B.J. and Gresshoff, P.M. 1991a. DNA amplification fingerprinting using very short arbitrary oligonucleotide primers. *Bio/Technology* 9: 553-57.

Caetano-Anolles, G., Bassam, B.J. and Gresshoff, P.M. 1991b. DNA amplification fingerprinting: A strategy for genome analysis. *Plant Molecular Biology Reporter* 9: 294-307.

Dickson, E.E., Kresovich, S. and Weeden, N.F. 1991. Isozymes in North American *Malus* (*Rosacece*): Hybridisation and species differentiation. *Systematic Botany* 16: 363-75.

Edwards, A., Civitello, A., Hammond, H.A. and Caskey, C.T. 1991. DNA typing and genetic mapping with trimeric and tetrameric tandem repeats. *American J. Human Genetics* 9: 746-56.

Frankel, O.H. 1984. Genetic perspectives of germplasm conservation. In Arber, W., Llimensee, K., Peacock, W.J. and Starlinger, P. (eds) *Genetic Manipulation: Impact on Man and Society*. Cambridge, UK: Cambridge University Press.

Jacob, H.J., Lindpainter, K., Lincoln, S.E., Kusumi, K., Bunker, R.K., Mao, Y.P., Ganten, D., Dzau, V.J. and Lander E.S. 1991. Genetic mapping of the gene causing hypertension in the stroke-prone hypersensitive rat. *Cell* 67: 213-24.

Jeffreys, A.J., MacLeod, A., Tamaki, K., Neil, D.L. and Monckton, D.G. 1991. Minisatellite repeat coding as a digital approach to DNA typing. *Nature* 354: 204-09.

Kresovich, S. 1992. Plant genetic resources conservation and utilisation: An evolving paradigm. *Field Crops Research* 29: 183-84.

Kresovich, S. and McFerson, J.R. 1992. Assessment of plant genetic diversity: Considerations of intra- and inter-specific variation. *Field Crops Research* 29: 185-204.

Kresovich, S., Williams, J.G.K., McFerson, J.R., Routman, E.J. and Schaal, B.A. 1992. Characterisation of genetic identities and relationships of *Brassica oleracea* L. via random polymorphic DNA assay. *Theoretical and Applied Genetics* 85: 190-96.

Kresovich, S., Lamboy, W.F., Szewc-McFadden, A.K., McFerson, J.R. and Forsline, P.L. 1993. Molecular diagnostics and plant genetic resources conservation. *AgBiotech News and Information* 5: 555-58.

Lamboy, W.F., McFerson, J.R., Westman, A.L. and Kresovich, S. 1993. Use of isozyme information in the management of the United States *Brassica oleracea* L. genetic resources collection. *Genetic Resources and Crop Domestication.* (in press).

Litt, M. and Luty, J.A. 1989. A hypervariable microsatellite revealed by *in vitro* amplification of a dinucleotide repeat within the cardiac muscle actin gene. *American J. Human Genetics* 44: 397-401.

Nickerson, D.A., Kaiser, R., Lappin, S., Stewart, J., Hood, L. and Landegren, U. 1990. Automated DNA diagnostics using an ELISA-based oligonucleotide ligation assay. *Proc. National Academy of Science, USA* 87:8923-27.

Silva-Dias, J.C., Monteiro, A.A. and Kresovich, S. 1994. Genetic variation of Portuguese Tronchuda cabbage and Galega kale landraces using isozyme analysis. *Euphytica* (in press).

Welsh, J. and McClelland, M. 1990. Fingerprinting genomes using PCR with arbitrary primers. *Nucleic Acids Research* 18: 7213-18.

Welsh, J. and McClelland, M. 1991. Genomic fingerprinting using arbitrary primed PCR and a matrix of pairwise combinations of primers. *Nucleic Acids Research* 19: 5275-79.

Weining, S. and Langridge, P. 1991. Identification and mapping of polymorphisms in cereals based on the polymerase chain reaction. *Theoretical and Applied Genetics* 82: 209-16.

Williams, J.G.K., Kubelik, A.R., Livak, K.J., Rafalski, J.A. and Tingey, S.V. 1990. DNA polymorphisms amplified by arbitrary primers are useful as genetic markers. *Nucleic Acids Research* 18: 6531-35.

Williams, J.G.K., Hanafey, M.K., Rafalski, J.A. and Tingey, S.V. 1993. Genetic analysis using RAPD markers. *Methods in Enzymology* 218: 704-40.

Core Collections of Plant Genetic Resources
Edited by T. Hodgkin, A.H.D. Brown, Th.J.L. van Hintum and E.A.V. Morales

A Wiley-Sayce Publication

3.6

Towards a Brazilian core collection of cassava

C.M.T. Cordeiro, E.A.V. Morales, P. Ferreira, D.M.S. Rocha, I.R.S. Costa, A.C.C. Valois and S. Silva

Abstract

Guidelines for the construction of a Brazilian core collection for cassava are presented in this chapter. Three criteria are proposed for the stratification of accessions: category (landraces or improved materials); origin (grouping accessions according to agroecological classification of origins); and characters of importance to the breeder (selected according to importance, heritability and reliability of data). The accessions are classified, according to origin, into nine Brazilian ecogeographic zones. A validation exercise of the proposed zoning, based on characterisation and evaluation data and using multivariate statistical techniques, is described. Recommendations are made on sampling methodology to build the core and on enlarging it to include wild materials.

The impressive growth of gene banks and the difficulties in obtaining appropriate funding for their operation have highlighted the need to devise strategies for the efficient management and more intensive use of collections. A major constraint in the germplasm collections of developing countries is the low level of characterisation and evaluation, mainly because of financial difficulties. The idea of concentrating efforts on a representative sample of the total reserve collection (Frankel, 1984; Brown, 1989a, b), to be distinguished as a core collection which is accessible to plant breeders and other users, has a major role to play in this context. Of necessity, it involves setting priorities for activities and resource allocation within a gene bank. When dealing with vegetatively propagated materials maintained in the field, conservation is expensive and risky, and thus special consideration should be given to planning activities such as evaluation, *in vitro* conservation, hybridisation, and duplication of accessions for other gene banks. This underlines the importance of the construction of a core collection.

THE BRAZILIAN CASSAVA COLLECTION

Cassava is a staple food for millions of people in the neotropics and paleotropics. The crop is grown almost throughout Brazil, and in terms of area under cultivation it is one of the most important crops in the country. The total Brazilian cassava root production is approximately 24 million tons per year; production in the north-eastern region accounts for almost 50% of the total (EMBRAPA, 1991).

Cassava is an important component of low-input agriculture because it is able to produce good yields under unfavourable conditions such as low soil fertility, acid soils or drought. The major constraints to production are pests and diseases. The most important uses of the crop in Brazil are roots and flour for human consumption, biomass, roots, chips and pellets for animal consumption, and roots for industrial uses, including starch production. The cultivation of cassava can be traced back 3000 to 7000 years, when it was grown by tribes in the lowland humid tropics (Hershey, 1991). There are many centres of origin and diversity for *Manihot*. Spath (1973) suggested four separate areas of origin: Guatemala and Mexico; the coastal savannas of north-western South America; eastern Bolivia and north-western Argentina; and eastern Brazil. Within the genus *Manihot*, 98 species are widely distributed throughout the lowland tropics of the Americas.

Although there has been considerable progress in cassava breeding in the past 20 years, there is much room for improvement. Floral biology and pollination habit make cassava a predominantly allogamous and highly heterozygous species. Thus hybridisation of selected parents creates source populations with high genetic variability (Bueno, 1991). Since cassava is propagated vegetatively through stem cutting, superior genotypes identified in breeding programmes may be maintained indefinitely.

Breeding programmes are carried out by a number of institutes in Brazil. Work on cassava improvement can be traced back to the end of the 19th century when Brazil's first cassava collection was initiated at Sao Bento das Lajes (Zehnter, 1919). Some of these materials were used for pioneer breeding work in Africa and Asia.

Instability of institutional policies, at a national level, and insufficient budgets have had a strong negative impact on conservation activities and, in particular, on those crops which require vegetative propagation, such as cassava. Periodic loss of materials, or even of small collections, points to the need for action in terms of establishing priorities and defining optimum strategies. In this context, the idea of a core collection of cassava offers a promising approach.

The Centro Nacional de Pesquisa em Recursos Genéticos e Biotecnologia (CENARGEN) has the national mandate for germplasm maintenance and research activities. In particular, it coordinates a national cassava conservation network, based on a central data bank. The Brazilian collections of cassava (*M. esculenta* Crantz) are maintained in the field by various research institutes and universities (*see* Table 1). A few *in vitro* collections are also maintained by several institutes; among these is the CENARGEN *in vitro* collection, which holds about 800 accessions.

A small number of additional collections are not included in Table 1 because no information on the number of accessions in them was readily available in the CENARGEN data bank. These collections hold a significant percentage of the materials from the Amazonian and Cerrado regions.

ESTABLISHING THE PARAMETERS OF A CORE CASSAVA COLLECTION

The objective of this study is to suggest guidelines for the construction of a Brazilian core collection of cassava. Basic information was obtained by analysing the inventory of all Brazilian cassava collections. Characterisation and agronomic data are available for about 40% of the whole collection. Also, some information is available on stress tolerance and on resistance to major pathogens or pests. Information on origin (passport data) has been obtained to the level of municipality for almost 80% of the accessions.

Table 1 **Number of accessions of *Manihot esculenta* Crantz in Brazilian gene banks, based on data from the CENARGEN data bank, 1992**

Institution	Location	No. of accessions
EMBRAPA-CNPMF	Cruz das Almas-Bahia	1454
EMBRAPA-CPAA	Manaus-Amazonas	78
EMBRAPA-CPATSA	Petrolina-Pernambuco	86
EMBRAPA-CPATU	Belém-Pará	145
EMAPA	Bacabal-Maranhao	136
EMCAPA	Linhares-Espírito Santo	255
EPACE-BARBALHA	Barbalha-Ceará	75
EPACE-TIANGUA	Tianguá-Ceará	25
EPACE-PACAJUS	Pacajús-Ceará	64
EPAGRI	Itajaí-Santa Catarina	282
IPA-ITAMBE	Itambé-Pernambuco	89
IPA-ITAPIREMA	Itapirema-Pernambuco	226
IPA-ARARIPINA	Araripina-Pernambuco	113
IAC	Ubatuba-Sao Paulo	328
IAPAR	Londrina-Paraná	487
IPAGRO	Taquarí-R. G. do Sul	289
Total		4132[a]

Note: a Approximately 1200 redundancies (duplications held in different collections) are known

Several requirements are considered as crucial. The first one is representativeness of the core as a sample of the whole collection. A second requirement is the inclusion of genotypes adapted to relevant environmental conditions defined according to the agroecological zones of Brazil. However, since some zones have a low representation in the whole collection, it is acknowledged that the final goal of representing the variability and diversity of cassava througout the country will be difficult to attain.

In the case of *M. esculenta* we are dealing with a vegetatively propagated and domesticated species. Human selection and multiplication of adapted genotypes for specific environmental conditions has played a strong role in determining the present distribution of this species. In this sense, the situation with *Manihot* might be compared to that of asexual species, such as banana and yam, where conservation of genotypes and phenotypic expressions in specific environments are crucial.

In addition, the conceptual background used for the definition of the Brazilian core for cassava places the emphasis on pragmatic demands. It is assumed that better use of collections depends upon the criteria for the selection of entries. The selection of genotypes on the basis of such factors as agronomic information (that is, adaptability, stability, resistance to pests, stress tolerance, yield and other characters of interest to breeders) should play a leading part in developing the core. The goal is to give priority to characters known to be useful. However, much of this crucial information is lacking, and this hinders the agronomic definition of a core collection.

The dilemma in selecting a strategy for constructing a core collection has been aptly expressed by Frankel (1985): 'The distinction between plant and gene introduction is significant for the strategy of a collection. If its objective is genotypes rather than genes, then the collection should consist of diverse and representative cultivars or ecotypes, the main criterion for inclusion being ecological relevance. This would result in a substantial rationalisation of collections which are now inflated by a plethora of accessions unlikely ever to be used, while preserving a broad representation of the genetic heritage.' With regard to the choice of a strategy for a number of crops, Frankel points out that 'the question is which of these crops is now, or is likely to be in the forseeable future, subject to plant breeding other than the selection of individuals or populations.'

In the case of *Manihot* most of the breeding work until now has been made through selection, but hybridisation will soon have high priority. Crossing is quite easily achieved in most cases, segregation is high, and thus the emphasis on the conservation or representativeness of alleles instead of genotypes is logical. This is particularly true in the case of landraces or wild *Manihot* material.

In an attempt to address these two important points raised by Frankel (1985), improved cultivars with characters of interest in cultivation have been treated separately from the landraces, which may also have useful combinations of genes. The improved materials are classified into specific strata, separated from landraces. Breeding materials are separated from those obtained through formal selection of landraces by breeders, thus creating two strata. Here a pragmatic strategy is stressed, the goal being representativeness of elite materials. The composition of these strata will be more flexible and dynamic. As advances are made in *Manihot* breeding work in Brazil, new materials will be included to replace obsolete ones. It is understood that breeders' needs change with time and that conflicting priorities may arise. Useful variation in terms of current breeding needs may not satisfy tomorrow's requirements (Chapman, 1989). Furthermore, landrace strata are conceptualised as highly stable, although the removal of materials to delete duplicates or the inclusion of accessions as a result of collecting or evaluating activities are expected. The aim here is to represent materials showing adaptability to different environmental conditions.

Wild relatives are not considered in this first attempt to define a core collection. However, these materials are an important source of genetic diversity and should be taken into account in future attempts to enrich the core. In particular, they have been used by breeders as sources of genetic resistance (Bueno, 1991). Brazilian wild relatives are an important part of the genetic diversity of *Manihot*. According to Rogers (1963), there are two major concentrations of *Manihot* species, one in Mexico and the other in Brazil. The species are predominantly South American; in South America, the greatest number are found in eastern-central Brazil, mainly in the states of Goias (including the new state of Tocantins), Minas Gerais and the interior of Bahia (Rogers and Appan, 1973).

METHODOLOGY

Criteria for the stratification of the collection

The first criterion for the hierarchical classification of the accessions is the category of material: landraces or improved material (resulting from formal breeding procedures, such as crossing and selection, and outstanding landraces obtained from formal selection procedures).

The second criterion, applicable only to landraces, is region of origin, which should be determined in terms of climatic and ecological characteristics.The Brazilian collection contains material from different ecosystems, which should be appropriately represented in the core. Those of unknown origin constitute a separate stratum.

The third criterion relates to characterisation and evaluation data. This study considers the most important source of information to be data on strongly inherited characters of importance to the breeder.

Ecogeographical zones

Nine ecogeographical zones have been defined in Brazil: Amazon, Cerrado, Caatinga, Agreste, South Littoral, North Littoral, Subdecidous Tropical Forest (or Humid Interior), South and Pantanal. These zones were defined by combining information from vegetation maps (Alonso, 1977a, b; Kuhlmann, 1977a, b; Santos et al., 1977; Nimer and Brandao, 1989) and climatic maps based on dry-season duration (Nimer, 1989). The zones constitute the strata used to classify the materials of the Brazilian collection and to sample the landraces selected to build the core collection. A detailed description of these regions is given in Table 2.

Some of the ecogeographical zones coincide with possible centres of origin of *Manihot*, as defined by Spath (1973), and with centres of diversity of the wild species identified by Nassar (1978). These are central Brazil, western Mexico, north-eastern Brazil, western Mato Grosso and eastern Bolivia.

Further information on diversity distribution was provided by Gulik et al. (1983) who distinguished three primary diversity zones for *M. esculenta*. These are also zones where the crop is widely and traditionally cultivated. The first zone includes parts of north-eastern, eastern-central, south-eastern and southern Brazil, as well as Paraguay. The second zone includes eastern Colombia, southern Venezuela and northern Brazil, extending southwards to the Amazon river and along the Orinoco river. The third zone is centered in Nicaragua and extends to Panama and Honduras. Other areas of secondary diversity, where cassava is an important or moderately important crop but appears to be less genetically variable, are Bolivia, a large part of the Amazon basin, southern Mexico and areas near the coast of north-eastern Brazil. Thus, there is high genetic diversity of cassava in north-eastern Brazil and moderate genetic diversity in the coastal region of the north-east.

Stratification of the collection

According to the first two criteria for the stratification of the collection, 11 strata were defined. The sizes of these strata are shown in Table 3. The third criterion applies only to characterised and evaluated accessions, which at present constitute 40% of the Brazilian collection. This criterion will be used when sampling landraces, aiming at optimum representation of diversity within strata.

Sampling of core entries

Marshall and Brown (1975, 1983) suggested classifying alleles into the following categories: rare and common (according to their frequency within populations); and widespread and localised (according to their presence in many or a few populations). When discussing the subject of core collections Brown (1989a, b) adapted these concepts and defined a common allele as one whose frequency within any one accession is greater than 10%. In the case of collections of clones, an accession is constituted by identical individuals and thus the definitions of 'rare and common' alleles, as proposed by Brown (1989a, b) are not useful. What really matters in this case is whether alleles are present or absent in a specific clone and how the frequency of clones carrying alleles of interest behaves in the whole collection and in the core collection.

Table 2 **Brazilian agroecological zones on which the stratification of landraces is based**

Zone	Description
Amazon	*Climatic types:* humid, super-humid (1-3 months dry); tropical humid (1-2 months dry); equatorial super-humid (without a dry season); tropical semi-humid (4-5 months dry) (Nimer, 1989). *Vegetation types:* evergreen periodically or permanently flooded forest; evergreen hylean amazonic rainforest; subdeciduous amazonic forest; Roraima complex (Kuhlmann, 1977b).
Cerrado	*Climatic type:* semi-humid (4-5 months dry) (Nimer, 1989). *Vegetation type:* Cerrado, a savanna-like vegetation (Rizzini, 1976; Nimer and Brandao, 1989).
Caatinga	*Climatic type:* semi-arid (6-11 months dry) (Nimer, 1989). *Vegetation type:* Caatinga, a deciduous thorny forest (Kuhlmann, 1977a).
Agreste	*Climatic type:* semi-humid (4-5 months dry) (Nimer, 1989). *Vegetational type:* deciduous non-thorny forest (Kuhlmann, 1977a). The Agreste vegetation is considered to be a subdivision of Caatinga by Rizzini (1976), where the humidity is greater because of its proximity to the sea. The soils are deeper and the vegetation is taller and closer than typical Caatinga vegetation. Because of the difference in dry-season duration allied to vegetation, this zone is treated as a separate region.
Littoral (north and south)	*Climatic types:* humid, super-humid (1-3 months dry); semi-humid (4-5 months dry) (Nimer, 1989). *Vegetational types:* evergreen coastal rain forest; evergreen bahiana hylean forest (Alonso, 1977a, b; Kuhlmann, 1977a, b). This region is divided into north (from southern Espirito Santo northwards) and south (from northern Rio de Janeiro southwards to Rio Grande do Sul) because of differences in winter temperatures and great differences in species compostion in the coastal vegetation of Espirito Santo compared with the composition of the southern floras.
Subdeciduous Tropical Forest	*Climatic type:* humid, super-humid (1-3 months dry) (Nimer, 1989). *Vegetational type:* subdeciduous tropical forest (Alonso, 1977a).
South	*Climatic type:* humid, super-humid (1-3 months dry). *Vegetational types:* subdeciduous subtropical forest; subdeciduous subtropical forest with *Araucaria*; grasslands (Alonso, 1977b).
Pantanal complex	*Climatic types:* humid, super-humid (1-3 months dry); semi-humid (4-5 months dry). *Vegetational type:* Pantanal, a complex of several types of vegetation affected by floods. The effect of the periodic floods of the Paraguai river differ across the region, which results the differences in vegetation. There are permanently damp areas, temporarily flooded areas and areas unaffected by the floods (Santos et al., 1977).

Table 3 **Number of accessions contained in each of the 11 strata of the Brazilian collection of cassava (*Manihot esculenta* Crantz)**

Stratum	Number of accessions
Agreste	228
Amazon	179
Caatinga	279
Cerrado	95
Subdeciduous Tropical Forest	465
North Littoral	309
South Littoral	348
South	200
Breeding materials	274
Selected landraces	65
Unknown origin	558

When working with vegetatively propagated material and focusing on adapted genotypes, the recovery of clones possessing the allelic structures, or allelic blocks, conferring adaptability characteristics is a crucial goal. Because of early and wide dispersal of the crop and relatively low levels of genetic exchange among regions, many distinct and locally adapted gene pools have evolved (Hershey, 1991). Grouping accessions according to an agroecological criterion enhances the possibility of recovering alleles responsible for local adaptability (localised alleles). An appropriate stratification should place localised alleles in a specific stratum. Inappropriate stratification might lead to the dispersion of a localised allele through different strata. Widespread alleles should also be considered when selecting the core. These alleles will be present in different environments.

Preliminary studies show that the four sampling methods suggested by Brown (1989b) behave similarly for the recovery of dispersed allelic blocks. In addition, constant allocation is recommended when recovering alleles concentrated in small strata, while proportional allocation is the best procedure where concentration occurs into large strata. The logarithmic strategy behaves as an intermediate strategy (Ferreira and Cordeiro, 1994).

If stratification is successful in concentrating alleles of interest, then the logarithmic allocation procedure is a good way of recovering alleles presenting different distribution patterns. In the case of an unsuccessful stratification, the three strategies are equally appropriate and slightly less efficient than simple random sampling.

The size of the core collection must be such that it ensures the safe recovery (90% of probability) of at least one copy of an allele present in the collection with a frequency of at least 5%. Analyses of the four sampling methods show that a 15-20% sample size proportionally allocated to the log of stratum size is sufficient to attain this goal and that smaller samples may lead to the loss of material of interest. The study is based on probabilistic models and makes no restrictive assumptions about the genetic structure of the collection. Therefore, 300-400 landraces will be selected for inclusion. This size is sufficient to recover localised and widespread alleles with a frequency of at least 5% in the whole collection.

Although a proportional allocation allows the utilisation of an inferior sampling rate in order to reach the same recovery level, the logarithmic allocation is recommended for the specific case of the

Brazilian cassava collection because areas of high diversity for this crop (Cerrado and Amazon) are poorly represented in the whole collection. It is worth noting that there is no theoretical background on which to base the proportion of improved materials that should be included in the core. Pragmatic decisions should be made in order to avoid overweighting these strata. About 80 introductions are expected to be included in the core from this group.

This approach is similar to that considered by Brown (1989b). He argued that, according to the sampling theory of neutral alleles, the allocation of stratified random sampling (STRS) should be made proportional to the logarithm of strata sizes. Other alternatives considered by the same author are proportionality to strata sizes, constant allocation and simple random sampling (SRS). In this case, the size of the core collection was determined according to the need to recover at least 70% of the class of widespread and rare alleles.

Selection of landraces from ecogeographical strata

A cluster analysis of landraces within each of the edaphoclimatic zones is recommended in order to define the final strata. Random sampling of landraces from these strata is proposed, following the rule of proportionality to the logarithm of strata sizes.

To perform the analysis, 18 variables (seven discrete, 11 continuous) were chosen based on their interest to breeders (*see* Table 4). The discrete characteristics were checked in terms of of their reliability (that is, small variation of data over the years). The continuous characteristics were also chosen according to their heritabilities.

Refinements of the core structure employing further agroecological parameters should be considered in the future because such parameters may vary greatly within the above-defined Brazilian edaphoclimatic zones.

Selection of improved germplasm

Cassava is usually grown under stress conditions, such as little or no fertilizer, no irrigation, no chemical control of diseases or pests, and poor soils. Thus, one of the objectives when selecting

Table 4 **Recommended variables to be used in the cluster analysis of cassava landraces within edaphoclimatic zones**

Discrete variables	Continuous variables	Heritabilities
HCN content	Plant height	($h^2 = 0.43$)
Root surface	Height of first branch	($h^2 = 0.47$)
Phelloderm colour	Starch percentage	($h^2 = 0.50$)
	Root flesh colour/root weight	($h^2 = 0.43$)
Branching pattern	Yield index (root/biomass weight)	($h^2 = 0.52$)
Root pedicle	Storey length	($h^2 = 0.44$)
Leaf pubescence	Root length	($h^2 = 0.47$)
	Median lobe length	($h^2 = 0.62$)
	Median lobe width	($h^2 = 0.70$)
	Number of lobes	($h^2 = 0.47$)
	Petiole length	($h^2 = 0.62$)

breeding entries for the core is to represent genotypes with the ability to thrive under low-input, marginal conditions. There has also been increasing interest recently in stability and integrated pest-and-disease management, based on host-plant resistance (Hershey, 1991).

Selection from breeding populations should employ several additional criteria. The traits to be considered include general adaptation, resistance to pests and diseases, tolerance to abiotic factors, ability to thrive under low-input, marginal conditions, stability, plant architecture, yield and root quality. Possible uses, such as consumption of fresh cassava, flour production, starch production or other industrial applications, also deserve consideration.

About 260 Brazilian breeding materials have been identified. Of these, 10% will be included in the core as a stratum. Some 60 selected landraces will constitute a separate stratum.

VALIDATION STUDY USING CHARACTERISATION AND EVALUATION DATA

A validation study of the proposed ecogeographical stratification of landraces was made using characterisation and evaluation data. The basic tool of this study was a cluster analysis of a subset of the Brazilian cassava collection. Two additional multivariate techniques (a correspondence analysis and multidimensional scaling) were used to 'validate' the cluster and the ecogeographic classification.

Data on the characterisation and evaluation of 389 landrace accessions in the collection held at the Centro Nacional de Pesquisa de Mandioca e Fruticultura (CNPMF), obtained over a period of 14 years (1977-90), were employed in this study. These landraces were selected because their information on origin was reliable and available. They were representatives of the Amazon, Cerrado, Caatinga, North Littoral, Agreste and Subdeciduous Tropical Forest agroecological zones.

To perform the clustering of the Brazilian collection, 29 characteristics were selected, 17 being categorical and 12 being numerical (continuous and discrete). The criterion used to select the categorical variables was consistency of estimates over a 2-year period. All available numerical variables were used, apart from those showing a considerable amount of missing data.

Two distance matrices, one for the numerical variables and the other for the categorical variables, were constructed and weighted according to the number of variables considered, thus producing a combined distance matrix. The distance matrix for the numerical variables (x_i, i = 1,..., p) was constructed using the following distance measure (derived from Gower's scoring system for quantitative data; *see* Wishart, 1986):

$$d(a,b) = \frac{1}{p}\Sigma |x_i(a) - x_i(b)| / \text{Range}(xi)$$

where:

$x_i(a) - x_i(b)$ = values of variables i for landraces a and b

The distance matrix for the categorical variables was obtained as a complement of a similarity matrix based on the following scoring system:

$$I_i(a, b) = \begin{cases} +1 \text{ if a and b coincide for the } i^{th} \text{ character} \\ +0 \text{ otherwise} \end{cases}$$

where the similarity is a weighted average of scores, and weights are proportional to the logarithm of the number of categories of each variable.

The combined distance matrix was used as a starting point for the use of the Ward clustering method. On the basis of the pseudo-t^2 statistic, six clusters were considered as the optimum partition (SAS, 1990).

The clusters/regions contingency table was assessed by a correspondence analysis using the 'Corresp' procedure devised by the SAS (1988). The graphical joint display of this analysis enables one to visualise the association among clusters and regions. Angles between vectors corresponding to ecogeographical regions and clusters measure levels of association. Furthermore, each individual display (looking separately at the cloud of points of the ecogeographical regions or clusters) enables one to associate distances between points with similarities and dispersion (*see* Figure 1). This representation provides an approximate view of the data because the first two axes retained 75% of the total inertia.

The analysis shows a clear relationship between the Amazon and North Littoral zones, with two specific clusters. Two other clusters appear as intermediate between the Caatinga and Agreste zones and the Amazon and Cerrado zones. It is worth noting that these two pairs of regions are pairs of geographically neighbouring regions.

The ecological transition from the Littoral to the Agreste and Caatinga zones, an east-west geographical transition, is clearly reflected in the spatial configuration of the corresponding vectors

Figure 1 **Display of a correspondence analysis for ecogeographical regions vs. clusters**

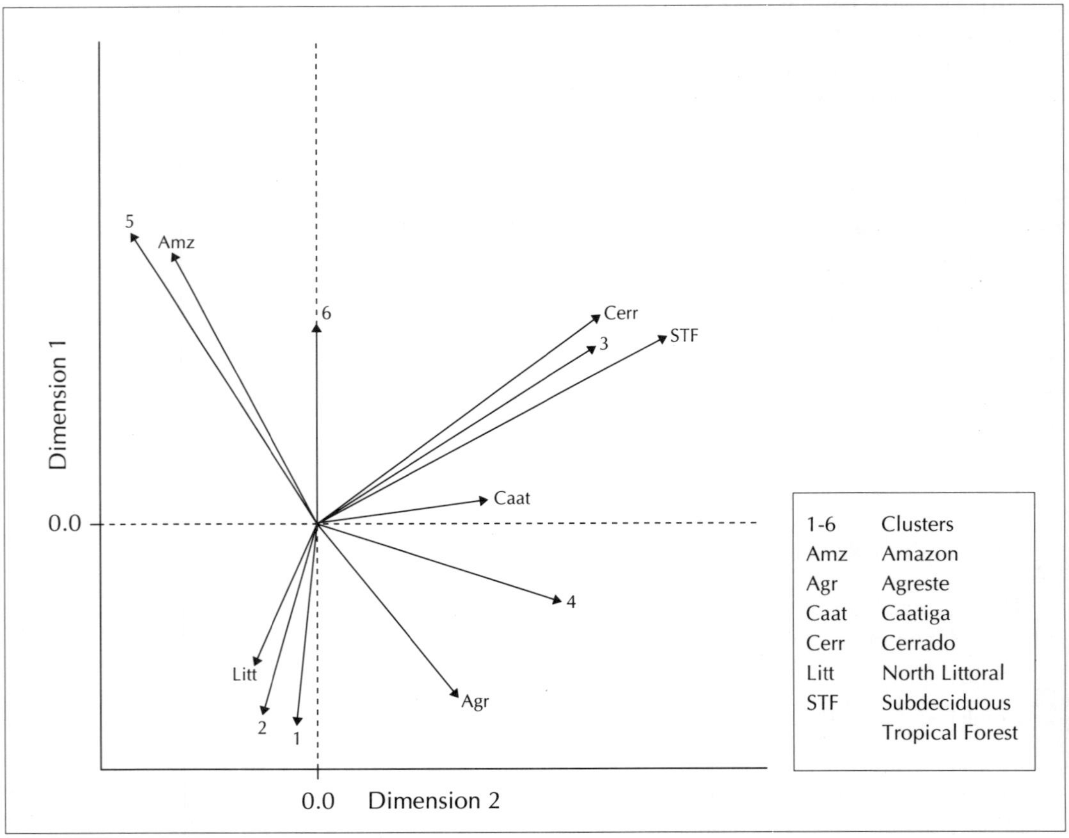

in the graphical display. The Cerrado and Subdecidous Tropical Forest zones, which are geographically neighbouring regions, appear in close association in the correspondence analysis and are isolated from other groups.

The vertical axis is a bipolar dimension comparing Amazon and Cerrado materials with those of the Littoral and Agreste zones, a west vs. east dimension. The horizontal axis compares the Amazon

Figure 2 **Display of 389 accessions of cassava according to the first two dimensions of a multidimensional scaling**

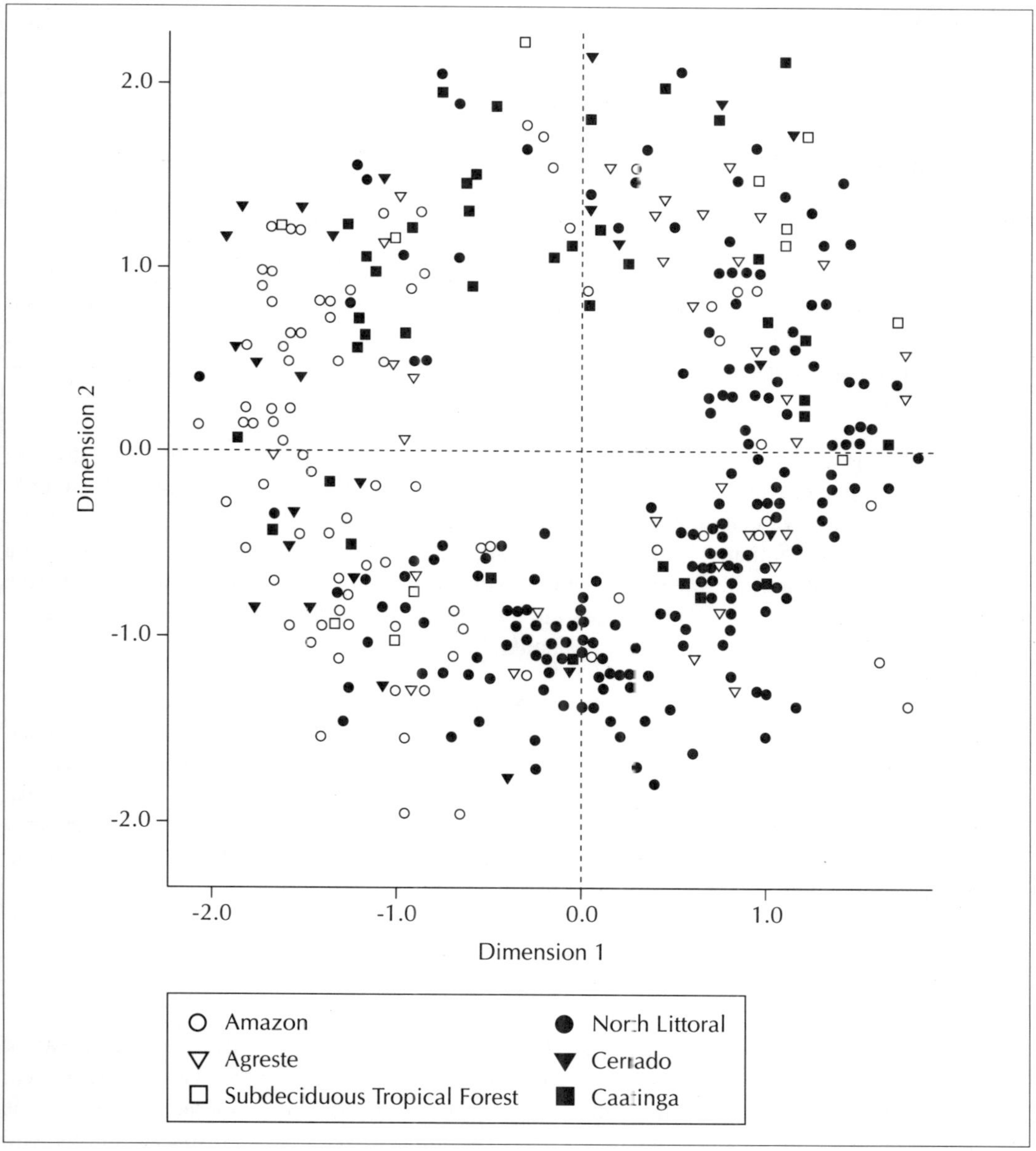

materials with the rest, excluding the Littoral. These bipolar dimensions should be further analysed in terms of the characteristics of their corresponding materials.

A representation of the distance matrix computed for the 389 accessions was made using a multidimensional scaling procedure (Alscal procedure; *see* SAS, 1986). A graphical display of the first two dimensions is presented in Figure 2, and shows a clear separation between the North Littoral and the Amazon accessions. In addition, material from the Agreste zone tends to be mixed up with accessions from the neighbouring North Littoral zone.

The validation exercise showed that the stratification of accessions according to agroecological information was successful because it bears close links with the classification based on phenotypic information provided by characterisation and evaluation data.

RECOMMENDATIONS

Other validation procedures

Some suggestions are given here for the further validation and study of the core collection:

- Experiments should be implemented to evaluate segregation using crosses between accessions selected from different strata of the core. These genetic parameters could be compared with similar quantities evaluated from pairs of accessions arising from the same stratum. This technique was used by Peeters and Martinelli (1989) as a tool to evaluate an hierarchical cluster analysis of barley germplasm.

- The use of genetic markers, such as isozyme, polymerase chain reaction (PCR), restriction fragment length polymorphism (RFLP) and protein, in samples obtained from each strata is recommended as a validation tool. The question to be answered is whether there is a quantitative or qualitative differentiation of the genetic variability among ecogeographical strata. These data could provide valuable information about the genetic variability of each stratum, as well as a comparison of between- and within-strata variation. Measuring variability may provide interesting information on the relative importance of strata.

- In the proposals outlined in this chapter, environmental factors play a crucial role in defining the strata of the core collection. However, information on genotype x environment interaction obtained from national trials of the collection's materials should be used to validate the stratification. Locations can be classified according to the similarity of their interactions with the core materials. Using such a methodology, locations could be classified into regions which are not necessarily contiguous, such that the interactions between entries and locations within regions are small (Abou-El-Fittouh et al., 1969; Cordeiro and Silva, 1980).

Future selection of wild material for inclusion in the core collection

The proposal put forward by Harlan and de Wet (1971) provides a convenient and useful way of defining what should make up a germplasm collection. They proposed a biological species concept with regard to applied plant breeding where the degree of kinship between related gene pools is ascertained in direct proportion to the degree of crossability between the crop species and its wild relatives. Thus, they label a wild gene pool readily crossing with the cultigen and forming fertile offspring a primary gene pool (GP1). Taxa crossing with difficulty but giving some results make up the GP2, and so on. We recommend that the inclusion of materials in the core should concentrate first on strategic folk varieties and then expand to cover the wild GP1 of cassava represented by the South American species *M. flabellifolia* and *M. peruviana*. The GP2 represented by section Glaziovianae in north-eastern Brazil is a close second priority. This proposal, suggested by A.C.Allem, is similar to the approach suggested by Allem and Hahn (1991) for the collection of plant materials.

Acknowledgements

The authors gratefully acknowledge the advice and helpful comments received from their colleagues A.C.Allem, R. Vencovsky, E. Paterniani, T. Valle and H. Cardoso on the draft version of this chapter.

References

Abou-El-Fittouh, H.A., Rawlings, J.O. and Miller, P.A. 1969. Classification of environments to control genotype by environment interactions with an application to cotton. *Crop Science* 9: 135-40.

Allem, A.C. and Hahn, S.K. 1991. Cassava germplasm strategies for Africa. In Ng, N.Q., Perrino, P. Attere, F. and Zedam, H. (eds). *Crop Genetic Resources of Africa.* (vol. 2) Rome, Italy: IBPGR.

Alonso, M.T.A. 1977a. Vegetaçao. In Fundaçao Instituto Brasileiro de Geografia e Estatística (Rio de Janeiro, RJ). *Geografia do Brasil — Regiao Sudeste (Rio de Janeiro)* 3.

Alonso, M.T.A. 1977b. Vegetaçao. In Fundaçao Instituto Brasileiro de Geografia e Estatística (Rio de Janeiro, RJ). *Geografia do Brasil — Regiao Sudeste (Rio de Janeiro)* 5.

Brown, A.H.D. 1989a. The case for core collections. In Brown, A.H.D., Frankel, O.H., Marshall, D.R. and Williams, J.T. (eds). *The Use of Plant Genetic Resources.* Cambridge, UK: Cambridge University Press.

Brown, A.H.D. 1989b. Core collections: A practical approach to genetic resources management. *Genome* 31: 818-24.

Bueno, A. 1991. Hybridisation and breeding methodologies appropriate to cassava. In Hershey, C. (ed) *Cassava Breeding: A Multidisciplinary Review.* Cali, Colombia: CIAT.

Chapman, C.G.D. 1989. Collection strategies for the wild relatives of field crops. In Brown, A.H.D., Frankel, O.H., Marshall, D.R. and Williams, J.T. (eds). *The Use of Plant Genetic Resources.* Cambridge, UK: Cambridge University Press.

Cordeiro, C.M.T. and da Silva, J.G.C. 1980. Estudo do zoneamento da regiao Centro-Sul do Brasil para a cultura de milho. *Pesquisa Agropecuaria Brasileira* 15: 191-205.

EMBRAPA. 1991. Programa Nacional de pesquisa de mandioca. In *Relatório Técnico do Centro Nacional de Pesquisa de Mandioca e Fruticultura Tropical (1987-1990).* Rio de Janeiro, Brazil: EMBRAPA/CNPMF.

Ferreira, P. and Cordeiro, C.M.T. 1994. Comparing sampling methods for the construction of core collections. In Salazar, R. (ed). *Proc. 1993 Scientific Week.* Turrialba, Costa Rica: CATIE.

Frankel, O.H. 1984. Genetic perspectives of germplasm conservation. In Arber, W., Llimensee, K., Peacock, W.J. and Starlinger, P. (eds). *Genetic Manipulation: Impact on Man and Society.* Cambridge, UK: Cambridge University Press.

Frankel, O.H. 1985. Into the second decade: Genetic resources and the plant breeder. In Mehra, K.L. and Sastrapradja, S. (eds). *Proc. of the International Symposium of SE Asian Plant Genetic Resources* . Bogor, Indonesia: Lembaja Biology Nacional.

Harlan, J.R. and de Wet, J.M.J. 1971. Toward a rational classification of cultivated plants. *Taxon* 20: 509-17.

Hershey, C. 1991. Cassava germplasm resources. In Hershey, C. (ed) *Cassava Breeding: A Multidisciplinary Review*. Cali, Colombia: CIAT.

Kuhlmann, E. 1977a. Vegetaçao. In Fundaçao Instituto de Geografia e Estatítica (Rio de Janeiro, RJ). *Geografia do Brasil — Regiao Sudeste* (*Rio de Janeiro*) 1.

Kuhlmann, E. 1977b. Vegetaçao. In Fundaçao Instituto de Geografia e Estatística (Rio de Janeiro, RJ). *Geografia do Brasil — Regiao Sudeste* (*Rio de Janeiro*) 2.

Marshall, D.R. and Brown, A.H.D. 1975. Optimum sampling strategies in genetic conservation. In Frankel, O.H. and Hawkes, J.G. (eds). *Genetic Resources for Today and Tomorrow*. Cambridge, UK: Cambridge University Press.

Marshall, D.R. and Brown, A.H.D. 1983. Theory of forage plant collection. In McIver, J.C. and Bray, R.A. (eds). *Genetic Resources in Forage Plants*. Melbourne, Australia: CSIRO.

Nassar, N.M.A. 1978. Conservation of genetic resources of cassava (*Manihot esculenta*): Determination of wild species location with emphasis on probable origin. *Economic Botany* 32: 311-20.

Nimer, E. 1989. *Climatologia do Brasil*. (2nd edn). Rio de Janeiro, Brazil: IBGE.

Nimer, E. and Brandao, A.M.P.M. 1989. *Balanço Hídrico e Clima da Regiao dos Cerrados*. Rio de Janeiro, Brazil: IBGE.

Peeters, J.P. and Martinelli, J.A. 1989. Hierarchical cluster analysis as a tool to manage variation in germplasm collections. *Theoretical and Applied Genetics* 78: 42-48.

Rizzini, C.T. 1976. *Tratado de Fitogeografia do Brasil — Aspectos Sociológicos e Florísticos*. Sao Paulo, Brazil: HUCITEC/EDUSP.

Rogers, D.J. 1963. Studies in *Manihot esculenta* Crantz and related species. *Bull. Torrey Botanical Club* 90: 43-54.

Rogers, D.J. and Appan, S.G. 1973. Manihot *and Manihotides (Euphorbiaceae). (Flora Neotropica, 13).* New York, USA: Hafner Press.

Santos, L.B. dos, Inocêncio, N.R. and Guimaraes M.R.S. 1977. Vegetaçao. In Fundaçao Instituto Brasileiro de Geografia e Estatística (Rio de Janeiro, RJ). *Geografia do Brasil — Regiao Sudeste* (*Rio de Janeiro*) 4.

SAS. 1986. *SUGI Supplemental Library User's Guide*. (5th edn). Cary, North Carolina, USA: SAS Institute.

SAS. 1988. *SAS Technical Report*. Cary, North Carolina, USA: SAS Institute.

SAS. 1990. *SAS/STAT User's Guide. Version 6*. (4th edn). Cary, North Carolina, USA: SAS Institute.

Spath, C.D. 1973. Plant domestication: A case of *Manihot esculenta*. *J. Steward Anthropological Society* 5: 45-67.

Wishart, D. 1986. Hierarchical cluster analysis with messy data. In Gaul, W. and Shader, M. (eds). *Classification as a Tool for Research*. New York, USA: Elsevier.

Zehntner, L. 1919. *Estudos de Algumas Variedades de Mandioca brasileira*. Rio de Janeiro, Brazil: Sociedade Nacional de Agricultura.

Part 4

MANAGING AND TESTING CORE COLLECTIONS

Core Collections of Plant Genetic Resources
Edited by T. Hodgkin, A.H.D. Brown, Th.J.L. van Hintum and E.A.V. Morales

A Wiley-Sayce Publication

4.1

The Barley Core Collection: An international effort

H. Knüpffer and Th.J.L. van Hintum

Abstract

The Barley Core Collection (BCC) started as a collaborative initiative of the European Cooperative Programme for the Conservation and Exchange of Crop Genetic Resources (ECP/GR) in 1989. Its concept was elaborated by a working group which was reconstituted as an international committee in 1992. The BCC is a limited set of accessions selected from gene bank collections to represent optimally the genetic diversity of barley. The objectives of the BCC are to increase the efficiency of germplasm evaluation and utilisation, to provide a manageable and representative set of barley accessions for use in research and breeding, and to provide standardised material for scientific investigations. It includes five main categories: cultivars, landraces, *Hordeum spontaneum*, other wild *Hordeum* species, and genetic stocks and reference material. The structure of the gene pool and of the BCC is hierarchical and can be described by a dendrogram. Procedures and criteria to partition the diversity and select accessions are described. Experts for specific sections of the gene pool will be involved in the actual selection of BCC accessions. To ensure continued integrity, accessions will be, as far as possible, homozygous and homogeneous lines, developed from gene bank accessions by such methods as single-seed descent or doubled haploids. Advantages and disadvantages of this approach are discussed. The BCC is expected to become operational in 1995. Aspects of creating and maintaining the collection, including international cooperation, coordination of activities, and documentation, are discussed.

Barley is an economically important and genetically well-studied crop. The cultivated forms belong to one diploid species, *Hordeum vulgare* L. s.l., comprising the primary gene pool along with the *spontaneum* complex. The species *H. bulbosum* forms the secondary gene pool, while the tertiary gene pool consists of the remaining wild species (about 30) of the genus *Hordeum* Bothmer et al., 1991). Breeding began early and, together with intensive collecting of landraces and the creation of genetic stocks, led to the accumulation of large numbers of accessions in *ex situ* collections such as gene banks and breeders' collections worldwide. The total number of barley accessions is estimated to be about 280 000 (Plucknett et al., 1987), probably with a high degree of duplication (IBPGR, 1989a).

A recent survey of ongoing work on core collections, carried out by the International Board for Plant Genetic Resources (IBPGR), revealed more than 20 such projects and four established core collections (Hodgkin, 1991). These projects differed widely in scope, in the differentiation of groups of accessions from which the core was selected, in the sampling procedures used and in the practical aspects of developing and maintaining a core.

Compared with other core collections, the Barley Core Collection (BCC) approach has several distinct features. The main differences are that it is not a selection from the germplasm collection of a single institution but from the entire gene pool of a crop, as far as it is represented in germplasm collections, and that it is not a part of an existing gene bank collection but a collection created on the basis of gene bank collections and maintained separately.

THE BARLEY CORE COLLECTION WORKING GROUP

Within the European Cooperative Programme for the Conservation and Exchange of Crop Genetic Resources (ECP/GR), a collaborative core collection approach was initially discussed for *Beta* (Hintum, 1989). In 1989, the Barley Working Group of the ECP/GR considered the usefulness of setting up a barley core collection on the basis of European collections (IBPGR, 1989a). The group recognised that creating one synthetic core collection instead of several independent core collections in larger gene banks would facilitate the coordination of efforts and sharing of responsibilities in genetic resources activities. The objectives of creating the BCC were to increase knowledge about the barley gene pool, and thus to use the existing germplasm collections more efficiently and to provide standards for studies of genetic diversity.

The use of the BCC in research and evaluation would lead to the accumulation of a large amount of data for a limited standard set of accessions. When looking for a new trait, the researcher could always start from a relatively small sample covering a considerable part of the whole diversity of the gene pool, and thus avoid expensive screenings of large collections containing duplicated material.

The Barley Working Group set up an *ad hoc* task force which put forward various recommendations (IBPGR, 1989a). The International Barley Working Session (IBPGR, 1989b) endorsed the establishment of a European Barley Core Collection and considered it as a pilot project to explore possible methodologies and to provide a first subset of an international core collection. The network aspect of the BCC was stressed from the outset. Accessions for the European BCC should be selected using the European Barley Database (Knüpffer, 1988) which provides information on about 55 000 accessions held in European collections.

The working group met three times (Anon., 1989, 1990; Bothmer et al., 1990). The report of the second meeting was distributed for comment to nearly 100 colleagues in many parts of the world. The report of the third meeting summarised the results and reflected the controversies and compromises. Short notes on the subject were published in appropriate journals and newsletters (for example, Hintum et al., 1990). The subject was discussed at two events before a wide international audience during the 6th International Barley Genetics Symposium in Sweden: the International Barley Genetic Resources Workshop (46 participants from 28 countries; IBPGR, 1992), and the Barley Core Collection Workshop (51 participants; Hintum, 1992). Both workshops endorsed the earlier work and recommended its continuation and worldwide extension, with the aim of setting up a BCC with continued IBPGR support within 3 years. Subsequently, an international BCC committee was formed which also included non-European specialists (Anon., 1992).

CONCEPT OF THE BARLEY CORE COLLECTION

Definition

To avoid confusion in the use of terminology, the BCC is defined as a selected and limited set of accessions, optimally representing the genetic diversity of cultivated barley (*H. vulgare* L. s.l.) and wild species of *Hordeum*, and providing well-known genetic standards (Bothmer et al., 1990). Its objectives are to:

- increase the efficiency of evaluation and thus of utilisation of existing collections
- provide a manageable and representative selection of the available barley germplasm for use in research and breeding
- provide adequate material for the needs of standardisation in scientific work with barley

The BCC is based on the world's barley holdings and comprises the whole genus *Hordeum*, including the secondary and tertiary gene pools, genetic stocks and reference material. The original concept of a core collection was that of a subset selected from a particular gene bank collection and marked in a database. The collection would thus be divided into the core collection and the 'reserve' collection (Brown, *Chapter 1.1, this volume*). The BCC, however, will be created as a physically separate, 'synthetic' collection (Brown, *Chapter 1.1*), selected from existing gene bank collections. It is not intended to replace gene bank collections and does not make them superfluous, but it will improve their accessibility. This also implies that the well-known concepts of base collections, safety duplicate collections and active collections will have to be applied to the BCC.

Within the BCC, subsets can be selected for specific purposes. An algorithm will have to be developed which will always give the same subset on the same request. For example, if one needs 50 BCC accessions of two-rowed landraces, the algorithm should indicate which accessions out of the possibly 200 meeting the criteria form the desired subgroup.

The composition of the BCC should be kept as stable as possible. With new scientific insights, however, it may become necessary to add accessions to, or delete accessions from, the BCC. Such suggestions will be accumulated and addressed on a regular basis.

Structure and size

Genetic variation of the barley gene pool is considered to be classifiable in a hierarchical structure which can be described by a dendrogram (Hintum, *Chapter 2.1*). In order to keep the collection manageable, the size of the BCC should not exceed 2000 accessions. Accordingly, the BCC will include the following categories:

- cultivars — about 500 accessions
- landraces (that is, 'all cultivated material except that resulting from continuous commercial breeding activities'; Anon., 1992) — about 800 accessions
- *H. spontaneum* (including the *agriocrithon* complex and the products of introgression between *H. spontaneum* and cultivated barley) — 150-200 accessions (two-thirds from central areas of distribution and one-third from marginal areas)

- other wild *Hordeum* species — 60-100 accessions (about two per species)
- genetic stocks and reference material — a maximum of 200 accessions

Coordinators for these categories and some of their major ecogeographical subgroups have been nominated from the members of the committee (Anon., 1992) (*see* Table 1).

Table 1 The Barley Core Collection subgroups and their coordinators

Subgroup	Coordinator
Ethiopia (cultivars and landraces)	(open)
West Asia and North Africa (cultivars and landraces)	J. Valkoun
South and East Asia (cultivars and landraces)	T. Konishi
Europe (cultivars and landraces)	G. Fischbeck
North and South America (cultivars and landraces)	H.E. Bockelman
Other areas (cultivars)	A.A. Jaradat
Hordeum spontaneum	A.A. Jaradat
Other wild species	R. von Bothmer
Genetic stocks and reference material	(open)

Source: Anon (1992)

A further ecogeographical subdivision was recommended for the first three categories. Tentative sizes of the regional subgroups of the cultivar and landrace groups are given in Table 2. This subdivision will be refined by the coordinators in close collaboration with other experts for the particular group; it will result in a dendrogram, the terminal groups of which will be represented by selected BCC accessions (*see* Figure 1). The classification concept for cultivated barley, both landraces and cultivars, proposed by Lukyanova et al. (1990) and comprising seven centres of diversity and 37 agroecological groups characterised in detail (*see* Table 3), could serve as a basis for the initial subdivision.

Table 2 Tentative sizes of regional subgroups of the cultivar and landrace groups

Region	Cultivars	Landraces
West Asia and North Africa	15	300
South and East Asia	80	300
North and South America	150	30
Ethiopia	5	100
Europe	200	80
Australia, New Zealand, Southern Africa and other regions	35	0
Total	485	810

Source: Anon (1992)

Figure 1 Part of the barley diversity tree

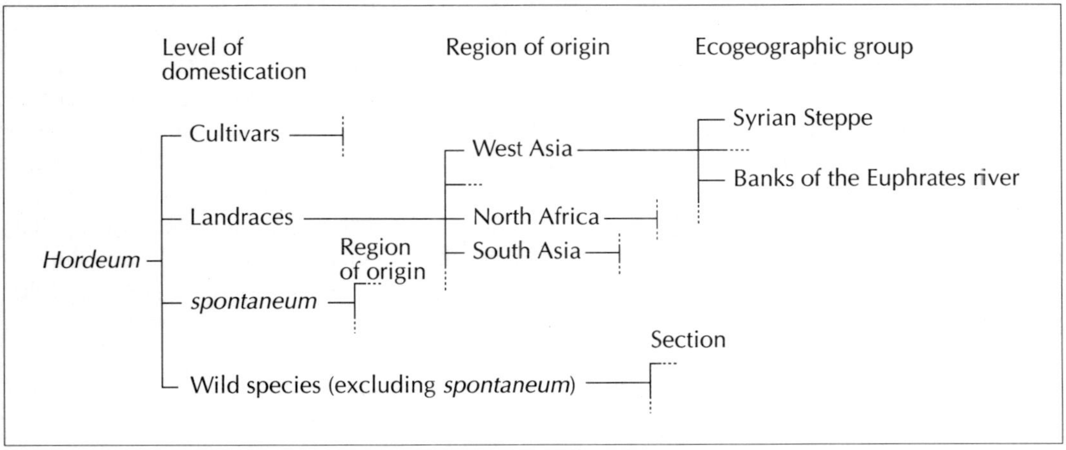

Table 3 Centres of diversity of cultivated barley

Centres of diversity	Area	Main characteristics of barley forms
Ethiopian	Ethiopia	Large diversity, many endemic forms
East Asiatic[a]	Eastern China, Korea, Japan and areas near eastern Tibet	High diversity of endemic forms, short straw, dense short ear, small almost round grain, short awns or awnless
Near Eastern	Azerbaijan, Armenia, Georgia and Anatolia	Large ecological diversity of forms, area overlapping with that of *Hordeum spontaneum*
Mediterranean	Egypt, Algeria, Tunisia, Palestine, Syria, Greece and the Greek islands, Spain, Italy; and parts of western and south-western Anatolia	Waxy forms, drought resistant, resistant to several diseases
Central Asiatic	Afghanistan, western Tien-Shan, Tadjikistan and Uzbekistan	Heat and drought resistant, mainly forage types
European-Siberian	Western Europe, Ukraine, northern Caucasus, and western and eastern Siberia	High level of breeding, tolerant to soil acidity
New World	North, South and Central America	Lodging resistance, earliness, resistance to diseases

Note: a Also known as the Chinese-Japanese centre of diversity, after Zhukovsky's classification
Source: Lukyanova et al. (1990)

The decisions made during this repeated stratification process will be carefully documented so that a branching criterion can be associated with each node of the branching tree. The combination of all decisions leading to the formation of a particular small group will describe this group and the BCC accessions representing it in terms such as 'two-rowed Danish malting barley cultivar of spring type'.

Further characteristics that might be used in constructing the dendrogram for cultivars (category 1) are, for example, phylogenetic group (occidental or oriental), growth habit (winter or spring barley), climatic zone of cultivation, ear type and pedigree data. For landraces (category 2), important characteristics include use, ecogeographical data and the agricultural system practised. For *H. spontaneum* (category 3), mainly ecogeographical data will be used to represent the diversity. The representatives of wild species (category 4) will be chosen within each species according to ecogeographical and/or morphological criteria. For genetic stocks (category 5), experts in barley genetics will be approached to propose appropriate BCC accessions. Characters for which reference material should be included in the BCC will also have to be selected.

Accessions for the international BCC might well be selected from existing core collections created for a specific purpose or region (for example, an Ethiopian core collection of landraces).

Homogeneity of accessions

In order to ensure the continued integrity of the BCC, the accessions will be, as far as possible, homozygous and homogeneous lines. Carefully chosen methods, such as single-seed descent or doubled haploids, have been suggested for obtaining homozygous lines wherever initial accessions are found to be heterozygous. Heterogeneous or heterozygous accessions will be included in the BCC only in the case of the two strictly outcrossing *Hordeum* species, and possibly for particular accessions in the category of genetic stocks, such as those containing lethal alleles. In these cases, other appropriate measures will be taken to ensure identical reproduction.

The decision to use homogeneous accessions was taken after thorough discussion (Hintum, 1992). The decisive advantages of using homozygous and homogeneous accessions are that identical multiplication over generations and locations is possible, and that correspondence between information and material can be guaranteed. A high stability of the accessions and reliability of the documentation is considered very important; if a scientist uses a BCC accession, he must be sure that the seeds he analyses are genetically exactly the same as those previously analysed by others.

The major disadvantage of using homogeneous accessions is that variation within landraces is not reflected in the BCC accessions (Brown, 1992). A landrace will be represented in the BCC by a single line; however, it is understood that such a line is not identical to the landrace and can never be a substitute for it. To study the variation within a landrace, one should refer to the original gene bank accession from which the BCC accession was derived. For the choice of these landraces the BCC can be a useful guide. Another disadvantage is that the number of allele combinations in the core is considerably reduced compared to the normal, often heterogeneous, gene bank accessions.

For modern cultivars it can be expected that the accessions will already be homozygous lines. Local experts will be needed to verify the identity of the gene bank accessions and, in the case of contamination or heterogeneity, to decide on the line best fitting the variety description.

Choosing the line to represent landraces and wild relatives is a more difficult task. It is presumed that a single line contains the genetic background common to the material it represents. Local experts will be asked to choose these lines (for example, a line optimally representing the barley grown in a given river valley in Pakistan, or the two lines representing *Hordeum procerum* in the BCC).

Phases of development

The phases in the establishment and operation of the core collection (Hintum, 1992) are:

- *First phase:* development and presentation of the concepts (1989-92)
- *Second phase:* establishment of the collection (1992-94)
- *Third phase:* BCC becoming operational (1995-)

It was recommended that IBPGR be involved from the outset (IBPGR, 1992). An important criterion for selecting entries for the BCC is the availability of reliable passport data which are as complete as possible. To this end, an International Barley Germplasm Database as proposed by the International Barley Genetic Resources Workshop (Knüpffer and Perry, 1992) would greatly facilitate the selection of the accessions for the BCC.

Future work

Once the BCC accessions have been selected, the coordinators for the subgroups will oversee the preparation of the material. They will also coordinate the multiplication of the accessions to the extent required to meet the requests of the users. The base collection of the BCC will be maintained at the International Center for Agriculture in the Dry Areas (ICARDA) and duplicated in Canada. Active collections should be kept in major research centres. In each of these collections the standard minimum recommended was 4000 seeds per accession, with possible exceptions for material that is difficult to multiply. Normally, only small samples would be distributed.

A well-organised documentation system will be of vital importance, preferably coordinated with the International Barley Germplasm Database (Knüpffer and Perry, 1992). Researchers will be asked to provide their results on BCC accessions to this documentation system. Publications of investigations of BCC material should always refer to the BCC accession numbers. Reference material for BCC accessions (such as herbarium or spike samples) will be needed for regular checking of authenticity during multiplication.

The maintenance of the core collection will require additional funds and efforts. The BCC committee considers that the benefits resulting from the BCC will more than compensate for the costs. The committee anticipates meeting regularly to coordinate activities. Although additional funds will be needed for special activities, the implementation of BCC activities are considered to be the responsibility of the participating gene banks. It is expected that highly valuable systematic information about the genetic diversity in barley will accumulate during the compilation of the BCC.

CONCLUSION

The BCC can be considered to be the first large-scale synthetic core collection. Its success will determine, to a large extent, the future of this type of international cooperation. So far, scientists and curators have shown great willingness to contribute and cooperate, possibly because of the commonly felt need to make barley germplasm collections more accessible and to reduce redundancy in gene bank activities. The BCC is an example of an international effort to make genetic resources management more efficient. As such, the approach could be applied to other large decentralised gene bank crops.

Acknowledgements

This chapter draws on the results of the meetings of the BCC committee and the contributions of numerous barley researchers, curators and breeders. We wish to acknowledge all those who were involved. We also wish to thank colleagues who reviewed the draft version of this chapter for their valuable comments and suggestions.

References

Anon. 1989. The Barley Core Collection. Report of the First Meeting of the BCC Working Group, Wageningen, Netherlands, September 1989. Mimeograph. BCC Working Group.

Anon. 1990. The Barley Core Collection. Report of the Second Meeting of the BCC Working Group, Gatersleben, Germany, March 1990. Mimeograph. BCC Working Group.

Anon. 1992. Report of the First Meeting of the Barley Core Collection Committee, Aleppo, Syria, May 1992. Mimeograph. BCC Committee.

Bothmer, R. von, Fischbeck, G., Hintum, Th.J.L. van, Hodgkin, T. and Knüpffer, H. 1990. The Barley Core Collection. Report of the BCC Working Group, Weihenstephan, Germany, September 1990. Mimeograph. BCC Working Group.

Bothmer, R. von, Jacobsen, N., Baden, C., Jørgensen, R.B. and Linde-Laursen, I. 1991. *An Ecogeographical Study of the Genus* Hordeum. *Systematic and Ecogeographic Studies on Crop Gene Pools. No. 7.* Rome, Italy: IBPGR.

Brown, A.H.D. 1992. Genetic variation and resources in cultivated barley and wild *Hordeum.* In Munck, L. (ed) *Barley Genetics VI.* Copenhagen, Denmark: Munksgaard.

Hintum, Th.J.L. van. 1989. Strategies for selecting subsets within collections. In *Report of an International Workshop on* Beta *Genetic Resources.* International Crop Network Series No. 3. Rome, Italy: IBPGR.

Hintum, Th.J.L. van. 1992. The Barley Core Collection workshop summary. In Munck, L. (ed) *Barley Genetics VI.* Copenhagen, Denmark: Munksgaard.

Hintum, Th.J.L. van, Bothmer, R. von, Fischbeck, G. and Knüpffer, H. 1990. The establishment of the Barley Core Collection. *Barley Newsletter* 34: 41-42.

Hodgkin, T. 1991. The core collection concept. In Hintum, Th.J.L. van, Frese, L. and Perret, P. (eds) *Crop Networks. Searching for New Concepts for Collaborative Genetic Resources Management.* International Crop Network Series No. 4. Rome, Italy: IBPGR.

IBPGR. 1989a. *Report on a Working Group on Barley (Third Meeting), Gatersleben, Germany, April 1989.* Rome, Italy: IBPGR.

IBPGR. 1989b. *Report of an International Barley Working Session, Gatersleben, Germany, April 1989.* International Crop Network Series No. 1. Rome, Italy: IBPGR.

IBPGR. 1992. *Barley Genetic Resources. Report of an International Barley Genetic Resources Workshop, Helsingborg , Sweden, July 1991.* International Crop Network Series No. 9. Rome, Italy: IBPGR.

Knüpffer, H. 1988. The European Barley Database of the ECP/GR: An introduction. *Kulturpflanze* 36: 133-62.

Knüpffer, H. and Perry, M. 1992. An international barley documentation system — concepts and strategies. In *Barley Genetic Resources. Report of an International Barley Genetic Resources Workshop, Helsingborg, Sweden, July 1991.* Rome, Italy: IBPGR.

Lukyanova, M.V., Trofimovskaya, A.Y., Gudkova, G.N., Terentyeva, I.A. and Yarosh, N.P. 1990. *Kul'turnaya Flora SSSR. Tom II, Chast' 2. Yachmen' (Flora of Cultivated Plants of USSR. vol. II, part 2. Barley).* Leningrad, USSR: Agropromizdat.

Plucknett, D.L., Smith, N.J.H., Williams, J.T. and Murthi Anishetty, N. 1987. *Gene Banks and the World's Food.* Princeton, New Jersey, USA: Princeton University Press.

Core Collections of Plant Genetic Resources
Edited by T. Hodgkin, A.H.D. Brown, Th.J.L. van Hintum and E.A.V. Morales

A Wiley-Sayce Publication

4.2

The dynamics of a core collection

A.A. Jaradat

Abstract

If a core collection is to represent a living record of a species or a genus, and contains as much genetic variation as possible, it should evolve over time. Alterations in size and content of the collection should be a relatively slow process in order to build up information on the accessions. However, some alterations may be necessary in the short term to: maximise the variation in the collection; provide additional sources to meet new needs; identify traits for which useful variability is limited in the source collection; include the potential usefulness in untested materials; reduce changes resulting from contamination; and clarify the structure of genetic diversity in a species and refine the structure of the core collection. This chapter discusses these alterations and the circumstances which require a change in the composition of a core collection. Such circumstances include cases where: little is known about the species/crop and its population structure, and more information on its gene pool is available; new breeding methods are applied to a crop, and new genetic variability identified/created; required variability is incorporated into desirable genetic background; mutant stocks undergo changes; expanding cultivation and changes in an ecosystem call for new genes and greater diversity; genetically contrasting landrace genotypes are identified; the genetic base of the collection's cultivar component becomes narrow; accessions are received from new areas or represent new taxa; and accessions with questionable authenticity can be replaced with new ones from comparable sources. The chapter concludes with a list of indicators to guide curators in making the necessary changes to a core collection.

The development of core collections to provide an adequate sample of a species range, to streamline germplasm evaluation and to devise a global strategy for the management and more effective use of variation in germplasm collections is receiving much attention. Core collections of winter wheat (MacKey, 1989), okra (Hamon and van Sloten, 1989), wild *Glycine* species (Brown, 1989a), lentil (Erskine and Muehlbauer, 1991), peanuts (Holbrook et al., 1992), barley (ICARDA, 1992) and wild emmer wheat (Jaradat, 1994) have already been described. A survey conducted by the International Board for Plant Genetic Resources (IBPGR) (Hodgkin, 1990) pointed out that if a core collection is to represent a living record of, and optimise the genetic diversity in, a species or a genus, it should be a dynamic rather than a static set of accessions. Alteration in size and content of a core collection may be desirable to optimise its genetic diversity; however, theoretical and practical considerations suggest that changes in size and content may become counter-productive. This chapter discusses the advantages and disadvantages of altering a core collection, outlines criteria to guide managers as to when, why and how a core collection should be changed and puts forward some ideas on how these criteria may differ for different core collections.

A CORE COLLECTION: STATIC OR DYNAMIC?

An important measure of the effectiveness of a core collection is the extent to which it contains the genetic diversity present in gene bank holdings. However, material in gene banks is often unrepresentative of the total diversity of a crop or a species (ICARDA, 1992). For the major crops, this diversity includes obsolete cultivars and genetic stocks, landraces and primitive cultivars, and genetically related wild and weedy species.

It has been long recognised (Frankel and Brown, 1984) that the collection and collation of characterisation and evaluation data describing genetic diversity is a first requirement for the effective management and use of plant genetic resources. However, a better understanding of the way in which the genetic diversity of a species is distributed in gene banks and in nature is also becoming important. Since the core collection concept focuses on the extent and distribution of genetic diversity in a species, the collection should evolve over time in size and content in order to provide optimal representation of the available genetic diversity in that species (Brown, 1989b). However, the formation of a core collection aims at building up a body of information on its accessions and providing standard material for scientific work. This could lead, in the long run, to better information flow, allowing potential users to identify appropriate accessions for use in their programmes (Hodgkin, 1990). It has been suggested that too rapid a flux of accessions through the core collection would defeat these aims (Brown, 1989b).

Brown (1989a) suggested that altering a core collection, in the light of new information and/or availability of new accessions, may be necessary. However, a core collection should conserve only those additions which are distinct (Chang, 1989). The size of the collection should be related to comprehensiveness in genetic diversity and it should also be manageable (Frankel and Brown, 1984; Brown, 1989a, b).

Altering the size and composition of a core collection may be necessary for one or more of the following reasons:

- To maximise its variation, in terms of frequency and quality of alleles, relative to the collection from which it was derived. Smith and Duvick (1989) concluded that building a genetically diverse germplasm base is almost pointless unless it encompasses genes that are useful, either in themselves or in combination with other previously evaluated germplasm. Brown (1989b) stressed the importance of using the number of types or alleles per locus as a measure of diversity in core collections, and pointed out that the critical issue is whether any type or allele is present in the core, not its frequency. It is assumed (Brown, 1989a) that breeders, through crossing and selection, can recover desirable allels when required, and they need access to only one copy of such alleles.

- To provide additional genetic sources to take account of biotic and abiotic stresses, breeders' needs and new farming systems. For example, expanding cultivation to newly opened areas and dynamic changes in a 'crop' ecosystem will certainly call for new genes and greater diversity. Examples can be cited from the rice ecosystems in the Asian tropics, the Amazon basin in Brazil and many parts of Africa (Chang, 1989), where farmers are facing new problems associated with fauna, flora, soil and pests.

- To identify traits for which useful variability is limited in germplasm collections and, probably, to include some rare and valuable alleles. Genes for male sterility, dwarfness and disease resistance have been exploited by breeders mainly to develop crop varieties which are cultivated over large areas. The hazards associated with the use of such a narrow genetic base are well documented (Nevo et al., 1986) and call for new sources of variability for these traits.

- To include the potential usefulness of genetic resources in as yet untested wild and cultivated materials. Ladizinsky (1989) remarked that wild relatives of crop plants are under-represented in gene banks compared with cultivated germplasm. Wild relatives harbour genetic diversity which may be absent from cultivated material. Also, they display local and regional adaptive differentiation as a result of diversifying selection (Nevo et al., 1984, 1986). However, information is inadequate on the entire distributional range and the ecological niches in which the wild relatives of most crop plants grow, and thus it is likely that this will be a future source of material for a core collection.

- To reduce changes caused by contamination through mutations, foreign pollen or seed, and to minimise genetic drift by ensuring a sufficiently large sample size and reducing opportunities for natural selection. Palmer (1989), for example, pointed out that certain mutant stocks may undergo changes with time as a result of reversions, suppressions and secondary mutations. Therefore, stocks carrying cytological markers have to be characterised microscopically to ensure that the variant of interest has not been lost. Moreover, maintenance of small seed samples, as in the case of core collections of cross-pollinated crops, can lead to genetic drift (Gill, 1989).

- To clarify the structure of genetic diversity in a species and refine the structure and composition of the core collection. It is envisaged (Marshall, 1989) that the core collection will recieve priority in evaluation and characterisation, so that in time a large number of traits would be evaluated at different locations. This could lead to a better understanding of the genetic structure of a particular species and promote the distribution of information and material, thus facilitating use of the core collection.

GENETIC DIVERSITY AND DYNAMICS OF CORE COLLECTIONS

An understanding of the way genetic variation is partitioned among populations is of primary importance for the conservation of genetic diversity in general (Nevo et al., 1986) and for the formation and evolution of core collections in particular (Marshall, 1989). Surveys of biochemical (Nevo et al., 1986), molecular (Hamrick and Godt, 1990), morphological and developmental (Nevo et al., 1984) variation provide data that are critical for the establishment of strategies designed to preserve and optimally sample genetic diversity in plant species.

Variation within and among populations has been thoroughly examined (Nevo et al., 1986), but little information is available on variation at the species level (Hamrick and Godt, 1990; Brown, 1991). Two important issues have to be addressed during the formation and evolution phases of a core collection: the accuracy with which genetic variation at the population level reflects variation at the species level; and the relationships that exist between levels of genetic diversity and the characteristics of species.

A survey of allozyme variation in plant species (Britting and Goodman, 1989; Hamrick and Godt, 1990) indicated that genetic diversity within species is significantly influenced by phyletic group, life form, geographic range, breeding system and seed dispersal mechanism. Geographic range was found to be the best predictor of levels of allozyme variation within species. Species with widespread ranges had significantly higher levels of diversity than more narrowly distributed ones.

Within populations, differences existed in the categories of phyletic group, life form, geographic range, regional distribution, breeding system and seed dispersal mechanism. The highest level of variation within populations was associated with geographic range and life form of the species.

Significant differences in proportion of genetic diversity among populations (G_{ST}) occurred in the categories of phyletic group, life form, regional distribution, breeding system, seed dispersal mechanism and successional status. Breeding system and life form were most closely associated with the variation in G_{ST}; together, these categories accounted for 84% of the variation in all eight characteristics of species. Hamrick and Godt (1990) concluded that variation at the species level was positively and significantly associated with variation at the population level, and that both are influenced by different evolutionary processes. This finding has important implications for the development of germplasm sampling (Hamrick and Godt, 1990) and conservation (Britting and Goodman, 1989) strategies.

Changing the contents and composition of a core collection

The contents and composition of a core collection should be changed for one or more of the following reasons:

- If little is known about the species/crop and its population structure (Smith and Duvick, 1989), and if more detailed information on its gene pool is available. The new information about accessions from different sources may call for the revision of categories or affinities among accessions (Brown, 1989a). The lack of taxonomic verification, characterisation, documentation and description of germplasm collections, especially old ones, is an important technical constraint to their use (Frankel and Brown, 1984. More information on germplasm collections, especially from developing countries, can be made available if facilities for proper evaluation, characterisation and data storage, retrieval and dissemination are available (Chapman, 1989).

- As new breeding methods are applied to a crop, and new genetic variability is being identified or created. A wealth of increasingly useful genetic diversity is being generated through the diversity of germplasm, environments and selection schemes among plant breeders. New breeding methods being applied to sorghum are inbred selection for hybrid sorghum (Nath et al., 1984) and the tropical conversion programme (Prasada Rao et al., 1989) to introduce diverse alleles from tropical germplasm into dwarf, photoperiod-insensitive breeding lines.

- When required variability is being incorporated into desirable genetic background. Plant breeders will not be able to make efficient use of the required variability if it is not available in an agronomically desirable background (Hermsen, 1986). For example, some of the *Yr* and *Lr* genes for resistance to stripe rust and leaf rust in wheat, hiproly genes for high lysine and high protein in barley, and the *ph* mutant gene for induced homoeologous recombination in wheat are present in agronomically poor backgrounds. Similarly, genetic variability available in wild species must be transferred into backgrounds of cultivated species before it can be used for crop improvement (Nevo et al., 1984; Ladizinsky, 1989).

- When mutant stocks undergo changes over time as a result of reversions, suppression and secondary mutations (Palmer, 1989).

- If expanding cultivation and dynamic changes in an ecosystem call for new genes and greater diversity. The area of cultivation for a number of staple food crops is still expanding into new ecosystems (Chang, 1989). Intercropping and multiple cropping are also increasing. These situations will inevitably lead to changes in pest damage and edaphic and other ecological stressess. Core collections have to cope with these situations through acquiring and broadening the spectrum of genetic diversity. One possibility is to include sources of pest resistance from new environments where both host and pest have been introduced (Chapman, 1989).

- When genetically contrasting landrace genotypes are identified (Jaradat, 1992). A landrace, being composed of a mixture of homozygous genotypes in a self-pollinated crop, usually exhibits considerable variation. These landrace genotypes will facilitate selection for high levels of recombinations and thus the ability to generate more adapted genotypes.

- When the genetic base of the cultivars in the core collection, based on the genetic divergence between them, becomes narrow. Martin et al. (1991) reviewed several recent studies on allogamous crop species and concluded that average cultivar diversity tended to decrease over time in some crops (for example, soft red winter wheat and malting barley) and increase in others (for example, hard red winter wheat and oats). Known pedigree data to estimate coefficients of parentage (Souza and Sorrells, 1991a, b), which estimate the probability that two alleles chosen at random from each individual are identical by descent, can be used as an index of the relationship between pairs of cultivars. However, Souza and Sorells (1991b) suggested combining pedigree data and genetic marker information to produce a better summary for improving the estimate of the relationship among cultivars in a core collection.

- When new accessions are received from distinct or new areas or represent new taxa (Brown, 1989a).

- If the genetic basis for a desirable trait is different in various species or groups of species (for example, species evolved in different gene centres) (Ladizinsky, 1989), and if desirable traits are scattered among different accessions and even among different plants within an accession (such as, potato). Obviously it is important to include several sources of variation for the trait in question in the core collection.

- When accessions with questionable authenticity can be replaced with new ones from presumably comparable sources (Brown, 1989a). Information on collection sites of accessions is important for all phases of genetic resources work in general, and for the designation and evolution of core collections in particular (Frankel and Brown, 1984). The coordinates of collection sites of wild material and landraces may be lacking (Chapman, 1989), especially for older material collected from developing countries. Moreover, this information may be of limited circulation.

Indicators for altering the content and composition of a core collection

The indicators that could be used by curators for changing the content and composition of a core collection include:

- changes in allele frequencies over succeeding multiplications; mean number of alleles per locus and mean level of diversity could be used
- comparisons between the core and the source collection in terms of genetic diversity and allele frequency
- relative variance components, because of different sources of variation, in the analysis of diversity indices for a set of characters
- the non-random, partially adaptive distribution of certain molecular markers, their combinations and frequencies
- average genetic divergence between cultivars in a core collection based on diversity in polymorphic discrete characters

Small core collections (such as those maintained by plant breeders and experimental biologists) may be subjected to changes more frequently than large ones, for three reasons. The first is that rare and special alleles or biotypes could be important for a small but not a large core collection. Second, a small core collection should be more dynamic than a large one so that new genetic stocks and 'elite' strains possessing desirable genetic variability could be added more frequently. And third, plant breeders may wish to have all the different adaptive genes for their breeding work; these genes may not be all available in a core collection.

CONCLUSION

Recently, much attention has been paid to the development and use of core collections of plant species. The idea behind the core collection is that it should represent, with minimum redundancy, the genetic diversity of a crop species and its wild relatives. A core collection is expected to be a living record of, and to optimize the genetic diversity in, a species or a genus. Consequently, it has to be a dynamic, rather than a static, set of accessions.

The size and composition of a core collection could be altered in order to optimize its genetic diversity. This chapter has provided guidelines for gene bank managers on why, when and how to alter the size and composition of a particular core collection, depending upon the species in question and the size of the core. In addition, indicators for changing the content of a core collection were suggested. These include, but are not limited to: comparisions between the core and the source collections and allele frequencies over succeeding multiplications.

References

Britting, P.K. and Goodman, M.M. 1989. Genetic variation in crop plants and management of germplasm collections. In Stalker, H.T. and Chapman, C. (eds) *Scientific Management of Germplasm: Characterisation, Evaluation and Enhancement*. North Carolina, USA/Rome, Italy: North Carolina State University/IBPGR.

Brown, A.H.D. 1989a. The case for core collection. In Brown, A.H.D, Frankel, O.H., Marshal, D.R. and Williams, J.T. (eds) *The Use of Plant Genetic Resources*. Cambridge, UK: Cambridge University Press.

Brown, A.H.D. 1989b. Core collections: Practical approach to genetic resources management. *Genome* 31: 818-24.

Brown, A.H.D. 1991. Genetic variation and resources in cultivated barley and wild *Hordeum*. In Munk, L. (ed) *Barley Genetics VI*. (vol. II). Helsingborg, Sweden.

Chapman, C.G.D. 1989. Collection strategies for the wild relatives of field crops. In Brown, A.H.D, Frankel, O.H., Marshal, D.R. and Williams, J.T. (eds) *The Use of Plant Genetic Resources*. Cambridge, UK: Cambridge University Press.

Erskine, W. and Muehlbauer, F.J. 1991. Allozyme and morphological variability, outcrossing rate and core collection formation in lentil germplasm. *Theoretical and Applied Genetics* 83: 119-25.

Frankel, O.H. and Brown, A.H.D. 1984. Current plant genetic resources — a critical appraisal. In *Genetics: New Frontiers* (vol. VI). New Delhi, India: IBH Publishing.

Gill, K.S. 1989. Germplasm collections and the public plant breeder. In Brown, A.H.D, Frankel, O.H., Marshal, D.R. and Williams, J.T. (eds) *The Use of Plant Genetic Resources*. Cambridge, UK: Cambridge University Press.

Hamon, S. and van Sloten, D.H. 1989. Characterisation and evaluation of Okra. In Brown, A.H.D, Frankel, O.H., Marshal, D.R. and Williams, J.T. (eds) *The Use of Plant Genetic Resources*. Cambridge, UK: Cambridge University Press.

Hamrick, J.L. and Godt, M.J.W. 1990. Allozyme diversity in plant species. In Brown, A.H.D., Clegg, M.T., Kahler, A.L. and Weir, B.S. (eds) *Plant Population Genetics, Breeding and Genetic Resources*. Sunderland, Massachusetts, USA: Sinauer.

Hermsen, J.G.T. 1986. Effecient utilisation of wild and primitive species in potato breeding. In Jellis, G.J. and Richardson, D.E. (eds) *The Production of New Potato Varieties: Technological Advances*. Cambridge, UK: Cambridge University Press.

Hodgkin, T. 1990. The core collection concept. In Hintum, Th.J.L. van, Freese, L. and Perret, P.M. (eds) *Crop Networks: Searching for New Concepts for Collaborative Genetic Resources Management*. Rome, Italy: IBPGR.

Holbrook, C.C., Anderson, W.F. and Pitman, R.N. 1992. Selection of a core collection from the US germplasm collection of peanut (*Arachis hypogaea* L.). *Agronomy Abstracts*.

ICARDA. 1992. Report of the first meeting on the Barley Core Collection committee, May 1992. Mimeograph. Aleppo, Syria: ICARDA.

Jaradat, A.A. 1992. Estimate of phenotypic diversity and trait associations in *durum* wheat landraces from Jordan. *J Genetics and Breeding* 46: 69-76.

Jaradat, A.A. 1994. A core collection of wild emmer wheat from Jordan. *Crop Science* (submitted).

Ladizinsky, G. 1989. Ecological and genetic considerations in collecting and using wild relatives. In Brown, A.H.D, Frankel, O.H., Marshal, D.R. and Williams, J.T. (eds) *The Use of Plant Genetic Resources*. Cambridge, UK: Cambridge University Press.

MacKey, M.C. 1989. Utilising wheat genetic resources in Australia. In McLean, R. (ed) *Proc. 5th Assembly, Wheat Breeding Society of Australia*.

Marshal, D.R. 1990. Crop genetic resources: Current and emerging issues. In Brown, A.H.D., Clegg, M.T., Kahler, A.L. and Weir, B.S. (eds) *Plant Population Genetics, Breeding and Genetic Resources*. Sunderland, Massachusetts, USA: Sinauer.

Martin, J.M., Blake, T.K. and Hocket, E.A. 1991. Diversity among North American spring barley cultivars based on coefficients of parentage. *Crop Science* 31: 1131-37.

Nath, B., Omran, A.O. and House, L.R. 1984. Genetic divergence among non-restorer collection of sorghum (*Sorghum bicolor* L. Moench.) and its relationship to heterosis. *Euphytica* 34: 441-47.

Nevo, E., Beiles, A., Gutterman, Y., Storch, N. and Kaplan, D. 1984. Genetic resources of wild cereals in Israel and vicinity. I. Phenotypic variation within and between populations of wild wheat, *Triticum dicoccoides*. *Euphytica* 33: 717-35.

Nevo, E., Beiles, A. and Zohary, D. 1986. Genetic resources of wild barley in the Near East: Structure, evolution and application in breeding. *Biological J. Linnaean Society* 27: 355-80.

Palmer, R.G. 1989. Germplasm collections and the experimental biologist. In Brown, A.H.D, Frankel, O.H., Marshal, D.R. and Williams, J.T. (eds) *The Use of Plant Genetic Resources.* Cambridge, UK: Cambridge University Press.

Prasada Rao, K.E., Mengesha, M.H. and Gopal Reddy, V. 1989. International use of a sorghum germplasm collection. In Brown, A.H.D, Frankel, O.H., Marshal, D.R. and Williams, J.T. (eds) *The Use of Plant Genetic Resources.* Cambridge, UK: Cambridge University Press.

Smith, J.S.C. and Duvick, N. 1989. Germplasm collections and the private plant breeder. In Brown, A.H.D, Frankel, O.H., Marshal, D.R. and Williams, J.T. (eds) *The Use of Plant Genetic Resources.* Cambridge, UK: Cambridge University Press.

Souza, E. and Sorrells, M.E. 1991a. Relationships among 70 North American oat germplasm. I. Cluster analysis using quantitative characters. *Crop Science* 31: 599-605.

Souza, E. and Sorrells, M.E. 1991b. Relationships among 70 North American oat germplasm. II. Cluster analysis using qualitative characters. *Crop Science* 31: 605-12.

Core Collections of Plant Genetic Resources
Edited by T. Hodgkin, A.H.D. Brown, Th.J.L. van Hintum and E.A.V. Morales

A Wiley-Sayce Publication

4.3

Verifying and validating the representativeness of a core collection

N.W. GALWEY

Abstract

Two definitions of the representativeness of a core collection are proposed: how nearly it contains the full range of variation present in the whole collection; and how closely its pattern of variation resembles that of the whole collection. The effects on representativeness of different methods of selecting accessions for the core are examined using data from the Cambridge (UK) *Phaseolus* bean germplasm collection. Maximising representativeness (in terms of definition 1) for one descriptor did not consistently increase representativeness for other descriptors, and selection at regular intervals through the collection did not produce a more representative core (definition 2) than random selection. Maximising representativeness had little effect because associations between descriptors, although significant, were weak. Multivariate methods are usually used in the definition of groups of accessions for selection of a core collection, and the procedures for this situation are discussed. The effects of missing data have not been found to be serious in the *Phaseolus* collection except where a descriptor is missing for most accessions. The author considers ways in which descriptors too expensive to record in the whole collection could be used in the definition of the core, and notes that genetic diversity within accessions will often have to be treated on this basis in the formation of core collections, since information on this type of diversity will not be available for all accessions. The use of passport, taxonomic and ecogeographical data can help to ensure that the core collection is representative not only of the whole collection but also of the whole plant taxon under consideration. It is concluded that verification of representativeness, although important, is not an obstacle to the development of core collections.

The representativeness of a core collection can be defined in two ways: (1) how close it comes to including the full range of variation present in the whole collection; and (2) how closely the pattern of variation in the core resembles that in the whole collection. The usefulness of a core collection for plant breeding is related to its representativeness according to the first definition. However, definition (1) may not always be the most appropriate: a core collection which gives rare and unusual types equal prominence with the common forms of a crop may be misleading. In particular, the ease with which results from studies on the core collection can be applied to the whole collection will depend upon its representativeness according to the second definition. It is usually desirable to extrapolate further, and make inferences about the crop taxon (species, genus or other unit) as a whole. Their validity will, in turn, depend upon how closely the pattern of variation in the whole collection resembles that which exists (or formerly existed) in the field.

Much of the discussion of core collections has been in terms of the conservation of alleles (for example, Brown, 1989a, b). Although this is helpful for identifying and resolving the issues involved, in practice other approaches will be required when defining core collections, because information about individual alleles will be unavailable for many traits and, in any case, alleles do not occur independently. In the extreme case of an obligately clonally propagated organism, it is the whole genotype which is included or omitted, with no prospect of recombination of its component alleles. But even in obligate outbreeders, alleles occur in linkage groups which are difficult to recombine. Thus, even when the objective is the conservation of alleles, the assessment of the representativeness of the core collection should include the use of phenotypic characters, some of which are determined by many loci. Moreover, combinations of characters may be considered together: a particular character in one genetic background may be of much more use than the same character in another background. In addition to genotypic or phenotypic characteristics of the accession, the definition of a core collection may be based on the passport data of the accessions; indeed, good passport data probably have no rival as a concise, revealing and yet inexpensive source of information about accessions.

Phenotypic characters are represented in crop germplasm collections by descriptors (such as plant height and seed colour), each having a range of states (for example, 0.30-1.90 m; red, brown, black) (IBPGR, 1982). Descriptors and descriptor states, rather than loci and alleles, will therefore be considered here. Descriptors may be either continuous (having a range of numerical values, such as plant height) or discrete (having values which do not form a natural sequence, such as seed colour). Quasi-continuous descriptors, in which only integer values can occur, will be treated as continuous.

The measures which have been proposed for species diversity in ecosystems are reasonable candidates as indices of the diversity of a discrete descriptor. One such measure is:

$$Dv = \Sigma p_i \exp(-p_i)$$

where:
p_i = the proportion of accessions having the ith descriptor state

This index has the properties which are expected of a reasonable measure of diversity: it gives larger values when more descriptor states are present and, other things being equal, it gives larger values when descriptor states are equally common than when some are common and others rare. However, it is little affected by the inclusion or exclusion of rare descriptor states in the sample (Emlen, 1973). This is a desirable attribute in some contexts, but in a core collection rare descriptor states are of considerable importance. Nevertheless, it would be undesirable to use a measure of diversity which was strongly influenced by their chance inclusion. A more commonly used measure of diversity is the Shannon-Weaver information theoretic expression (Shannon and Weaver, 1963), often known as the Shannon index:

$$H' = -\Sigma p_i \log p_i$$

which has similar properties to Dv.

A third option is $1/\Sigma p_i$ (Emlen, 1973). The Shannon index is strictly appropriate when a random sample is taken from an infinite population. When a complete, finite population is studied, its diversity is given by Brillouin's measure (Poole, 1974):

$$H = \frac{c}{N}(\log_{10}N! - \Sigma\log_{10}Ni!)$$

where:

Ni = number of accessions having the ith descriptor state
N = total number of accessions
c = a constant for conversion of logarithms from base 10 to the base chosen for the measure

The sampling of a core collection lies between these two extremes.

Possible measures of the diversity of a continuous descriptor include its variance, its standard deviation, its range and the distance between its upper and lower quartiles. In the examples which follow, the Shannon index and the variance will be used; these two measures have the advantages of familiarity and simplicity.

COMPARISON OF STRATEGIES FOR THE DEFINITION OF A CORE COLLECTION

In practice, a core collection will be defined on the basis of several descriptors, but the principles involved can be examined using a single descriptor. A core collection may be defined by dividing the whole collection into groups and then selecting accessions from the groups according to one of the following strategies (Brown, 1989a):

- Strategy C (constant strategy): equal numbers of accessions selected from each group
- Strategy L (logarithmic strategy): an intermediate strategy whereby, for example, the number selected is proportional to the logarithm of the number in the group
- Strategy P (proportional strategy): the number selected is proportional to the number in the group

If the core collection is formed on the basis of a single normally distributed continuous descriptor, the equivalent of strategy C would be to over-represent, proportionally, the extreme values of the descriptor. Strategy C will tend to produce a core that is representative according to definition (1), whereas strategy P will produce a core that is more in accord with definition (2).

Once the core collection has been selected, its representativeness can be assessed by comparing its diversity with that of the whole collection for each individual descriptor. The consequences of applying different criteria in the selection of a core collection can be illustrated using data from the Cambridge *Phaseolus* bean germplasm collection in the UK. This collection was assembled between 1968 and 1979 and now contains 4939 accessions that were able to produce seed when grown out in the environment of Cambridge (52°N) (Anon., 1984). These accessions come from 73 countries, as far as their origins are known. Twenty-seven descriptors have been recorded in the collection, but many of these are missing for many accessions (*see* Table 1).

Table 1 Descriptors recorded in the Cambridge *Phaseolus* bean germplasm collection

Descriptor	% accessions for which descriptor present
Accesssion number	100.0
USDA PI number	25.9
Country of origin	96.2
Season in which grown	100.0
Seed weight	93.9
Seed shape	
longitudinal section	99.6
transverse section	99.6
Seed coat pattern	99.6
Seed coat colour	
base	99.6
mottles	25.7
Seedling pigmentation	60.9
Growth habit	72.9
Branching	56.1
Number of nodes	60.5
Height	45.9
Leaf shade	56.9
Leaf size	75.6
Flower colour	
standard	56.2
wing	56.2
Days to flowering	71.9
Days to maturity	61.3
Pod position	42.9
Number of pods	60.1
Pod length	44.5
Pod curvature	37.4
Pod fibre	43.4
Seeds per pod	57.3

A simple way of selecting a core collection would be to include accessions at regular intervals through the sequence of accession numbers (say, every 10th or 50th). If accessions with adjacent numbers are likely to have similar characteristics because they have similar origins, this approach may be expected to produce a representative collection (in terms of definition 2) more reliably than a random sample of the same size. This idea was tested, and the results are presented in Table 2, where the Shannon indices and variances of 10 randomly selected core collections are compared with those of 10 systematically selected ones. These core collections each contain between 96 and 100 accessions, as do all others discussed here. This is a much smaller number than would be sampled in practice.

Table 2 **Measures of diversity in randomly and systematically selected core collections**

Descriptor		Method of selection: Random	Method of selection: Systematic
Shannon indices:			
Seed coat pattern	min	0.306	0.333
	mean	0.384	0.385
	max	0.440	0.423
Seedling pigmentation	min	0.262	0.230
	mean	0.328	0.322
	max	0.427	0.421
Growth habit	min	0.239	0.206
	mean	0.257	0.259
	max	0.278	0.280
Flower standard colour	min	0.411	0.401
	mean	0.490	0.476
	max	0.552	0.554
Variances:			
Height	min	3315	2930
	mean	4463	4049
	max	5003	5282
Days to flowering	min	165	145
	mean	238	193
	max	357	235
Pod length	min	4.85	4.98
	mean	7.15	7.69
	max	9.91	12.96
Seeds per pod	min	1.58	1.49
	mean	2.01	2.11
	max	2.43	2.58

However, the use of a small sample size will highlight differences between the consequences of different selection methods, since the properties of large samples are similar to those of the populations from which they are drawn. If systematic sampling were more consistent in producing a representative core collection than random sampling, the measures of diversity would be expected to vary less among the systematic samples. In the present example this is not the case, but different results may be obtained in other collections.

The effect on other descriptors of maximising the diversity of the core collection for a particular descriptor can also be investigated. Core collections were selected by taking equal numbers of accessions with each seed coat pattern, or each growth habit, or equal numbers from each of three arbitrarily defined classes for number of days to flowering or pod length. The number of accessions available in each group for each descriptor is given in Table 3. It is when the numbers in the different groups are most unequal that there is the most scope for increasing diversity by an appropriate choice of selection strategy. The boundaries between the groups for the continuous descriptors were therefore placed at one standard deviation below the mean and one standard deviation above the mean, ensuring that small numbers of accessions would fall into the extreme groups. The results of selecting a core collection on the basis of these four criteria are compared with those of random selection in Table 4. In general, the different selection strategies have very similar outcomes. For example, selection on the basis of seed coat pattern increases the Shannon index for seed coat pattern itself, and hence representativeness in terms of definition (1), but selection on other bases does not produce values for this index outside the range of those obtained by random selection. Even among pairs of descriptors that are correlated, such as number of days to flowering and plant height, or pod length and number of seeds per pod (*see* Table 5), selection on the basis of one does not increase the variance of the other.

Table 3 **Numbers of accessions in groups defined on the basis of various descriptors**

Descriptor	Group	Number of accessions
Seed coat pattern	0	3648
	1	776
	2	123
	3	29
	4	151
	5	131
	6	40
	7	22
	missing	19
Growth habit	determinate	1016
	indeterminate	2586
	missing	1337
Days to flowering	$X \leq X - SD_X$	549
	$X - SD_X < X \leq X + SD_X$	2445
	$X > X + SD_X$	558
	missing	1387
Pod length	$X \leq X - SD_X$	353
	$X - SD_X < X \leq X + SD_X$	1508
	$X > X + SD_X$	335
	missing	2743

Table 4 Measures of diversity of core collections selected on the basis of various descriptors

Descriptor on which selection is based	Seed coat	Shannon index: Seedling pattern	Growth pigmentation	Flower habit	Standard colour
Random	min	0.306	0.262	0.239	0.411
	mean	0.384	0.328	0.257	0.490
	max	0.440	0.427	0.278	0.552
Seed coat pattern		0.903	0.314	0.255	0.563
Growth habit		0.421	0.361	0.301	0.523
Days to flowering		0.366	0.463	0.282	0.555
Pod length		0.315	0.374	0.192	0.509

Descriptor on which selection is based		Variance: Height	Days to flowering	Pod length	Seeds per pod
Random	min	3315	165	4.85	1.58
	mean	4463	238	7.15	2.01
	max	5003	357	9.91	2.43
Seed coat pattern		4856	182	6.99	2.58
Growth habit		4244	231	9.02	1.57
Days to flowering		4068	460	7.98	2.28
Pod length		4637	193	16.19	1.95

Table 5 Correlations between continuous descriptors

DF = 2082	Height	Days-to-flowering	Pod length
Days to flowering	0.333		
Pod length	0.327	-0.133	
Seeds per pod	0.352	0.096	0.355

Any effect of selection strategy on diversity would have to derive from associations between the descriptors, and the strengths of these associations are investigated in Tables 5, 6 and 7. Associations between continuous descriptors are assessed by correlation coefficients, those between continuous and discrete descriptors by analyses of variance, and those between discrete descriptors by calculation of the deviances from contingency tables of pairs of descriptors, which are distributed as chi-square. Since the sample sizes are very large, nearly all these statistics are significant, but this does not imply that the associations are strong. For example, the largest correlation coefficient (between pod length and seeds per pod) is only 0.355. Evidently, such associations are not strong enough for sampling on the basis of one descriptor to affect the diversity of another.

Table 6 Analyses of variance of continuous descriptors grouped by discrete descriptors

Discrete descriptor	DF	F ratio Height	Days to flowering	Pod length	Seeds per pod
Seed coat pattern	7 1494	2.66	2.77	13.00	3.24
Seedling pigmentation	5 1489	1.95	3.52	4.32	5.48
Growth habit	1 1498	1096.17	243.92	7.15	212.60
Flower standard colour	5 1493	11.25	10.20	35.42	18.13

Table 7 Degrees of freedom and deviances[a] for associations between discrete characters

	Seed coat pattern DF	Seed coat pattern deviance	Seedling pigmentation DF	Seedling pigmentation deviance	Growth habit DF	Growth habit deviance
Seedling pigmentation	49	166.5				
Growth habit	7	69.3	7	13.3		
Flower standard colour	35	365.2	35	1124	5	146.8

Note: a Distributed as chi-square

DEFINITION OF THE CORE COLLECTION BY MULTIVARIATE METHODS

In practice, the core collection will be selected on the basis of several descriptors, perhaps by using them to define clusters of accessions and then applying one of the three selection strategies to the clusters. If the levels of diversity within the different clusters are not approximately equal, the selection strategies may be modified so as to take more accessions from the more diverse clusters. Another approach, which would avoid this requirement, would be to ignore any natural groupings among the accessions and to divide the multivariate space which they occupy into several equal volumes, applying one of the selection criteria to the accessions in each volume. Once the core collection has been defined, its representativeness can be assessed by comparing its diversity with that of the whole collection for each individual descriptor.

It will not usually be possible to include the whole collection in a multivariate cluster analysis, and this means that the choice of accessions for the cluster analysis to some extent affects the definition of the core collection. However, it is desirable that accessions not used in the cluster analysis should nevertheless be candidates for inclusion in the core collection. It will therefore be necessary to develop methods to determine to which cluster each of the remaining accessions belongs. The purpose of doing this is to make possible the selection of a core collection more representative than the sample of accessions on which the multivariate analysis was based. This can be achieved for definition (1) of representativeness by adopting strategy C or strategy L, but it is not clear that it can be achieved for definition (2).

FURTHER CONSIDERATIONS

Missing data

Missing data are likely to pose a problem in the definition of a core collection, as the extent of available data is likely to be very uneven among accessions. Methods such as hierarchical cluster analysis cannot deal with this, at least as implemented in standard statistical packages; they generally require a complete, rectangular data matrix. All other things being equal, accessions with complete data should be preferred for inclusion in the core collection, but total exclusion of accessions with incomplete data is likely to reduce the diversity of the core collection considerably.

The extent of this problem in the Cambridge *Phaseolus* bean collection is shown in Table 1, and the effects of missing data are explored in Table 8. The latter table shows the results of eliminating all

Table 8 Effects on measures of diversity of omitting accessions with missing descriptors[a]

Missing descriptor	% acc. retained	SCP		SP		GH		FSC	
		% kept	index	% kept	index	% kept	index	% kept	index
None	100.0	100.0	0.392	100.0	0.371	100.0	0.258	100.0	0.522
Country	96.2	96.2	0.396	97.6	0.371	97.0	0.259	97.8	0.522
PI number	25.9	26.0	0.286	25.7	0.514	27.5	0.256	25.9	0.517
SP	60.8	61.1	0.347	100.0	0.371	78.1	0.250	99.2	0.521
GH	72.9	73.2	0.396	93.5	0.357	100.0	0.258	97.0	0.522
DF	71.9	72.1	0.393	91.9	0.349	96.1	0.261	99.6	0.522
S/P	57.3	57.5	0.398	71.6	0.332	78.1	0.258	77.8	0.527

Missing descriptor	H		DF		PL		S/P	
	% kept	var.	% kept	var.	% kept	var.	% kept	var.
None	100.0	4407	100.0	221	100.0	7.13	100.0	1.92
Country	98.2	4417	97.1	223	98.3	7.16	97.4	1.94
PI number	24.2	2235	27.7	232	23.8	4.22	26.5	1.90
SP	99.0	4402	77.8	247	99.0	7.14	76.1	1.54
GH	99.6	4405	97.5	220	99.5	7.14	99.5	1.92
DF	98.5	4379	100.0	221	99.4	7.10	99.5	1.91
S/P	95.6	4391	79.3	181	98.9	7.14	100.0	1.92

Note: a acc. = accessions; SCP = seed coat pattern; SP = seedling pigmentation; GH = growth habit; FSC = flower standard colour; H = height; DF = days to flowering; PL = pod length; S/P = seeds/pod accessions for which a particular descriptor is missing, and then calculating diversity measures for the other descriptors. Elimination of the accessions without United States Department of Agriculture (USDA) Plant Introduction (PI) numbers leaves only about a quarter of the collection and results in a considerable reduction in diversity for seed coat pattern and plant height, although not for the other descriptors. On the other hand, elimination of the accessions for which the number of days to flowering is missing leaves about 70% of the collection, but in general the accessions with missing days to flowering also had missing plant heights, and thus over 98% of those whose heights were known remain. Consequently, there is little effect on the variance of height. Except in the extreme case of PI numbers, the effect of eliminating accessions with missing values is minimal, but very different results might be obtained in a germplasm collection in which the distribution of missing values was different.

Use of additional descriptors

A further test of the representativeness of the core collection would be to measure some new variable in the whole collection, and then to compare the diversity of the core collection with that of the whole collection for this variable. However, it would then be appropriate to re-define the core collection, taking the new information into account. A partial way out of this dilemma is as follows. A new variable, perhaps too expensive to be determined on the whole collection (a molecular marker, for example, or an agronomic trait influenced by the environment) could be measured in the core collection. If it were found that a group of accessions uniform for the variables so far considered was diverse for the new variable, more accessions of this group could be included in the core collection, and could be presumed to contribute to genetic diversity although the new variable had not been measured on them.

Genetic diversity within accessions

Genetic diversity within accessions is certain to be widespread, particularly in outbreeding seed crops, and strategies for selecting a core collection can be evaluated in terms of their success in capturing rare but widespread alleles (that is, alleles which occur at low frequencies in a large proportion of accessions) (Brown, 1989b). However, the constraints of germplasm catalogues often require that there is only one state of each descriptor for each accession. Hence, when the selection of a core collection is based on catalogue information, it may be impossible to take account of within-accession genetic diversity. This strengthens the case for considering broad phenotypic categories, rather than alleles, as the units to be conserved. The extent to which this distorts reality will depend upon the relative magnitude of between-accession and within-accession variation. However, genetic diversity within accessions could be used as an 'additional descriptor' in the manner described above.

Validation: assessment of representativeness beyond the collection

So far, the discussion has been simplified by considering the representativeness of the core collection relative to the whole collection from which it is selected. This means that relatively straightforward and objective criteria for representativeness can be defined. However, a much more difficult objective — representative sampling of the plant taxon under consideration — must eventually be tackled. The term 'verification' may be used to indicate the assessment of representativeness relative to the whole collection, and 'validation' to indicate assessment relative to external criteria. Two types of evidence are relevant to the validation of a core collection:

- passport data for accessions in the collection
- taxonomic and ecogeographical studies of the plant, external to the collection

Passport data of high quality will indicate, for example, the range of locations, altitudes, soil types and cropping systems covered by the whole collection. It is probably more important that the core collection should be representative for these than for any other type of descriptor. Unfortunately, many otherwise valuable collections (including the Cambridge *Phaseolus* bean collection) lack good passport data. However, even very limited passport data can be of considerable value. For example, both country of origin and mean rainfall at the site of origin were found to be related to the salt tolerance of accessions in a barley germplasm collection, but of the two variables, country of origin was the more informative (Peeters et al., 1990). Moreover, lack of passport data can to some extent be compensated for by external taxonomic and ecogeographical information. For example, the common bean (*Phaseolus vulgaris*) comprises two gene pools, the Middle American and the Andean, that are quite distinct genetically and in their geographical origins (Gepts, 1988). If these two gene pools were considered to be of equal importance and interest, it might be decided that a core collection should comprise equal numbers of accessions of each. Most accessions could be unequivocally assigned to one or the other on the basis of seed size and other plant characteristics.

CONCLUSION

The discussion in this chapter suggests that the definition of a core collection should be an iterative process, with successive stages of refinement. It should not follow a mechanistic procedure: if a curator has a good knowledge of the collection, his or her discretion in including an accession or group of accessions will be a valuable supplement to any algorithm or formula. Since the stability of the composition of the core collection is important for the value of future work, this suggests that it should not be defined or re-defined hastily; nor should its definition be long delayed in the search for perfection. Even if the collection were selected randomly, it is likely that it would be sufficiently representative, and provide a sufficient improvement in manageability, to prove useful. In the studies of the Cambridge *Phaseolus* bean collection described here, it is striking how little difference the different methods of defining a core collection made to its diversity and representativeness. It cannot be assumed that other descriptors or stratification methods, or that the same methods applied to another collection, would have produced the same result. However, it is worth bearing in mind that a major motive for stratified sampling (for example, in market research or in conducting opinion polls) is the difficulty of obtaining a random sample. There is no such difficulty in the definition of a core collection.

Core collections usually contain about 1000 accessions, and this is a large enough sample to provide a representative description of a population. If it is concluded that core collections are a good idea in other respects, verifying and validating their representativeness, although important, should not pose a major obstacle to their development.

References

Anon. 1984. *The Cambridge* Phaseolus vulgaris *L. Germplasm Catalogue*. Cambridge, UK: Cambridge University.

Brown, A.H.D. 1989a. Core collections: A practical approach to genetic resources management. *Genome* 31: 818-24.

Brown, A.H.D. 1989b. The case for core collections. In Brown, A.H.D., Frankel, O.H., Marshall, D.R. and Williams, J.T. (ed) *The Use of Plant Genetic Resources*. Cambridge, UK: Cambridge University Press.

Emlen, J.M. 1973. *Ecology: An Evolutionary Approach*. Reading, Massachusetts, USA: Addison-Wesley.

Gepts, P. 1988. A Middle American and an Andean common bean gene pool. In Gepts, P. (ed) *Genetic Resources of* Phaseolus *Beans*. Dordrecht, Netherlands: Kluwer .

IBPGR. 1982. *Descriptors for* Phaseolus vulgaris. Rome, Italy: IBPGR.

Peeters, J.P., Wilkes, H.G. and Galwey, N.W. 1990. The use of ecogeographical data in the exploitation of variation from gene banks. *Theoretical and Applied Genetics* 80: 110-12.

Poole, R.W. 1974. *An Introduction to Quantitative Ecology*. Tokyo, Japan: McGraw-Hill Kogakusha.

Shannon, C.E. and Weaver, W. 1963. *The Mathematical Theory of Communication*. Urbana, Illinois, USA: University of Illinois Press.

Core Collections of Plant Genetic Resources
Edited by T. Hodgkin, A.H.D. Brown, Th.J.L. van Hintum and E.A.V. Morales

A Wiley-Sayce Publication

4.4

One core collection or many?

M.C. MACKAY

Abstract

The growth in the size of germplasm collections and in user demand during recent decades has caused concern about the effective management and use of these collections. The core collection, based on the selection of a small, representative sample from a large germplasm collection, has been proposed as a means of addressing this problem. There are at least two motives for establishing a core collection. First, it would assist in gene bank management; this has implications for policy and strategy at gene bank, national and international levels of plant genetic resources management. Second, it would contribute to the effective and efficient use of available germplasm; this is achieved by focusing attention on a sample of germplasm selected with representation of genetic variation as the basic criterion. These two motives are firmly linked: good gene bank management contributes to germplasm utilisation and hence to the requirements of germplasm users. These requirements may be simple requests for genotypes, giving accession number or name/designation, or they may be requests for a range of genetic variation for a particular attribute, often when such variation has not previously been reported. The latter type of request is best served by using the core collection principle to select the desired range of genetic variation. This chapter discusses the core collection with reference to its practical application and the possible inclusion of core collection principles into gene bank information management systems.

The proposal to divide *ex situ* plant germplasm collections into 'core collection' and 'reserve collection' components stemmed from the need to encourage germplasm evaluation, promote utilisation of germplasm and improve gene bank management (Frankel, 1984). The rationale was to reduce the effective size of a gene bank and thus encourage use of the representative genetic diversity in the core component.

The assumption that gene banks are not being effectively utilised is questionable. Marshall and Brown (1981) state that 'while it is true that evaluation is a necessary first step to utilisation, it does not necessarily follow that lack of information has been a major stumbling block to the greater utilisation of genetic resources in the past'. This puts the emphasis on information availability. It is immaterial whether it be passport, characterisation or evaluation information; what matters is that available information is accessible to germplasm managers and users so that the best use can be made of germplasm available. Emphasis has previously been placed on systematic evaluation (as a precursor to utilisation), resulting in confusion. Making the best use of available germplasm through sensible use of available information, whether it be evaluation data or some other type of information, is also a logical objective for core collections.

Various approaches to developing core collections for a range of objectives have been discussed elsewhere (Frankel and Brown, 1984; Mackay, 1986, 1990; Brown, 1989a, b; Hamon and van Sloten, 1989; von Bothmer et al., 1990; Beuselinck and Steiner, 1992). None of these authors has queried the premise that evaluation is a precursor to utilisation. They describe ways of extending or restricting the original proposal of the core collection (a set representing the variation held by a gene bank) and of selecting accessions for inclusion. Their ideas for making better use of available germplasm include the selection of sets of accessions with different objectives in mind. These might include sets that represent the broad genetic variation available for or in:

- a total crop genome or species
- a geographical region
- the accessions held by a gene bank
- one or more specific attributes or loci
- the alleles for a specific locus

Any one of these sets could be considered to be a core collection because it is representative of the genetic variation of a larger group of germplasm accessions. The major differences between the various approaches relate to the perceived expectations of users (Hodgkin, 1991) and to the choice of a method of selecting a set of accessions that will contain the genetic variation being sought. Nowhere has it been suggested that a core collection must be fixed, but rather that it should be flexible to facilitate the acquisition of knowledge or germplasm — there can be one core collection or many, and the objective in establishing a core collection can vary.

The word 'core', however, suggests that there can be only one such set and this has led to confusion. Brown (*Chapter 1.1, this volume*) reports on how distinctions have been made between core collections with different objectives by including an explanatory word. Examples include 'core sets', 'synthetic core' and 'specific purpose (attribute) core sets'; perhaps 'gene bank core' could be added to this list. The term 'core collection' has clearly become generic in that it now refers to a principle or concept rather than a single, finite set of accessions. This principle is to select a small, representative sample of genetic variation from a large germplasm collection to assist with germplasm management and utilisation. In this chapter, the term 'core collection' is used in the generic sense (*see* Figure 1).

The potential users of core collection sets fall into three main groups:

- plant breeders who wish to find and utilise germplasm
- germplasm specialists who wish to study germplasm or genetic variation
- curators who require assistance with germplasm management

Breeders require rapid identification of desirable attributes and immediate access to seed samples. Germplasm specialists attempt to understand genetic diversity more fully and require in-depth knowledge of how variation is distributed. Curators require some knowledge of this diversity and

Figure 1 **Varied uses of the representative genetic variation in a core collection set, ranging from a complete gene pool to a single genetic attribute and determining the size of the core set**

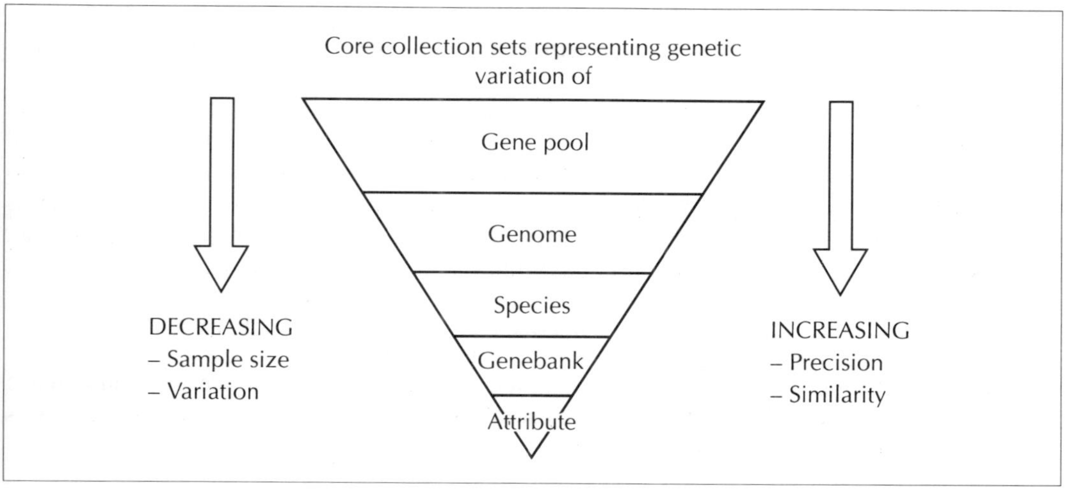

variation in order to meet the needs of breeders, and continually seek ways of improving the efficiency of gene bank management. Germplasm specialists and curators are concerned mainly with management issues, while breeders are concerned primarily with germplasm use. These different interests are not incompatible; in fact, advances in one will contribute to the success of the other (*see* Figure 2). Thus, while one can distinguish different objectives for a core collection, the basic principle seems to

Figure 2 **Complementarity between various types of core sets, resulting in improved gene bank management and germplasm use**

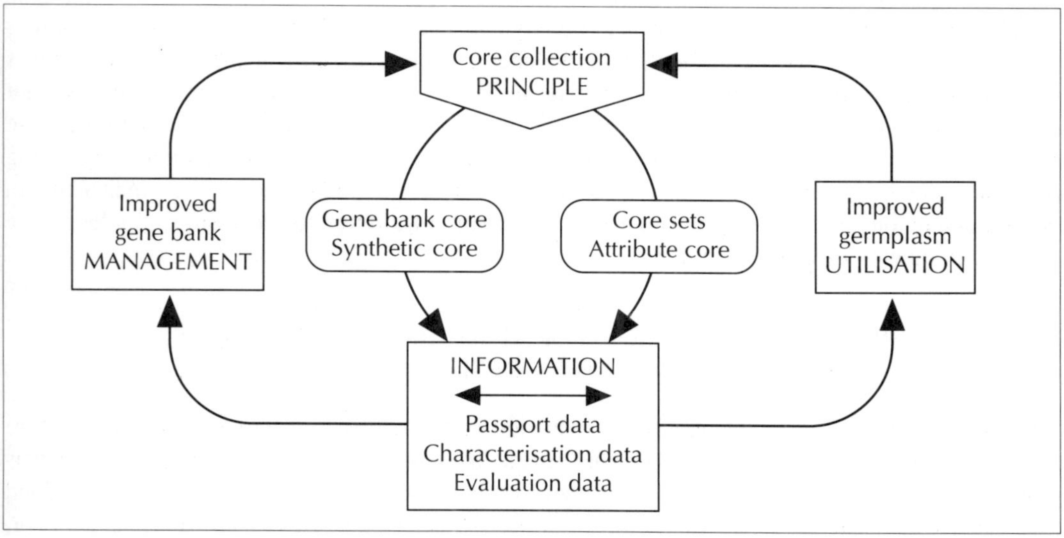

be sufficiently flexible to satisfy all users. It is suggested that the principle can be further extended to embrace innovative technologies, such as heuristic data models, and allow dynamic maintenance and upgrading of core collections.

EVALUATION — A PREREQUISITE TO UTILISATION?

The premise that the evaluation of germplasm is a prerequisite to utilisation merits some consideration. Evaluation was one of the basic tenets behind the core collection proposal and it is frequently advanced as a constraint to the effective use of plant genetic resources. In reviewing this premise, Marshall (1989) concluded that it was not necessarily valid. Different gene banks perform different functions, such as the management of active and base collections, and thus the importance and consequence of evaluation cannot be readily compared. The main constraints to utilisation identified by Marshall (1989) included:

- deficiencies in gene bank management that effectively restrict its accessibility to users
- limited information flow preventing users from acquiring available germplasm, or inadequate financial support to undertake basic gene bank activities; these are seen as resolvable by users alone, or by users in collaboration with gene banks

The significance of either type of constraint would differ from one gene bank to another, depending upon institutional and national priorities. Despite these differences, the various constraints highlight the need for effective gene bank management and accessible information. They do not promote systematic evaluation as a precursor to utilisation.

The availability of adequate genetic variation within the adapted germplasm held in breeders' working collections must also be considered as a possible constraint to the use of gene banks. In many cases, desirable attributes from cultivated species (such as single-gene disease resistance) are simply inherited and easily incorporated into appropriate genetic backgrounds by backcrossing. Desirable attributes in wild material are sometimes neglected because of the difficulty in transferring one allele (or a small number of alleles) into an adapted genotype. The lack of pre-breeding (transferring these attributes into suitable backgrounds) is another identified constraint to utilisation (Marshall, 1989).

The notion that evaluation is the primary prerequisite to the utilisation of plant genetic resources cannot and does not need to be substantiated. Evaluation received attention in the past because it was seen by users to restrict their access to germplasm. Systematic evaluation of germplasm does not represent the only or the most effective means of using available plant genetic resources. Prudent use of available information in selecting genotypes for evaluation is an attractive method for targeting desirable genetic variation. It addresses a range of issues, including gene bank management, information quality and availability and user involvement, in addition to addressing a model for the strategic evaluation of germplasm.

THE CORE COLLECTION FOR GENE BANK MANAGEMENT: ONE CORE COLLECTION

The paradigm for a core collection is the reduction in the size of a gene bank to make it more manageable and useable (Frankel, 1984). This gene bank management and accessibility strategy was proposed at a time when gene banks were growing rapidly (in numbers of accessions). This strained financial and physical resources. In addition, information management was usually done manually or, if computerised, was primitive compared with current systems. Information adds value to plant genetic resources — information availability and transfer are essential for effective and efficient utilisation.

For any assembly of germplasm, whether it be a gene bank or some other grouping, a core collection is representative of all or part of the genetic diversity of the original group. How this is achieved has been the subject of considerable discussion (Brown, 1989a; von Bothmer et al., 1990; and elsewhere in this volume). In essence, current thought has a 'single' core collection changing in content (as necessary) while providing a reasonably fixed set of germplasm for exhaustive evaluation.

The difficulties in handling large and growing numbers of accessions in gene banks in the decade up to 1984 were largely responsible for the core collection proposal. This constraint to utilisation is largely negated by modern database technology. Gene banks can now manipulate information in minutes compared to the days or even weeks previously required. Selecting a set of accessions for any purpose is not difficult if one has the knowledge, information and skill to exclude or include individual accessions according to the given goal. Some random sampling could also be considered legitimate in certain circumstances. Developing a procedure to make these selections for variation of a single genetic attribute is achievable, but certainly not simple. Applying similar procedures to multiple attributes or selecting one core collection representative of all genetic variation will obviously be more difficult. Attempts to develop such a core collection will undoubtedly contribute to the general understanding of genetic variation and its distribution in gene banks, which is a legitimate endeavour. What began as a proposal to assist in the management and use of large numbers of accessions is now focusing on methods of sampling genetic variation.

Developing a single core collection in order to encourage systematic evaluation and therefore utilisation is of questionable value to users in the short term. Such an approach would be useful in grouping accessions to enable, for example, a map of ecogeographical representation within a collection to be developed which would familiarise gene bank managers and users with deficiencies or excesses in a collection. There would be a flow-on value to users from improving the management or rationalising a gene bank in this way, which is obviously a legitimate goal.

The crucial component in developing a core collection is the means by which the set that contains the representative genetic variation is selected. It is not so crucial whether the objective is a set representative of a species or a single attribute. The objective, from the perspective of the user, is to increase the chance of finding the required attribute or attributes within a smaller number of accessions, thus making utilisation more effective and efficient.

A single core collection is legitimate if it facilitates the acquisition of practical knowledge and experience, and it must also contribute to the ease with which required variation is detected. This application of the core collection principle is focused more on gene bank and germplasm management than on utilisation *per se* (Brown, *Chapter 1.1*). Germplasm specialists and gene bank managers can use the information a core collection provides to gain a better understanding of plant genetic resources. Plant improvement, through the use of germplasm, remains a practical incentive and benefits from better management of genetic resources.

THE CORE COLLECTION FOR GERMPLASM USE: MANY CORE COLLECTIONS

One basis for considering multiple core collections is that germplasm users often request a set of accessions that is likely to contain a characteristic not previously described. Like most researchers, plant breeders will favour a technique that enables them to achieve their objective as quickly as possible. This implies a small number of genotypes containing a representative range of variation for the characteristic they are seeking. These non-specific requests have a similar, but narrower, objective than that of a single core collection. They differ from specific requests, which are requests for accessions that are readily identified, perhaps by name or designation, because of a reported characteristic or characteristics.

The non-specific request highlights the need for a method of 'targeting' germplasm or accessions that are likely to possess genetic variation for a particular attribute. When the extent of genetic variation is not known (for example, full host susceptibility, through various degrees of resistance to immunity for reactions to a fungal disease) the breeder will prefer a set of accessions likely to be representative of available variation. It should be noted that what is being sought is genetic variation at one or a few loci, not a set representing the genetic variation of an entire species, genome or gene bank.

The non-specific request is important because it represents the cutting edge of plant improvement — the identification of new genetic variation for overcoming constraints to productivity. Plant improvement in major crop species, such as wheat or maize, is a step-by-step process that is usually aimed at adding one or a few genes at a time to an adapted genetic background. Locating and identifying these genes is the objective of non-specific requests for germplasm.

An example could be that a breeder finds that a trace element toxicity is the cause of reduced productivity; variation for tolerance to the toxicity has not been reported and the breeder requests a range of genotypes that are likely to contain genetic variation for this attribute. A core collection of accessions, selected explicitly with a breeder's objectives in mind, provides an obvious means of efficiently identifying the desired genetic variation. By contrast, the single core collection would contain accessions that would be considered redundant to the breeder's objectives. It is both more effective and more efficient to base the accession selection or sampling strategy on the requirements of the breeder. This approach of generating many core collections that directly address the user's requirements complements the primary objectives of the 'single' core collection — to promote evaluation and use.

The cumulative information and experience gained in selecting small, multiple core collection sets will ultimately provide feedback benefits to germplasm specialists and gene bank managers. This, in turn, will benefit gene bank and genetic resources management, as well as facilitate utilisation. These two approaches to applying the core collection principle offer benefits that can be shared. The relationship between the two approaches is illustrated in Figure 2.

Examples of non-specific requests and core collections

Experience with non-specific requests at the Australian Winter Cereals Collection (AWCC) probably resembles that of other active gene banks. Prior to in-house computerised documentation systems it was difficult to sort or select accessions for a specific project. The AWCC initially provided wheat accessions at random for pre-harvest sprouting evaluation, and later was able to select accessions on the basis of ecogeographical data that had been previously held but was not readily available. Since 1986, the AWCC has encouraged breeders with non-specific requests to discuss their requirements with collection staff in an attempt to define the target accessions more precisely. Selection of sets of representative genetic variation for tolerance to toxic levels of boron in the soil and for resistance to the cereal cyst nematode (*Heterodera avenae*) are examples of the application of the core collection principle (Mackay, 1990).

Table 1 **Geographical distribution of boron-tolerant genotypes of wheat, based on visual assessment of tolerance**

Region or origin	Visual assessment score[a] VS	S	MS	MT	T	VT	Total	Frequency (%) of MT, T and VT groups
Asia/Asia Minor	22	82	89	105	49	10	357	47
Australia	5	31	45	8	1	—	90	10
Egypt	25	22	7	—	—	—	54	0
Ethiopia	—	—	—	—	—	—	—	—
Europe	18	98	98	26	7	3	250	14
Kenya/north-western Africa	5	10	9	5	2	—	31	23
Mexico	6	34	31	9	3	1	84	15
North America	11	40	27	5	—	2	85	8
South America	3	12	21	12	8	—	56	36
Unknown origin	90	220	192	46	18	3	569	12
Total	185	549	519	216	88	19	1576	
Frequency (%)	12	35	33	14	5	1		

Note: a VS = very susceptible; S = susceptible; MS = moderately susceptible; MT = moderately tolerant; T = tolerant; VT = very tolerant
Source: Moody et al. (1988)

More recent cases of using this approach include the selection of core sets of accessions to provide genetic variation for pasting quality of bread wheat, for the effect of heat shock on bread wheat quality and for spectral quality of anthocyanin across altitudes in a range of *Triticum* species, including wild and cultivated diploids, tetraploids and hexaploids. The possibility of using a single core collection (broadly representing the genetic variation of the gene bank in a set of about 3000 accessions) was raised with the user in each case. All three chose to use a set selected specifically for their purpose. Undoubtedly, if the only option available was a single core collection then it would have been used in preference to a random sample.

Another example of this more specific representation of genetic variation has been the recent search for genetic resistance to the Russian wheat aphid (*Diuraphis noxia*) in wheat. Initially, a number of sources of resistance were available in wild relatives of wheat. Resistant accessions among cultivated forms have only recently been found in germplasm from Asia Minor, whence the aphid's origins can be traced (CIMMYT, 1991). The AWCC supplied 42 accessions, representing a number of species, from this region for screening in 1989; one, an accession of *T. vavilovi*, provided useful resistance in a hexaploid form.

Wheat breeders at the Waite Agricultural Research Institute at the University of Adelaide, Australia, have submitted several non-specific requests for genetic diversity since 1968. Their requests were initially serviced with random sets of accessions. More recently, the core collection principle has been used to select representative sets of genetic variation for particular objectives (A.J. Rathjen, pers. comm.). The wheat breeders concede that a single core collection is, in the absence of any other information, a valid starting point. If, however, the option of using an attribute-biased core collection is available, then this is the preferred choice. Rathjen has suggested three reasons for avoiding the single core collection option if multiple core collections are available:

- it is unlikely that the breeder will be totally without information regarding the possible distribution of the attribute being sought
- the probability of identifying a useful accession within a single core collection is lower than it would be in a specially selected core collection because there will be very few genotypes from any area which experiences a biotic or abiotic stress that is limited in its distribution; severe expression is often localised
- the genetic diversity in a single core collection (representing, for example, the general variation of a single gene bank) could be far from representative of the whole range of genetic diversity in a species

In short, the Waite group believe that a germplasm user, in contrast to a gene bank manager or a germplasm specialist, would be more likely to use multiple core collections to locate desired genetic variation. This approach would make most efficient use of available resources, especially in terms of the number of accessions to evaluate and the potential success rate of locating the desired variation.

Thus, the core collection concept has merit in selecting representative diversity from a larger assembly of germplasm, but different germplasm users will have different objectives in sampling germplasm. A plant breeder's requirement for a range of genetic variation usually involves one or a few attributes at a time. Any number of multiple core collections can be constructed to service the non-specific requests of plant breeders. Such core sets should not be seen as a replacement for a single core collection. The attribute-biased core collection simply addresses the individual requirements of a germplasm user, and it can contribute to the longer-term success of single core collections. The single core collection has an important role to play in wider aspects of plant genetic resources policy development and in the management and planning of national and global programmes. It will also contribute to the effectiveness with which a collection can be subsampled for a specified range of genetic variation. The single and multiple core collection approaches, when viewed as complementary, open further horizons of germplasm management and use — the possibility of a dynamic system for effectively and efficiently locating and/or identifying a user-defined range of genetic variation. The user, in this case, can be a gene bank manager, a plant breeder, a population geneticist or an institutional policy maker.

FUTURE DEVELOPMENT

In achieving the stated objective of promoting the effective and efficient use of available plant genetic resources, curators, germplasm specialists and breeders must be mindful of changing technologies. Computerised gene bank information systems have contributed significantly to germplasm use in the past decade through improved information flow and availability. This technology has had a great impact on the current implementation of the core collection as an aid to gene bank management. The development of expert systems involving, for example, heuristics and pattern matching offers new dimensions in plant genetic resources management and use, particularly if such an approach is integrated into global crop networks. This technology is available and is being used in a variety of natural resource and agricultural fields. Examples involve scientific exploration with geographical information systems (Twery et al., 1991) and the use of an intelligent frame system for selecting cultivars (Bolte et al., 1991). Both these applications share common sampling practices involved in developing a core collection.

The development of such a system requires a re-definition of our understanding of information assembly and linkages. In a gene bank information management system there is a logical structure to the way in which discrete units of information are stored and retrieved. There are many discrete units of information, but they can be accessed only in ways that are defined by the system structure. For example, if one were seeking accessions with Russian wheat aphid resistance in a relational database model, a field for (or describing) this attribute would need to be defined. On the other hand, a pattern-matching process could be used to link the various units of information in less traditional ways. To expand on a previous example, the term 'RWA' (for Russian wheat aphid) might prompt a simile from another data file that links with a list of accession numbers, or with a geographical region where the insect is endemic (and perhaps how long the insect has been known to occur in that region). In this case, the traditional links between units of information have been replaced by new links which are less dependent upon traditional database structures but more dependent upon reasoning — by changing the rules for linking information units, the knowledge gained is new or different.

Such an application for a germplasm information management system would have elements that not only address the day-to-day operational aspects of a gene bank, but also provide procedures that allow previous experience and knowledge to be accumulated (heuristic learning process) and used as an aid in selecting germplasm for evaluation in the future. Thus, when a gene bank user requests accessions with a range of genetic variation for a particular attribute, the information management system could alert the breeder to the following types of information:

- other users have made a similar request (and the results obtained)
- the attribute being sought has already been identified
- literature references to variation studies on the subject

This model should also be used to request clarification from the breeder, such as:

- Are you seeking identified sources of variation?
- Are you seeking new sources of variation?
- Where is the variation you seek likely to occur in nature?

Such a system would result in the selection of a set of accessions on the basis of details determined from the interactive request session. The selection filter(s) generated (*see* Figure 3) would sieve through all available accessions and choose the final core sample of accessions; the filter(s) would be retained for future use or further refinement. This heuristic model would actually build upon knowledge and would minimise the need to re-define associated (or similar) requests each time they are made.

The intelligent frame system mentioned above is used to maintain cultivar information and select the optimal cultivar for specific sites or purposes (Bolte et al., 1991). It is not difficult to imagine how such an application can be extended to plant genetic resources management and the selection of accessions for specific purposes.

Figure 3 **Schematic representation of how a heuristic data model extends the core collection principle towards more precise selection of sets of accessions for evaluation and use**

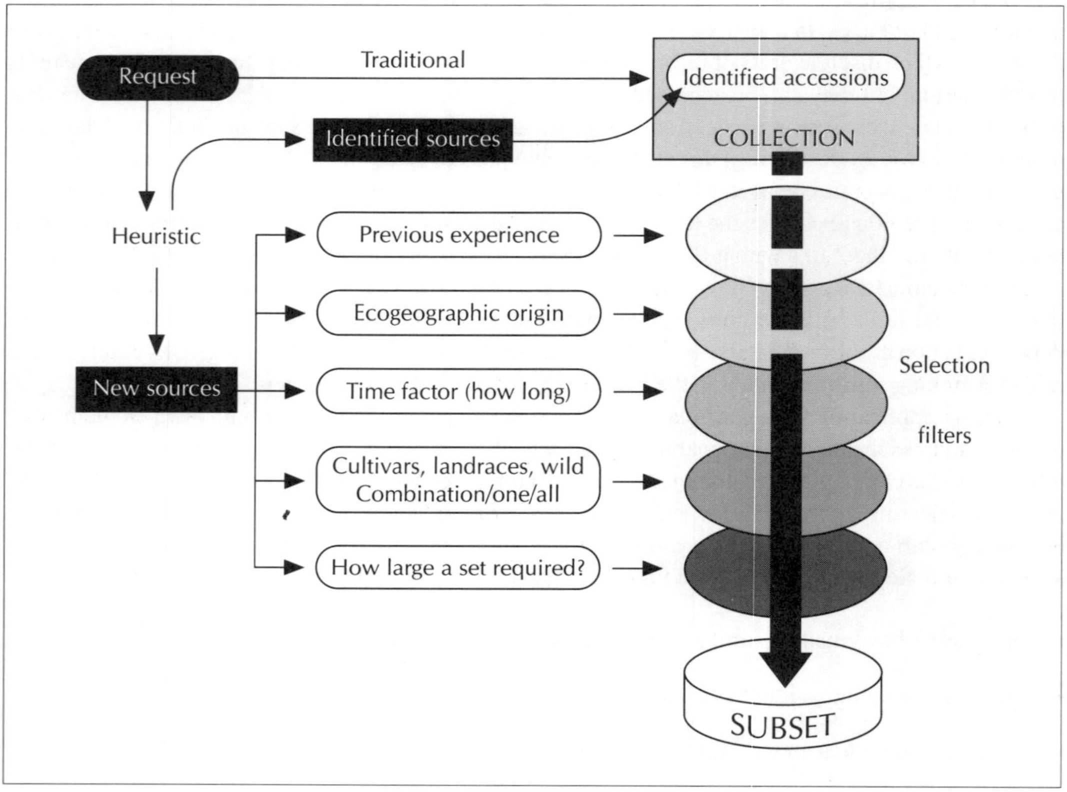

There are numerous possibilities for altering the way in which information is processed. The suggested heuristic model is designed to maximise the relevance of readily available information to give the inquirer, and add extra information of possible interest in order to 'extend' the person making the inquiry. It is seen as an efficient means of managing non-specific requests made to gene banks, while allowing a strategic set of evaluation data to be generated for the purposes of long-term management of genetic resources. The model would require minimal maintenance and would be able to accumulate knowledge as it is used. It would ideally be developed on an international basis so as to ensure continued relevancy and could well be an adjunct to global crop networks. It could possibly be considered as the 1990s replacement for the core collection concept in effectively and efficiently utilising plant genetic resources.

CONCLUSION

The core collection was proposed in 1984 as a means of helping gene banks manage large numbers of accessions. The proposal was put forward because of the need to promote the use of plant genetic resources. Extensive evaluation of a set of accessions that represented the general diversity in a gene bank was seen as an effective method of doing this. The core collection, by reducing the number of accessions available in the first instance, would foster greater use of genetic resources by plant breeders than large and cumbersome collections. While this had obvious appeal to gene bank users and managers, it also implied the need for further refinement to enhance the efficiency of germplasm use.

While evaluation can be used to promote utilisation, it is not the only means by which this can be achieved. Database technology that has become available since the core collection idea was first proposed has eliminated an important constraint to utilisation in the early 1980s — the lack of a means to manipulate large amounts of information.

There are two primary motives for establishing core collections: assisting germplasm management and enhancing germplasm use. Germplasm specialists seem more concerned with the management motive, while breeders are more concerned with usage. Curators have an interest in both motives but are (understandably) biased towards gene bank management. Establishing a core collection with one of these motives as the main objective does not necessarily serve the other. There are, however, joint benefits: the information obtained using either approach contributes to the success of both applications of the core collection. The two approaches are complementary. The beneficiaries are all those who have a stake in the conservation and judicious exploitation of plant genetic resources.

To maximise the long-term benefits of methods for making germplasm use more efficient and effective, a means of building on previous experience is necessary. This involves extending current gene bank information management technology towards a more heuristic design. The result will be that the cumulative knowledge or information, obtained by implementing either type of core collection, is fully used. In this way the full benefits of the core collection principle — improved germplasm management and use — can be achieved.

Acknowledgements

The author acknowledges Drs B. Skovmand and A.J. Rathjen for their suggestions and helpful discussion on the subject of the core collection concept, and R. Clark for his guidance on expert systems and heuristic information technology.

References

Beuselinck, P.R. and Steiner, J.J. 1992. A proposed framework for identifying, quantifying and utilising plant germplasm resources. *Field Crops Research* 29: 261-72.

Bolte, J.P., Hannaway, D.B., Shuler, P.E. and Ballerstedt, P.J. 1991. An intelligent frame system for cultivar selection. *AI Applications* 5(3): 21-31.

Brown, A.H.D. 1989a. Core collections: A practical approach to genetic resources management. *Genome* 31: 818-24.

Brown, A.H.D. 1989b. The case for core collections. In Brown, A.H.D., Marshall, D.R., Frankel, O.H. and Williams, J.T. (eds) *The Use of Plant Genetic Resources*. Cambridge, UK: Cambridge University Press.

Frankel, O.H. 1984. Genetic perspectives of germplasm conservation. In Arber, W., Llimensee, K., Peacock, W.J. and Starlinger, P. (eds) *Genetic Manipulation: Impact on Man and Society*. Cambridge, UK: Cambridge University Press.

Frankel, O.H. and Brown, A.H.D. 1984. Plant genetic resources today: A critical appraisal. In Holden, J.H.W. and Williams, J.T. (eds) *Crop Genetic Resources: Conservation and Evaluation*. London, UK: Allen and Unwin.

Hamon, S. and van Sloten, D.H. 1989. Characterisation and evaluation of okra. In Brown, A.H.D., Marshall, D.R., Frankel, O.H. and Williams, J.T. (eds) *The Use of Plant Genetic Resources*. Cambridge, UK: Cambridge University Press.

Hodgkin, T. 1991. The core collection concept. In Hintum, Th.J.L. van, Frese, L. and Perret, P.M. (eds) *Crop Networks. Searching for New Concepts for Collaborative Genetic Resources Management: Proc. EUCARPIA/IBPGR Plant Genetic Resources Symposium, Wageningen, Netherlands, 1990*. Rome, Italy: IBPGR.

Mackay, M.C. 1986. Utilising wheat genetic resources in Australia. In *Proc. 5th Assembly of the Wheat Breeding Society of Australia, Perth/Merredin, Western Australia*. Perth, Australia: Western Australian Department of Agriculture.

Mackay, M.C. 1990. Strategic planning for effective evaluation of plant germplasm. In Srivastava, J.P. and Damania, A.B. (eds) *Wheat Genetic Resources: Meeting Diverse Needs*. Chichester, UK: John Wiley.

Marshall, D.R. 1989. Limitations to the use of germplasm collections. In Brown, A.H.D., Marshall, D.R., Frankel, O.H. and Williams, J.T. (eds) *The Use of Plant Genetic Resources*. Cambridge, UK: Cambridge University Press.

Marshall, D.R. and Brown, A.H.D. 1981. Wheat genetic resources. In Evans, L.T. and Peacock, J.W. (eds) *Wheat Science — Today and Tomorrow*. Cambridge, UK: Cambridge University Press.

Moody, D.B., Rathjen, A.J., Cartwright, B., Paull, J.G. and Lewis, J. 1988. Genetic diversity and geographical distribution of tolerance to high levels of soil boron. In Miller, T.E. and Koebner, R.M. (eds) *Proc. 7th International Wheat Genetics Symposium, Cambridge, UK*. Cambridge, UK: Institute of Plant Science Research.

Twery, M.J., Elmes, G.A. and Yuill, C.B. 1991. Scientific exploration with an intelligent GIS: Predicting species composition from topography. *AI Applications* 5(2): 45-53.

von Bothmer, R., Fischbeck, G., Hintum, Th.J.L. van, Hodgkin, T. and Knüpffer, H. 1990. The Barley Core Collection. Final Report of the BCC Working Group. Mimeograph. BCC Working Group.

Part 5

USING CORE COLLECTIONS

Core Collections of Plant Genetic Resources
Edited by T. Hodgkin, A.H.D. Brown, Th.J.L. van Hintum and E.A.V. Morales

A Wiley-Sayce Publication

5.1

The durum wheat core collection and the plant breeder

P.L. Spagnoletti Zeuli and C.O. Qualset

Abstract

The choice of the exotic germplasm to use as parents is a major problem for plant breeders. Yet long-term progress can take place only if unrelated parents are introduced into breeding programmes. The issues considered in this chapter are: the estimate of genetic diversity within a core collection; the evaluation of the potential of exotic lines for use in a breeding programme; and the use of information obtained from genetic evaluation of a core collection to exploit the whole collection more fully. The role of plant breeders in the evaluation of phenotypic or genetic diversity among accessions is stressed. The testcross procedure can be used to assess the combining ability with improved varieties which may reflect diversity among exotic accessions. The differential in combining ability with available germplasm could form the basis for classification. The choice and number of testers to be used should be based on the degree of genetic similarity in the gene pool best adapted to the environment at which the breeding programme is aimed. A method based on the mating of a constant tester for selecting promising parents and estimating the genetic components of variance is illustrated. A multivariate analysis approach is presented as a means of identifying sources of 'new' yield-enhancing genes. Results illustrating approaches adopted in a programme initiated to develop greater understanding of a germplasm collection of durum wheat are presented.

The genetic potential of germplasm collections has been exploited only in a minor way. Breeders, confronted with a large number of entries in gene banks, do not yet have a good basis for selecting parental materials that will maximise potential genetic gain from derived hybrid populations. The use of these populations for breeding purposes could be greatly increased if there was more information on the amount and kind of variation available, but in most cases the resources needed to characterise fully, both phenotypically and genetically, thousands of accessions may not be available. This task could be more easily accomplished if core collections were established (Frankel and Brown, 1984; Brown, 1989).

How and to what extent plant breeders will use core collections will depend on a number of factors: the crop species; the objectives of their programme; how well the accessions are characterised and documented; and how much breeders know about the collection. This means that these collections should be characterised and evaluated primarily for characters which are 'useful' for breeding and that this information should be effectively transferred to the breeder. Clearly, these requirements would be met if breeders participated in the evaluation effort themselves. In this chapter we outline and discuss some approaches a breeder might take to evaluate and use a core collection. The issues involved include:

- estimating the diversity within a core collection
- evaluating the breeding potential of exotic lines
- determining methods that can be used to identify promising parents
- using information obtained from core genetic evaluation to exploit the reserve collections more fully

As an example, we discuss the results of some studies conducted on a large section (about 3000 accessions) of the United States Department of Agriculture (USDA) world collection of durum wheat (*Triticum turgidum* L.). This collection has been extensively studied (Qualset and Puri, 1974; Jain et al., 1975; Spagnoletti Zeuli and Qualset, 1987, 1990) and a core collection of 496 accessions was extracted on the basis of a multivariate analysis of spike characters and other yield components (Spagnoletti Zeuli and Qualset, 1987). For most of the measured characters, the variance in the core sample was larger than in the collection. This was attributable mainly to the relative increase of frequencies in the least represented classes and to a parallel reduction in the most represented classes (Spagnoletti Zeuli and Qualset, 1993). To assess the amount and kind of genetic variation, we have used a testcross design involving mating a common parent with a random sample of accessions from the core collection. Two improved high-yielding cultivars, Creso and Modoc, have been crossed with accessions representing most countries where this species is grown (Spagnoletti Zeuli, 1993).

Wheat is an autogamous species and is already at an advanced breeding stage, but most conclusions can be extended to other self-pollinated crop species.

EVALUATION OF CORE SUBSAMPLES AND ESTIMATION OF DIVERSITY

Phenotypic evaluation

To use a germplasm collection, breeders must know that it exists (Peeters and Williams, 1984), and to formulate an educated request of germplasm they should know the amount and kind of variation the germplasm collection contains.

As pointed out by Prasada Rao and Ramanatha Rao (*Chapter 3.2, this volume*), many breeders are not familiar with the kind and amount of variation available in gene banks. Thus, the very first task for the curator of a core collection would be to promote germplasm collections among those for whose benefit they were established in the first place. The core collection should be distributed as widely as possible to plant breeders. Because of the relatively small size of these collections, breeders will be more inclined to devote some effort to growing the core collection as part of their breeding programme and to consider some sort of strategy for evaluation and characterisation to familiarise themselves with this 'exotic' germplasm.

A quick and simple visual inspection of plants in the field will give most breeders an idea of how much better these collections could benefit their work than any computerised database. Many required descriptors are either complex to score or environmentally sensitive (Peeters and Galwey, 1988). While descriptions of the test site and of evaluation procedures may have indicative value, in general they will not replace tests in the breeder's own environment (Frankel, 1989).

When the world collection of durum wheat was evaluated at the University of California, Davis, USA some years ago, some accessions with very large kernels (kernel weight > 80 mg) were identified. Evaluation data for this multigenic character were the means from five random spikes per accession, from an unreplicated experiment, but kernel weight is well known to be fairly stable. In a study of three environments, the correlation between environments was $r > 0.7$ (Spagnoletti Zeuli and Qualset, 1993) and heritability was high ($h^2 = 0.57$) (Lebsock and Amaya, 1969; Spagnoletti Zeuli, 1993). The evaluation data were made available to the US National Plant Germplasm database (GRIN) and shortlists of accessions showing extreme values for this and other plant and spike characters were published (Spagnoletti Zeuli et al., 1986; Spagnoletti Zeuli and Qualset, 1990). However, further evaluation and use of these accessions apparently occurred only at Davis. Of the many possible reasons for the relatively low use of evaluation data, one was that breeders did not know that this germplasm was available (Peeters and Galwey, 1988) and another was they did not trust evaluation data. These kernel weight figures, for example, are hard to believe when looked at on a computer printout, even if extensive information on how and where the measurements were obtained are provided.

Evaluation for high yield

In most species and for most environments phenotypic evaluation would rarely identify new germplasm with yield comparable to any cultivar already grown in the target environment. However, for crops that have been little improved or where the target environment has been the subject of little breeding effort, it is possible that even a rough evaluation of the core sample for adaptation would identify promising accessions to be included in the breeder's population or even to be released as improved varieties (Kenworthy, 1980; Ceccarelli et al., 1987). This would be the case, for example, when breeding for extreme environments where improved germplasm performs poorly or does not survive at all. Experienced breeders will also be able to identify sources of variation not thought to be available and may discover characters not yet described. Breeding aims change rapidly with time and differ between individual breeders (Frankel, 1989).

Ceccarelli et al. (1987) conducted a study to assess the value of landraces for breeding barley for environments where environmental stresses such as drought, cold, heat and salinity are common; under these agroclimatic conditions, local landraces or selections from landraces were considered to be more reliable and often outyielded improved cultivars. They evaluated a small sample from 7000 barley lines collected in Jordan and Syria for yield and other characters, and 15 lines outyielding the tester landrace were identified. They concluded that some of the high-yielding lines had the potential to be released as pure line varieties and that those superior lines with such adapted background could be also used effectively in the crossing programme.

In general, however, yield performance of exotic germplasm may be a poor indicator of breeding potential because of adaptation, but at least some accessions might have genetic potential as parents. Some type of pre-breeding effort will be needed before germplasm not adapted to a given environment could be even looked at (Marshall, 1989). Given the dimension of recombination potential for multigenic traits (MacKey and Qualset, 1986), the genetic variation in the segregating population must be narrowed to increase the chances of recovering whatever useful gene was contributed by the exotic parent. In wheat, the following combination of crosses can be used (A x B) x C, where B is the exotic germplasm and A and C are two well-adapted varieties. This incomplete backcross scheme has proved to be efficient in overcoming the problems encountered during initial attempts to widen the gene pool (MacKey and Qualset, 1986).

Evaluation of breeding potential of exotic lines using the testcross design

A major advantage of using a core collection is that a more accurate evaluation of quantitative characters is possible. In fact, many traits of interest for breeding are multigenic, and thus environmentally labile and replicated trials are needed. As Frankel (1989) pointed out, information about interaction between accessions and locally adapted cultivars is relevant to plant breeders. Because of its obvious limitations, evaluation on a phenotypic basis should not be considered the end point.

The performance of exotic strains gives no indication of genetic differences between the strains and current cultivars. The genetic basis of any observed phenotypic diversity can be studied by adopting several possible mating designs. The amount and accuracy of the genetic information gained is unfortunately directly proportional to the amount of resources required. It is also inversely proportional to the number of genotypes that can be evaluated and thus a balance between 'how well' and 'how many' must be found.

Testcross design

The mating of a constant parent with a sample of accessions (core) from the germplasm collection provides a good compromise for ranking tested material on the basis of combining ability with improved germplasm. This (testcross) approach has been used in a few self-pollinated species, including soybean (Reese et al., 1988; Sweeney and St Martin, 1989), barley (Gebredikidan and Rasmusson, 1970), peanuts (Isleib and Wynne, 1983), and wheat (Qualset, 1979), and more extensively in cross-pollinated species such as maize (reviewed by Hallauer and Miranda, 1988).

Three issues need to be considered:

- What is the best tester genotype?
- How many testers should be used?
- How should testcross progenies be studied?

Choice of tester genotype

The choice of the tester genotype is critical because it should ensure the correct classification of the relative performance and discriminate efficiently among accessions being tested. The tester should:

- have the ability to maximise variance among testcrosses
- be distantly related to most tested accessions

- carry a low frequency of favourable alleles for the trait of interest
- be able to compensate for genotype x environment interactions
- be, as far as possible, homozygous recessive at all loci

Different tester lines could be used to rank the tested germplasm accessions in different orders since the degree of dominance could change for different sets of tested lines. The best tester would be homozygous recessive at all loci. In this case, the regression of offspring on non-recurring parent would be a maximum. Although a tester should be weak, it should also be adapted to the target environment. Good combining ability with available germplasm should also be a criterion, and thus the testers should be selected from a group of locally adapted varieties.

An important issue is how many testers should be used. We have compared the results of crossing the same 51 accessions to two improved varieties: Modoc from California, and Creso from Italy. The degree of phenotypic correlation between the tested accessions and their testcrosses will also indicate the value of the performance of tested accessions in predicting the potential for selecting superior lines from a cross. Despite the difference in pedigree between the two testers, the pattern of correlations (*see* Table 1) between the F_1s and the non-recurring parents for eight spike characters was very similar in the two sets of crosses. Correlation coefficients for six characters were highly significant and in most cases were larger than 0.45, while the correlation for number of kernels and kernel weight were lower or not significant. Thus, for example, in both sets of crosses for spike length, the performance *per se* is of some value in predicting the potential for selecting desired lines from exotic unadapted germplasm, but this is not the case for the other two important yield characters. High correlations for yield between tester and selected testcross lines was also observed in soybean (St Martin and Aslam, 1986; Sweeney and St Martin, 1989) while, in barley, testcross selection compared with selection based on parental performance did not give a real advantage (Gebrekidan and Rasmusson, 1970).

If the accession x tester interaction is small, the number of tester varieties can be kept to a minimum. In durum wheat, although significant, the accession x tester component of variance was for most

Table 1 **Correlation coefficients for eight durum wheat characters between the non-recurrent parent mean and its F_1 offspring mean in two sets of 51 crosses with Modoc and Creso cultivars**

Character	Recurring parent: Creso	Recurring parent: Modoc
Spike length	0.50**	0.49**
Awn length	0.45**	0.51**
Number of kernels/spike	0.30*	0.22 ns
Kernel weight/spike	0.25 ns	0.25 ns
Awn/spike length	0.74**	0.46**
Kernel weight	0.63**	0.56**
Number of spikelets	0.35**	0.45**
Rachis internode length	0.74**	0.60**

Note: * = $P < 0.05$; ** = $P < 0.01$; ns = not significant

characters smaller than the testcross component (P.L. Spagnoletti Zeuli and C.O. Qualset, unpubl.). In soybean, St Martin and Aslam (1986) found a lack of interaction between parents and testers. They concluded that the choice of the tester could be made on the basis of convenience and that a single tester would adequately evaluate a parent's potential for producing a high-yielding line.

Thus, in self-pollinated species, the number and the choice of tester varieties should be determined on a case-by-case basis. The number should be kept as low as possible to allow for a larger number of lines to be tested. The definition of the preliminary criterion of choice should probably be based on the degree of genetic similarity among the varieties best adapted to the environment at which the breeding programme is aimed. If the genetic base of the improved gene pool is narrow, the ranking of tested exotic lines should be largely similar and only one or few testers would be needed.

Selection of testcrosses

The selection of testcrosses is actually the selection of half-sib families and could be performed in either F_1 or subsequent generations. Variance among testcrosses provides an estimate of variance for combining ability. Although an inbred line is a narrow-base tester, phenotypic selection of testcrosses leads to selection for genes with additive effects (Falconer, 1981). In general, if a broad-based tester is used, the estimate would be mainly for general combining ability. If the tester is narrow based, the estimate would be for specific combining ability. Since it is difficult to distinguish general vs. specific combining ability, the term 'combining ability' should be used in a broader sense (Hallauer and Miranda, 1988).

For a single locus, the combining ability ($C_i = T_i - T.$), where C_i is combining ability effect for line i, T_i is the mean of the testcross progeny with the line i, and T. is the mean of all testcrosses with the same tester variety) is independent from dominance effects only for a gene frequency of 0.5 in the tester or for dominance of zero (h = 0). For multigenic traits, the combining ability variance (V_c) must be summed over all loci, and when inbred lines are used as testers the variance will depend upon the balance of the number of loci with a frequency 0 or 1. An excess of important loci with a frequency of 0 in the tester will be always advantageous in increasing the variance and the efficiency of selection among testcrosses. Based on simple models at the single-locus level, the size of variance among testcrosses can be predicted (Hallauer and Miranda, 1988). The variance among testcrosses will be at

Table 2 Phenotypic variances for six spike characters among 80 durum wheat accessions and among their F_1s obtained with the cultivar Creso

Characters	V_P	V_{F_1}	F^a
Spike length	1.52	0.67	2.27**
Number of kernels/spike	816.35	1434.68	1.76**
Kernel weight/spike	521.10	659.82	1.26 ns
Kernel weight	0.071	0.026	2.71**
Number of spikelets	33.14	43.82	1.32 ns
Rachis internode length	4.68	1.32	3.54**

Note: a The F value tests the difference between variances; ** = $P < 0.01$; ns = not significant its maximum when the tester is fully homozygous recessive for the character of interest; if it is fully dominant the variance would be zero. Pseudo-overdominance for a given character can be caused by the presence of dominant genes at loci different in the tested line and the improved variety. When the tester is an improved variety this would seldom be the case. In fact, it can be expected that a high proportion of favourable genes have already been accumulated through breeding and the variance among testcrosses will be lower than among their non-recurring parents. Therefore the results could be quite different, depending upon the genotypes being tested, the character measured and how this character is genetically controlled. This is illustrated by the data in Table 2, where we show the phenotypic variances among non-recurring parents and their F_1s with Creso for six spike characters. Variance among parents was significantly larger for spike length weight per kernel and rachis internode length but smaller for number of kernels per spike, and not significantly different per kernel weight per spike and number of spikelets.

Identification of accessions with good combining ability for a desired trait

When the number of parents that are potential donors of a specific trait is rather large, the choice of the germplasm accessions to use as parents is a major problem for plant breeders. In this case, the regression of F_1 hybrids on the non-recurring parent (Qualset, 1979) can be used to choose a small sample of accessions for further study. In Figure 1 the results of such an analysis are shown for two spike characters observed among 80 durum wheat accessions and their F_1s with the Italian cultivar Creso. The offspring and non-recurring parent means are presented as a two-dimensional distribution and the observed regression is shown. The dotted diagonal line shows the hypothetical perfect correlation of the F_1s and non-recurring parents. This method readily identifies parents with heritable variation or sources of dominance effects. It has already been used in a breeding programme for improving the protein content in bread wheat, in which it appears that parents with heritable variation for grain lysine content were identified (Qualset, 1979).

Genetic analysis

It is evident that the elaboration of efficient screening methods that will make the gene pool more usable must be based on genetic knowledge of the traits of interest (MacKey and Qualset, 1986). The single array analysis can provide genetic information on the degree of dominance and an estimate of heritability of the traits scored in the set of core accessions in a specific environment. Table 3 gives estimates of additive and dominance genetic variance components, the degree of dominance and the heritability obtained using the method described by Griffing (1950). The two genetic components of variance, directly estimated from the analysis of variance from the replicated experiments, are the variance components among non-recurring parents and F_1-midparents, respectively. The degree of dominance is the square root of the ratio of dominance and additive components. Over-dominance is indicated when the dominance variance component is larger than the additive variance component. Additive genetic variances were significant for all traits and were larger than dominance variances for six of the eight characters. Dominance variance was evident for all traits, but was especially important for the fitness-related characters (number and weight of kernels per spike).

These estimates confirmed what the phenotypic correlations had suggested and what had been indicated previously for durum wheat by Porceddu and Scarascia Mugnozza (1983) and Spagnoletti Zeuli et al. (1983b,1985) and for other self-pollinated species (Matzinger, 1963).

Figure 1 **Joint distribution of parental values and their F_1s for 80 durum wheat accessions crossed with the cultivar Creso**

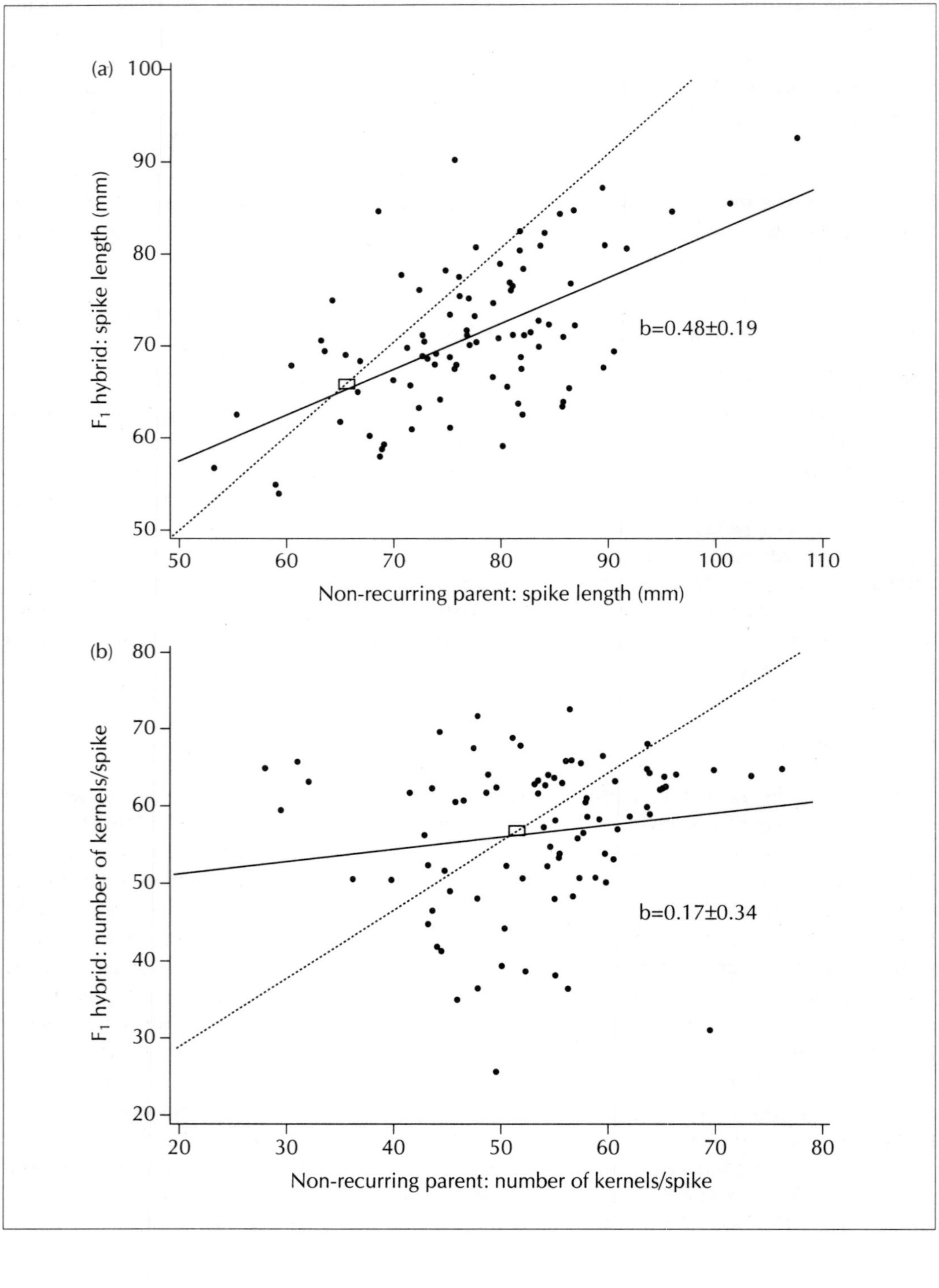

Table 3 **Genetic components of variance, degree of dominance (h) and heritability (h^2) estimated from a set of crosses between 80 durum wheat entries and the cultivar Creso**

Character	$\sigma^2 g$	$\sigma^2 h$	h	h^2
Spike length	84.11**[a]	31.43**	0.61	0.63
Awn length	508.70**	177.37**	0.59	0.66
Awn/spike length	0.12**	0.02**	0.38	0.70
Number of kernels/spike	58.86**	107.90**	1.35	0.42
Kernel weight/spike	38.70**	47.42**	1.11	0.47
Kernel weight	55.19**	8.78**	0.40	0.56
Number of spikelets	2.52**	2.29**	0.95	0.49
Rachis internode length	0.38**	0.02**	0.25	0.72

Note: a ** = $P < 0.01$

IDENTIFICATION OF 'NEW' YIELD-ENHANCING GENES

To achieve short-term objectives, breeders concentrate on locally adapted high-yielding varieties as potential parents to be crossed. When a yield 'plateau' is reached, a narrow genetic base is often suspected. Long-term progress can be made, however, by augmenting the flow of distantly related parents into the programme, thereby widening the gene pool of the breeder's population. 'Exotic' germplasm, available for many species in the gene banks, can be a source of additional genes for yield or environmental adaptation that are not available in improved cultivars or hybrids with a very narrow genetic base (Smith and Duvick, 1989).

The search would be for 'new' yield-enhancing genes that cannot be precisely identified. To maximise the potential genetic gain from subsequent hybrid populations, breeders will need to identify parental materials that will increase genetic diversity. In the genetic resources of cereals, phenotypic and genetic divergence has been linked mainly to geographic origin (Jain et al., 1975; Tolbert et al., 1979; Spagnoletti Zeuli and Qualset, 1987, 1990; Peeters et al., 1990). However, not all accessions carry reliable passport data (Spagnoletti Zeuli and Qualset, 1987; Sweeney and St Martin, 1989).

Evaluation and characterisation data based on environmentally insensitive characters or strongly inherited characters can further improve the identification of clusters within broad geographic groups. Multigenic characters that are environmentally labile provide a more serious problem in evaluation. Nevertheless, good reasons exist for computing estimates of genetic distances from morphological data showing continuous variation. Natural selection affects morphological traits linked to adaptation. Thus genetic distances from these traits could help in identifying patterns of adaptation and co-adaptation in a germplasm collection (Ceccarelli et al., 1987).

If evaluation is based on a large number of different characters, the ranking of tested accessions could be based on multivariate statistical techniques. For each character, an estimate of the genetic variance and the frequency of transgressive segregates in individual F_2 bulks could be also obtained.

Choice of characters

Since the multicharacter approach will involve the use of a considerable amount of resources, the characters should be easy to score and informative of yield potential. Yield components and characters revealing adaptation to the target environment could be used. A multigenic trait needs replicated trials for accuracy, and scores will be most valuable if obtained in the specific environmental conditions at which the breeder's efforts are aimed. An example is provided by the spike characters in wheat:

- they are easy to score
- they are multigenic, such that the genes could be considered representative of the whole genome
- two out of three yield components can be observed on single heads, thus providing useful information for breeding purposes
- they probably played a major role during wheat evolution, as conscious or unconscious human selection seems to have operated on these characters rather than on other traits

Alternatively, characters that are more affected by natural than human selection have contributed more to diversification among groups (Spagnoletti Zeuli and Qualset, 1990). It might be expected that when diversity for spike characters is present, diversity should also be present for other traits.

Multivariate analysis

Estimates of phenotypic or genetic distances can be obtained from qualitative and quantitative data (Goodman, 1973; Lefort-Buson and de Vienne, 1985). Accessions can be selected on the basis of a diversity matrix (Peeters and Martinelli, 1989) or cluster analysis of observed or transformed variables (Crossa et al., *Chapter 2.4*). Multivariate statistical analysis of measurements of quantitative characters provides estimates of morphological and genetic divergence among accessions and genotypes, which helps in planning crosses among genotypes belonging to different clusters (Whitehouse, 1969; Bhatt, 1970; Sneath, 1976; Camussi et al., 1983; Spagnoletti Zeuli et al., 1985). Several mating designs have been used (Hanson and Casas, 1968; Whitehouse, 1969; Cervantes et al., 1978; Qualset, 1979; Camussi and Ottaviano, 1981), and Camussi et al. (1985) have developed a general method for estimating and testing genetic distances, irrespective of the complexity of the genetic design.

Six durum wheat lines of differing geographical origin (Ethiopia, Algeria and Italy) and showing diversity in spike characters (Spagnoletti Zeuli et al., 1984, 1985) were crossed in diallel. The multivariate analysis of estimates for additive effects showed correspondence between geographic diversity and genetic divergence and other yield components. Most of the above studies have been restricted to inter-racial differences (for example, maize) or a small number of geographic groups (for example, durum wheat). It was obvious that the results extended only to the populations of origin because of the restricted number of genotypes included in the crossing design.

The degree of heterosis between populations, which reflects differences in gene frequencies (Falconer, 1981), is positively related to their genetic divergence. However, a reduction in heterosis is expected if the parents are too divergent (Moll et al., 1965; Cress, 1966). When the additive components account for a major proportion of total genetic variance, the phenotypic distances among accessions and the genetic distances based on additive effects are very similar.

A multivariate approach can be used to help plan crosses among genotypes in different groups or to select testcrosses. Distances among non-recurring parents and relevant testcrosses can be compared using multivariate statistical techniques. This would indicate how well observed overall phenotypic differences among germplasm accessions correlate with the performance of their progeny testcrosses.

We have conducted canonical discriminant analysis (Seal, 1964; Pimentel, 1979) separately for the set of non-recurring parents and of their F_1s. Canonical variate transformation permits an optimal visualisation of differences between populations through a reduction of dimensions that preserve most biological information. This technique had previously been applied to the whole world collection of durum wheat (Spagnoletti Zeuli and Qualset, 1987). Table 4 gives the results of such analysis calculated separately for the non-recurring parents and their F_1s. In the parent sets, diversity among accessions is attributable mainly to characters measuring lengths, as shown from their correlations to the first and second canonical variable. Kernel weight, one of the yield components, is associated with the third canonical variable. In the F_1 set, the picture is quite different. The first canonical variable, which accounts for 41% of the total variance, is strongly associated with yield-related characters. Most other characters are associated with the second transformed variable. This result reflects the larger contribution of these characters to total variance in the F_1 set than in the parental set.

Table 4 Correlation between canonical and observed variables for (a) parental accessions and (b) F_1s crossed with the cultivar Creso

	CAN1		CAN2		CAN3		CAN4	
Character	(a)	(b)	(a)	(b)	(a)	(b)	(a)	(b)
Spike length	0.75	-0.15	-0.35	0.73	0.15	0.41	0.44	-0.45
Awn length	0.48	-0.01	0.73	0.81	-0.13	-0.53	0.44	0.02
Awn/spike length	-0.20	0.15	0.91	-0.02	-0.15	-0.81	-0.06	0.49
Number of kernels/spike	-0.25	0.74	0.26	0.56	0.02	0.22	0.46	-0.15
Kernel weight/spike	-0.16	0.60	0.49	0.65	0.46	0.28	0.35	0.05
Kernel weight	-0.08	-0.26	0.54	0.33	0.73	0.24	0.06	0.58
Number of spikelets	-0.28	0.15	-0.35	0.69	0.41	0.51	0.71	-0.06
Rachis internode length	0.95	-0.40	-0.07	-0.06	-0.15	-0.17	-0.20	0.65
Eigenvalue	3.12	3.64	2.48	1.50	1.64	1.11	1.36	1.01
Proportion of total variance	0.28	0.41	0.23	0.17	0.15	0.12	0.12	0.11
Cumulative proportion of total variance			0.51	0.58	0.66	0.70	0.78	0.81

The plots of the non-recurring parents and their testcrosses, according to the first and second canonical variables, are shown in Figure 2. The plots provide a synthesis of information from several characters in describing the variability patterns and the phenotypic differences. The groups obtained through cluster analysis are also identified. The Ward method was used for the cluster analysis (Romesburg, 1984) of the transformed variables. Despite some overlapping because of the variation within groups, the parental accessions from different geographical regions form distinct groups. The two improved recurring parent cultivars occur in the same cluster. For the F_1s, three clusters are identified. Most of the testcrosses are grouped within values -1 and +2 of the first canonical variable and only one cluster is well separated from the others. The characters that were more highly correlated to the first canonical variable were number and weight of kernels per spike (positively) and kernel weight (negatively). They show a high degree of dominance (*see* Table 3).

Figure 2 **Scatter diagram of the first two canonical variable means for (a) 80 durum wheat accessions of different geographical origin and (b) their F_1s crossed with the cultivar Creso**

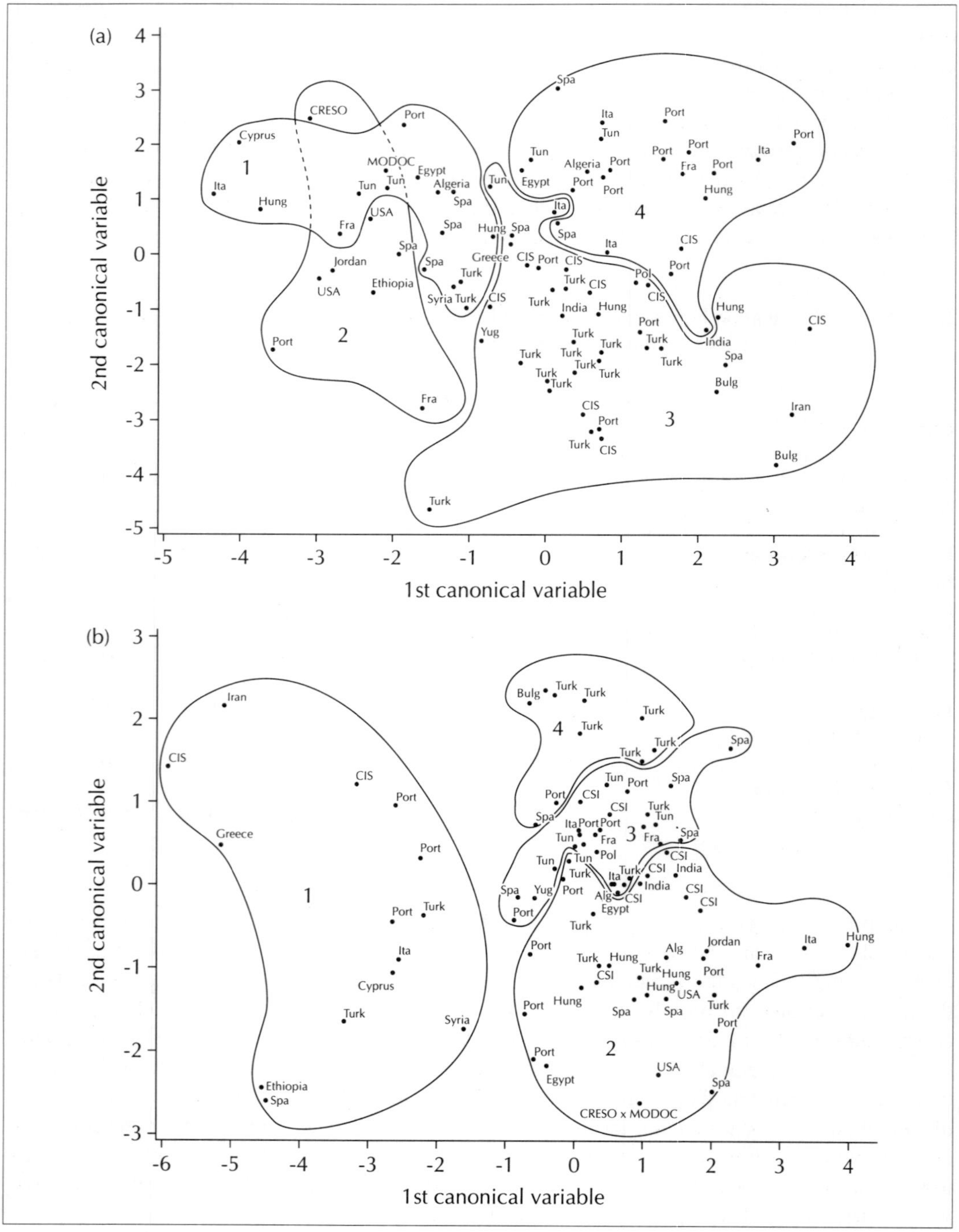

While most testcrosses cluster according to the country of origin of their non-recurring parent, cluster 1 is atypical in that it includes accessions from different geographical areas. Cluster analysis separated this group from the others. It shows the largest distance between this and all other accessions along the first canonical axis. The diversity among the F_1s along the first canonical axis is probably attributable to dominance at the loci controlling these characters. The degree of dominance was lower when the hybrids from cluster 1 were excluded from the analysis. It should be observed that only one accession (from Ethiopia) out of 14 clustered in the same group as Creso in the analysis that includes the parents. Thus, the probability of different gene frequency as measured by heterosis effect is more likely to occur among hybrids from different clusters. Most F_1s occur closer to each other, between values -1 and +2 of the first canonical axis, than is the case with the parents. This was expected from the observed smaller variances among the testcrosses. Distances among the other three clusters are much smaller, as are distances among parental clusters. In wheat, as in other autogamous species, genetic variance is expected to derive mainly from additive effects (Matzinger, 1963). Heterosis in the F_1 may not be of direct interest but heterotic crosses could produce desirable transgressive segregants.

The F_1s in the first cluster probably have unique genetic characteristics. Differences in gene frequencies are probably larger between these accessions and the cultivars used as testers. The phenotypic analysis of testcrosses on a multivariate basis seems to identify a few highly diverse accessions. Experimental evidence is needed, especially from the analysis of F_2 and following generations. Our study is one of the few attempts to establish the value of overall combining ability between exotic and improved germplasm, especially in a self-pollinated species.

The preliminary analysis of durum wheat F_2 data tends to support our findings. We also estimated genetic variances in F_2 testcross populations, which is expected to be larger when more diverse parents are crossed. Almost all accessions whose F_1s are in cluster 1 are not significantly different from Creso for kernel number and weight, but their F_2 variances were significantly large. Genetic diversity between parents in this case would be caused by dominance at different loci affecting the same multigenic characters. The second cluster includes the cross between the two improved varieties. The same characters generally showed F_2 variances as significant as in the F_2s where parents differed significantly. Here, genetic divergence between parents would be caused mainly by additive genetic effects.

In comparisons between advanced and primitive cultivars from Italy and Ethiopia, both the phenotypic and the genotypic variances were significantly larger in the segregating generation of crosses between accessions from those countries than from the same country (Spagnoletti Zeuli et al., 1983a). In general, the same results were obtained in the present study with this larger sample of accessions. Accessions from different countries differed significantly; crosses between accessions from different clusters generally showed larger variances. Some phenotypically similar accessions were also identified that showed heterotic response large enough to discriminate them from the rest of the core.

CONCLUSION

In our discussion of the approaches that plant breeders might take to evaluate and make effective use of a core collection, several important issues emerged. These can be summarised thus:

- It is very important that plant breeders contribute to the evaluation of core collections.
- Core collections should be extensively distributed among plant breeders.
- Core collections should be evaluated for multigenic characters useful for breeding purposes.

- Evaluation for combining ability with locally adapted cultivars is important in selecting exotic accessions as parental materials.
- The optimum testcross design rests on a compromise between the amount of genetic information that can be obtained and the number of accessions that can be evaluated.
- The choice of tester cultivars should be based on the degree of relatedness within the improved gene pool and their number should be small.
- The constant parent method that allows the screening of a large number of accessions permits the genetic evaluation of core collections; genetically diverse accessions could be also identified on the basis of heterotic effect.
- In durum wheat, phenotypic diversity among germplasm accessions indicates genetic differences that may be useful for breeding purposes; in fact, the additive genetic effects account for a large proportion of genetic diversity for most characters.
- The regression of offspring on non-recurring parent for characters of interest can be used to identify desired accessions.
- Multivariate analysis can be effective in identifying a group of accessions showing heterotic response for yield-related characters, and new yield-enhancing genes may be identified within a highly diverse group of accessions.
- The characters to be preferred should be: easy to score; multigenic, to represent the whole genome; useful for breeding purposes; and relevant to the evolution of the species.
- Phenotypic diversity among parents results in significant genetic variance in the F_2; if dominant gene action is significant, large F_2 variances arise from crosses of phenotypically similar parents.

References

Bhatt, G.M. 1970. Multivariate analysis approach to selection of parents for hybridisation aiming at yield improvement in self-pollinated crops. *Australian J. Agricultural Research* 21: 1-7

Brown, A.H.D. 1989. The case for core collections. In Brown, A.H.D., Frankel, O.H., Marshall, D.R. and Williams, J.T. (eds) *The Use of the Plant Genetic Resources*. Cambridge, UK: Cambridge University Press.

Camussi, A. and Ottaviano, E. 1981. Use of genetic effects of quantitative traits for the multivariate analysis of differences between populations. In Gallais, A. (ed) *Quantitative Genetics and Breeding Methods*. Paris, France: Institut National de la Recherche Agronomique.

Camussi, C., Spagnoletti Zeuli, P.L. and Melchiorre, P. 1983. Numerical taxonomy of Italian maize populations: Genetic distances on the basic of heterotic effects. *Maydica* 28: 411-24.

Camussi, A., Ottaviano, E., Calinsky, T. and Kaczmarek, Z. 1985. Genetic distances based on quantitative traits. *Genetics* 111: 945-62

Ceccarelli, S., Grando, S. and Van Leur, J.A.G. 1987. Genetic diversity in barley landraces from Syria and Jordan. *Euphytica* 36: 389-405

Cervantes, T.S., Goodman, M.M., Casas, E. and Rawlings, J.O. 1978. Use of genetic effects and genotype by environmental interactions for the classification of Mexican races of maize. *Genetics* 90: 339-48.

Cress, C.E. 1966. Heterosis of hybrid related to gene frequencies between the populations. *Genetics* 53: 269-74

Falconer, D.S. 1981. *Introduction to Quantitative Genetics.* New York, USA: Longman.

Frankel, O.H. 1989. Principles and strategies of evaluation. In Brown, A.H.D., Frankel, O.H., Marshall, D.R. and Williams, J.T. (eds) *The Use of Plant Genetic Resources.* Cambridge, UK: Cambridge University Press.

Frankel, O.H. and Brown, A.H.D. 1984. Plant genetic resources today: A critical appraisal. In Holden, J.H.W. and Williams, J.T. (eds) *Crop Genetic Resources: Conservation and Evaluation.* Winchester, Massachusetts, USA: Allen and Unwin.

Gebrekidan, B. and Rasmusson, D.C. 1970. Evaluating parental cultivars for use in hybrids and heterosis in barley. *Crop Science* 10: 500-02.

Goodman, M.M. 1973. Genetic distances: Measuring dissimilarity among populations. *Yearbook of Physical Anthropology* 17: 1-38.

Griffing, B. 1950. Analysis of quantitative gene action by constant parent regression and related techniques. *Genetics* 35: 303-22.

Hallauer, A.R. and Miranda, J.B. 1988. *Quantitative Genetics and Plant Breeding.* Ames, Iowa, USA: Iowa State University Press.

Hanson, W.D. and Casas, E. 1968. Spatial relationships among eight races of *Zea mays* L. utilising information from a diallel mating design. *Biometrics* 24: 867-80.

Isleib, T.G. and Wynne, T.C. 1983. F_4 bulk testing in testcrosses of 27 exotic peanut cultivars. *Crop Science* 23: 841-46.

Jain, S.K., Qualset, C.O., Bhatt, G.M. and Wu., K.K. 1975. Geographical patterns of phenotypic diversity in a world collection of durum wheats. *Crop Science* 15: 700-04.

Kenworthy, W.J. 1980. Strategies for introgressing exotic germplasm in breeding programs. In Corbin, F.T. (ed) *Proc. World Soybean Research Conference II, Raleigh, North Carolina, USA.* Boulder, Colorado, USA: Westview Press.

Lebsock, K.C. and Amaya, A. 1969. Variation and covariation of agronomic traits in durum wheat. *Crop Science* 9: 312-15.

Lefort-Buson, M. and de Vienne D. 1985. *Les Distances Genetiques: Estimations et Applications.* Paris, France: Institut National de la Recherche Agronomique.

MacKey, J. and Qualset, C.O. 1986. Conventional methods of wheat breeding. In *Genetic Improvement in Yield Wheat.* CSSA Special Publication No 13. Washington DC, USA: CSSA.

Marshall, D.R. 1989. Limitations to the use of collections. In Brown, A.H.D., Frankel, O.H., Marshall, D.R. and Williams, J.T. (eds) *The Use of Plant Genetic Resources.* Cambridge, UK: Cambridge University Press.

Matzinger, D.F. 1963. Experimental estimates of genetic parameters and their application in self-fertilizing plants. In *Statistical Genetics and Plant Breeding.* NAS-NCR Publication 82. Washington DC, USA: NAS.

Moll, R.H., Lonnquist, J.H., Fortuno J. and Johnson, E. C. 1965. The relationships of heterosis and genetic divergence in maize. *Genetics* 52: 139-44.

Peeters, J.P. and Williams, J.T. 1984. Towards better use of gene banks with special reference to information. *Plant Genetic Resources Newsletter* 60: 22-32.

Peeters, J.P. and Galwey, N.J. 1988. Germplasm collections and breeding needs in Europe. *Economic Botany* 42: 503-21.

Peeters, J.P. and Martinelli, J.A. 1989. Hierarchical cluster analysis as a tool to manage variation in germplasm collections. *Theoretical and Applied Genetics* 78: 42-48.

Peeters, J.P., Wilkes, H.G. and Galwey, N.W. 1990. The use of ecogeographical data in the exploitation of variation from gene banks. *Theoretical and Applied Genetics* 80: 110-12.

Pimentel, R.A. 1979. *Morphometrics. The Multivariate Analysis of Biological Data.* Dubuque, Indiana, USA: Kendall/Hunt Publishing.

Porceddu, E. and Scarascia Mugnozza, G.T. 1983. Genetic variation in durum wheat. In Sakamoto, S. (ed) *Proc. 6th International Wheat Genetic Symposium, Kyoto, Japan*. Kyoto, Japan: Plant Germplasm Insitute.

Qualset, C.O. 1979. Mendelian genetics of quantitative characters with reference to adaptation and breeding in wheat. In *Proc. 5th International Wheat Genetic Symposium, New Delhi, India*. New Delhi, India: Indian Society of Genetics and Plant Breeding.

Qualset, C.O. and Puri, Y.P. 1974. Heading time in the world collection of durum wheat. Photo and thermal-sensitivity related to latitudinal response. In Scarascia Mugnozza, G.T. (ed) *Proc. Symposium on Genetics and Breeding of Durum Wheat, Bari, Italy*. Bari, Italy: University of Bari.

Reese, P.F. Jr, Kenworthy, W.J., Cregan, P.B. and Yocum, J.O. 1988. Comparison of selection systems for the identification of exotic soybean lines for use in germplasm development. *Crop Science* 28: 237-41.

Romesburg, H.C. 1984. *Cluster Analysis for Researchers*. Belmont, California, USA: Lifetime Learning Publishing.

Seal, H.L. 1964. *Multivariate Statistical Analysis for Biologists*. London, UK: Methuen.

Smith, J.S.C. and Duvick, D.N. 1989. Germplasm collection and the private breeder. In Brown, A.H.D., Frankel, O.H., Marshall, D.R. and Williams, J.T. (eds) *The Use of Plant Genetic Resources*. Cambridge, UK: Cambridge University Press.

Sneath, P.H.A. 1976. Some applications of numerical taxonomy to plant breeding. *Zeitschrift für Pflanzenzücthtung* 76: 19-45.

Spagnoletti Zeuli, P.L. 1993. Assessing and sampling genetic variation in a world collection of durum wheat (*Triticum turgidum* L. durum group). PhD dissertation, University of California, Davis, USA.

Spagnoletti Zeuli, P.L. and Qualset, C.O. 1987. Geographical diversity for quantitative spike characters in a world collection of durum wheat. *Crop Science* 27: 235-41.

Spagnoletti Zeuli, P.L. and Qualset, C.O. 1990. Variation for flag leaf dimensions in durum wheat germplasm from different geographic origins. *Plant Breeding* 105: 189-202.

Spagnoletti Zeuli, P.L. and Qualset, C.O. 1993. Evaluation of five strategies for obtaining a core subset from a large genetic resource of durum wheat. *Theoretical and Applied Genetics* 78: 295-304.

Spagnoletti Zeuli, P.L., De Pace, C., Benedettelli, S., Lafiandra, D. and Porceddu, E. 1983a. Variation in durum wheat populations from different geographical origins. IV. Comparison between advanced and primitive cultivars. In Porceddu E. (ed) *Proc. FAO/University of Tuscia Workshop on Breeding Methodologies in Durum Wheat and Triticale, Viterbo, Italy*. Viterbo, Italy: Institute of Agricultural Biology, University of Tuscia.

Spagnoletti Zeuli, P.L., De Pace, C., Porceddu, E., Scarascia Mugnozza, G.T. and Volpe, N. 1983b. Variation in durum wheat populations from different geographical origins. IV. Genetic analysis of correlated sequential characters. In Sakamoto, S. (ed) *Proc. 6th International Wheat Genetic Symposium, Kyoto, Japan*. Kyoto, Japan: Plant Germplasm Insitute.

Spagnoletti Zeuli, P.L., De Pace, C. and Porceddu, E. 1984. Variation in durum wheat populations from different geographical origins. I. Materials and spike characteristics. *Euphytica* 33: 563-75

Spagnoletti Zeuli, P.L., De Pace, C. and Porceddu, E. 1985. Variation in durum wheat populations from different geographical origins. III. Assessment of genetic diversity for breeding purposes. *Zeitschrift für Pflanzenzücthtung* 94: 177-91

Spagnoletti Zeuli, P.L., Qualset C.O. and Smith, D.H. 1986. Germplasm resources in durum wheat: Extreme variants for some quantitative spike characters in the USDA World Collections. *Wheat Information Service* 67: 30-34.

St Martin, S.K. and Aslam, M. 1986. Performance of progeny of adapted and plant introduction soybeans lines. *Crop Science* 26: 753-56.

Sweeney, P.M. and St Martin, S.K. 1989. Testcross evaluation of exotic soybean germplasm of different origins. *Crop Science* 29: 289-93.

Tolbert, D.M., Qualset, C.O., Jain, S.K. and Craddock, J.C. 1979. A diversity analysis of a world collection of barley. *Crop Science* 19: 789-94.

Whitehouse, R.N.H. 1969. An application of canonical analysis to plant breeding. *Genetica Agraria* 23: 61-69.

Core Collections of Plant Genetic Resources
Edited by T. Hodgkin, A.H.D. Brown, Th.J.L. van Hintum and E.A.V. Morales

A Wiley-Sayce Publication

5.2

The core as a guide to the whole collection

D.A. VAUGHAN and M.T. JACKSON

Abstract

The composition of a germplasm collection is dependent upon the way the collection has been accumulated and conserved. A curator's knowledge and good databases enable a collection to be systematically organised and patterns of variation revealed. However, deficiencies in the comprehensiveness and quality of information in the database need to be considered when selecting germplasm for a core collection. Users' knowledge of the germplasm collection varies; curators should therefore strive to make germplasm collections 'user friendly'. Evaluation has been conducted on a massive scale in some crops. Where a required genotype is not rare, evaluation can benefit from using a core collection. Examples from major crops in US agriculture with large germplasm collections reveal sustained yield gains over many decades despite the rather limited use of conserved germplasm. The rigorous selection during the breeding process suggests that the relatively few ancestral varieties contributing to released varieties should also form a portion of a core collection. Curators need to focus their attention on germplasm which will complement this in the composition of a core collection. The key to the development of core collections is reliable information on all the accessions in the whole collection. Old and new information can be used to sort germplasm and select accessions for a core collection. The use of core collections developed from the whole collection of cultivated and wild rice has enabled scientists to identify rapidly the germplasm they require.

Germplasm banks exist primarily to furnish germplasm and information on germplasm to current and future germplasm users. The conserved germplasm supplied to plant breeders for use in the development of improved varieties for farmers ultimately produces the economic return on investments made in germplasm collecting, preservation and evaluation.

The core collection, which is a part of the active collection, aims to reduce genetic duplication and redundancy while maintaining maximum genetic diversity (Frankel, 1984). There is no way at present or in the foreseeable future that genetically identical accessions in a gene bank can be identified with complete certainty. Thus, the responsibility for the long-term conservation of the whole collection, which lies with gene bank curators, will continue and will not be affected by the development of a core collection. The relationship between the whole (or base) collection and its core is the focus of this chapter. In particular, we draw on our experiences with the collection of rice maintained at the International Rice Research Institute (IRRI) in the Philippines, where pilot studies have been initiated to determine the value of the core collection concept.

STRUCTURE OF A GERMPLASM COLLECTION

Germplasm collections are organised in different ways, but they all consist of some or all of the following: landraces and selected lines from landraces, elite breeding lines, released varieties, wild and weedy relatives of the cultigen, and genetic stocks (*see* Figure 1). These components show differences in genetic diversity within and between accessions; often, there are fewer accessions of the genetically most diverse germplasm such as wild relatives of crops.

The phases in the growth of the rice germplasm collection at IRRI, which are probably similar to those of some other collections, can be summarised thus:

- 1960-72: Donations of germplasm from existing collections
- 1972-85: Active collecting for rice landraces
- 1987-93: Active collecting for wild rices

Figure 1 Components of crop gene pools

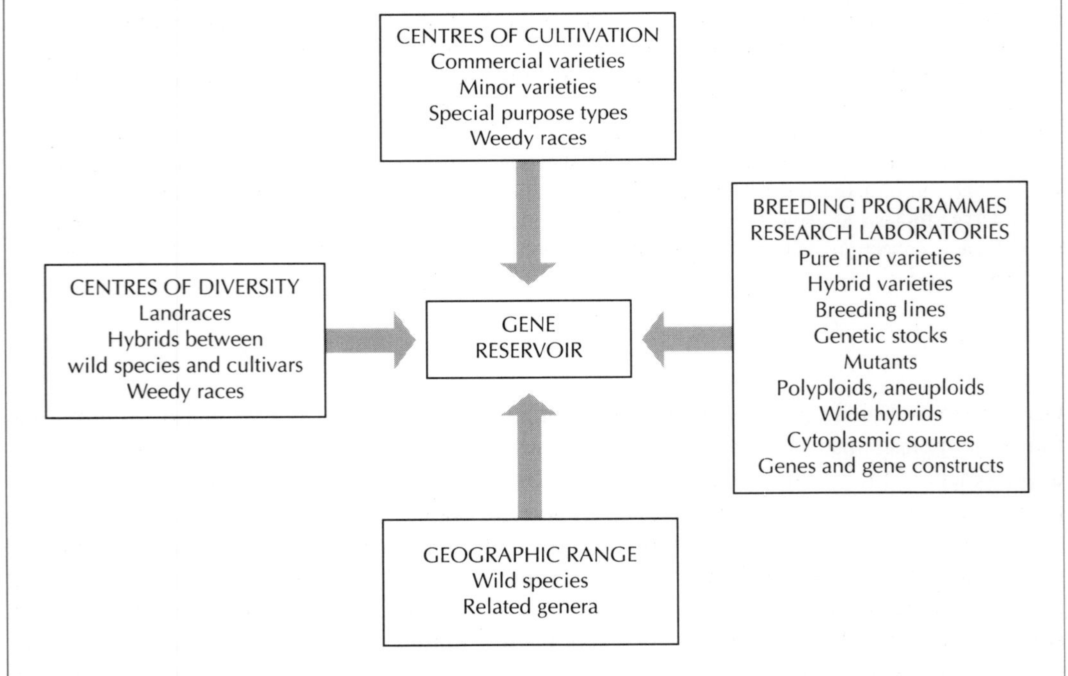

Source: Adapted from Chang (1985)

As a collection increases in size, so newly received germplasm requires careful checking for duplication of existing conserved germplasm before accepting it for long-term conservation. For seed-propagated species, this is a particularly difficult task.

It is often assumed that because a germplasm collection is very large, it is complete. The large size may hide the fact that germplasm from some areas may be over-represented, while that from other areas is unrepresented or under-represented. Although the germplasm collection at IRRI consists of about 80 000 accessions, it was only in 1992 that a comprehensive collection of rice cultivars was made in West Kalimantan, Indonesia. A four-volume work on the rice varieties of Mozambique (Gonçalves Valente, 1968) shows clearly that germplasm from Mozambique is under-represented in the collection (there are only eight Mozambique accessions in the collection). On the other hand, a collection of rice germplasm recently received from Cambodia had 59 variety names represented more than five times, suggesting considerable duplication during the collecting phase. The larger and more diverse the collection is, the more difficult it becomes to conserve germplasm in a routine way. Certain types of germplasm are more difficult to maintain than others, because of the environment in which they are grown or the inherent nature of the material, and the interaction between these factors (Chang, 1989). In selecting a core collection, these aspects should be taken into account to ensure that a representative sample of the whole collection is provided to germplasm users.

DATABASES

As germplasm collections increase in size, a good database is of fundamental importance in managing the collection and revealing patterns of variation. The database, coupled with a broad knowledge of the germplasm, becomes the basis for selecting accessions for the core collection.

Germplasm databases go through various developmental stages, usually starting with handwritten records before computerisation. The data obtained on an accession and incorporated into a database comprise three types: passport data (obtained at the collecting site), characterisation data (inherent characteristics of the germplasm) and evaluation data (germplasm screened in response to biotic and abiotic stresses). To this should be added management data which may or may not overlap with passport, characterisation and evaluation data.

New ideas, needs and concerns lead to improvements in germplasm databases from time to time. Thus, isozyme characterisation is now added to the characterisation database for rice. Isozyme classification increases our understanding of the genetic variation of varieties from different regions. In addition, varieties of certain isozyme groups have a shorter storage life; this information is helpful in managing the collection. Other biochemical and molecular characteristics of germplasm are likely to be incorporated into the existing germplasm databases in the future. Similarly, improved evaluation methods may stimulate re-evaluation of germplasm and evaluation for newly added descriptors (for example, tungro disease of rice is now known to be a complex of two viruses). New stresses arise requiring new evaluation methods (for example, evaluation is needed to identify germplasm resistant to parasitic plants of the genus *Striga*, which is increasing in importance in some African countries).

The database at IRRI has about 5 million items of information on the 80 000 conserved accessions of rice. However, the information available on each accession and on the germplasm collection as a whole depends upon many factors, including the relative importance of the information, the cost and the time taken to obtain information on the trait. Some traits may have priority for evaluation, but no reliable screening techniques have been developed to obtain reliable data on them. One example is the identification of germplasm adapted to different edaphic stresses. However, some stresses can be quickly, accurately and simply scored, such as the clipping method for evaluating the reaction of rice to bacterial blight (*Xanthomonas campestris* pv. *oryzae* ([Ishiyama] Dye) (Kauffman et al., 1973).

The accumulation of data in a well-organised database permits analysis which can provide new insights into the germplasm. For example, analysis of green leafhopper resistance revealed that a high proportion of resistant accessions are distributed in West Asia (Vaughan, 1991a). Similarly, a search for rice germplasm which has both good elongation ability and flood tolerance revealed that no single accession with both traits was available out of the 903 accessions screened. These two useful traits are probably mutually exclusive.

Good information on a germplasm collection is also essential to enable curators to respond appropriately to seed requests. Between January 1991 and June 1992, all 122 foreign seed requests for germplasm received at IRRI's germplasm bank were specific in nature (*see* Table 1). Germplasm users are often unaware of the core collection concept and the existence of core collections. When non-specific or broad seed requests are received, germplasm curators should promote accessions in a core collection as these would be a relevant starting point for experimental purposes.

Table 1 **Foreign seed requests for conserved germplasm received by the germplasm bank at the International Rice Research Institute, 1991 to mid-1992**

Total number of requests	122
Total number of countries	18
Type of request:	
Specific varieties/accessions	95
Specific trait/environment	19
Specific group/origin	3
Combination of variety, trait and site	5

USE OF THE WHOLE COLLECTION

Germplasm collections are used by scientists for basic, strategic, applied or adaptive research. Scientists vary in their knowledge of collections, and curators could help by making collections 'user friendly' for those least familiar with the collection. In general, the major use to date has been for strategic research, particularly evaluation of germplasm for specific traits for use in plant breeding. The use of exotic germplasm in basic research, such as biodiversity studies and adaptive research involving multilocational testing of germplasm, is likely to increase in the future. Results from the use of germplasm in basic and adaptive research should help in targeting new germplasm as potential parents for plant breeders.

Evaluation

Evaluation of some germplasm collections has been done on a large scale. In particular, the international agricultural research centres based in the tropics have taken a lead in the conservation, evaluation and use of their mandated crops. At the International Crops Research Institute for the Semi-

Arid Tropics (ICRISAT) in India, screening *Arachis* germplasm for various abiotic and biotic stresses has revealed many desirable accessions. One desirable accession was found, on average, in every 144 accessions evaluated for the 16 stresses described by Moss et al. (1989).

The multitude of evolving rice pests and diseases in the tropics has led IRRI to undertake large-scale evaluation of rice genetic resources. Over 20 000 accessions of rice have been evaluated for their reaction to 10 biotic and abiotic stresses (*see* Table 2). For all these stresses, many accessions showed resistance or moderate resistance. For some stresses, the resistant varieties were concentrated in a particular geographic region or regions; a good example is resistance to the brown planthopper in southern Indian and Sri Lankan germplasm (Khush, 1979). Clearly, a core collection approach could have helped in reducing the effort expended on evaluation. Finding many resistant accessions enables genetic analysis of these materials to determine whether or not the genetic basis for resistance is the same.

Table 2 **Number of rice accessions in the germplasm collection at the International Rice Research Institute tested for 10 stressses and found to be resistant or moderately resistant**

Stress	Number of germplasm accessions evaluated	Accessions with resistance or moderate resistance to the stress (score 1 to 3)
Brown planthopper	44 335	682
Green leafhopper	50 137	1 403
Rice whorl maggot	22 949	697
White-backed planthopper	52 042	871
Bacterial blight	49 752	5 512
Blast	36 634	9 616
Sheath blight	23 088	2 153
Drought resistance at early vegetative stage	28 319	4 288
Drought resistance at late vegetative stage	22 873	1 826
Recovery from drought stress	24 432	15 115

Lehmann (1984) reported that 11 accessions of barley in the Gatersleben collection out of 6000 tested were identified as being resistant to six isolates of leaf rust (*Puccinia hordei* Otth.). Since the variation among these accessions was limited, it was thought that the resistance was based on a single gene, *Pa7*. On the other hand, 71 strains of barley resistant to a new race of stripe rust (*P. striiformis* West.) suggested that several different genes contributed to the resistance. For biotic stresses caused by organisms that may rapidly evolve, seeking different resistance genes is an essential part of an evaluation programme.

To improve the quality of soybean, evaluation was undertaken in the USA to find germplasm lacking various chemical constituents of the seed. About 3300 accessions of wild and cultivated soybean were evaluated; of these, only two were found that lacked the kunitz trypsin inhibitor (Hymowitz, 1980). After testing 6499 soybean accessions, two were found that lacked the enzyme lipoxygenase-1 (Hildebrand and Hymowitz, 1981). A core collection approach is unlikely to have helped reveal these rare genotypes. When seeking rare genes, the relative costs of systematic screening of the entire collection compared with using a mutation approach to generate the desired trait, or even with introducing alien genes through genetic transformation, may need to be considered.

Use

Surveys of US crops indicated that the increase in yield as a result of genetic work was 0.5-0.8% per year during the 1960s and 1970s for wheat, cotton and soybean (Boerma, 1979; Meredith and Bridge, 1984; Schmidt, 1984; Specht and Williams, 1984). Hybrid maize in the USA showed yield gains of 1.4-1.78% per year between 1930 and 1980 (Duvick, 1984). However, gains in sorghum hybrids were about 2% per year in the 1970s (Miller and Kebede, 1984).

Despite these considerable and sustained gains, the number of plant introductions contributing to the genetic improvement of US crops has been rather limited. The total number of accessions in the US soybean collection is about 14 000. Delannay et al. (1983) reported that eight cultivars contributed about 65% of the genes of US soybean cultivars. Some 62 plant introductions, out of 30 000 in the total US wheat collection, contributed to breeding hexaploid wheat (*T. aestivum* L.) in the USA; of these, seven were landraces, 11 were other species and 44 were improved wheats (Cox, 1991). Frey (1991) stated that 'most of the oat germplasm utilised and developed in the USA until 1970 can be traced to seven landraces.' The US world oat collection consists of about 20 000 accessions, of which about 25% are wild and weedy species. The US rice collection consists of about 20 000 accessions but rice cultivars can be traced to only 45 plant introductions (Dilday, 1990).

The average annual growth rate of rice yield in Asia since 1965 has been 2.0-2.5% (David, 1991). Germplasm from the rice breeding programme at IRRI, which has been available to national programmes for direct use or use in their breeding programmes since the early 1960s, has been derived from 66 112 crosses involving 3985 parents not derived from IRRI lines. Of the parents used in crosses at IRRI, 124 varieties have each been used in over 100 crosses (*see* Table 3), and constitute the IRRI

Table 3 **Number of times non-IR varieties or their derivatives have been used in crosses at the International Rice Research Institute, 1962-92**

Number of times used in crosses	Number of varieties
>100	124
90-99	17
80-89	22
70-79	24
60-69	39
50-59	55
40-49	77
30-39	130
20-29	221
10-19	544
2-9	1919
1	813
Total	3985

breeders' collection (the breeders' active collection would include many of their own lines). From the thousands of crosses made at IRRI, less than 100 lines derived from these crosses have been released as varieties by national programmes.

The strong selection pressure that results in millions of lines from thousands of crosses being reduced to a few released varieties suggests that the ancestral parents of released varieties should also be a component of a core collection because of their unique genes, useful gene combinations and good combining ability. The germplasm curator has to determine which germplasm to include when developing a core collection. Particular attention may need to be given to newly received germplasm in the collection which has not undergone multiple testing over many years.

In a number of crops, such as oats and rice, one of the principal approaches is to infuse exotic germplasm to overcome stagnant or declining yields (Frey, 1991; Khush, 1991). Developing new plant types is an important approach to make a significant increase in yield potential if yields have reached a plateau.

The active role of germplasm curators in stimulating or participating in enhancing germplasm to develop agronomically acceptable, genetically desirable and diverse populations would overcome the primary constraint in the use of exotic germplasm for cultivar development. In addition, well-classified germplasm can immediately highlight segments of the whole collection or core collection suitable for use by the breeder.

DEVELOPMENT AND USE OF CORE COLLECTIONS, WITH PARTICULAR REFERENCE TO RICE

Information on the diversity of accessions in the whole collection is crucial to the development of core collections. Wild species collections tend to be relatively small, with good passport data, and if well classified taxonomically they can be readily incorporated into a core collection. Small germplasm collections permit comprehensive biodiversity studies, leading to better classification on which selection for a core collection can be based (Hamon and van Sloten, 1989).

With large collections, basic information such as passport data is often absent. For example, only about 25% of the cultivated rice germplasm at IRRI has some passport data, compared with 72% of the wild rice accessions. Efforts to develop core collections for major crops pose problems beyond the mere size of the collection. The advice given by Brown (1989b) to use a stratified sampling strategy to choose germplasm for a core collection has helped in the development of core collections for some major crops (Brown et al., 1989; Erskine and Muehlbauer, 1991; Vaughan, 1991a).

Problems encountered in developing and using core collections reflect those encountered with using the whole collection, such as inadequate coverage of certain geographic regions, lack of information on accessions and inadequate seed stocks of some accessions. The use of accessions in a core collection may be limited by its size and the cost of a particular experiment. For example, several thousand accessions may be analysed for isozyme diversity for the same cost as studying only scores or a few hundred using restriction fragment length polymorphisms (RFLPs). At IRRI, two small core collections have been developed to test the usefulness of the core collection approach. However, our principal interest in establishing a core collection is the safe duplication of accessions representing the broad diversity of the genus *Oryza* in several locations around the world; a duplication of the whole collection for this purpose would not be feasible.

The scientific basis for the two major rice varietal groups, *indica* and *japonica*, was established by Kato et al. (1928). Oka (1958) subdivided *japonica* varieties into temperate and tropical types. Using electrophoresis, Glaszmann (1987) surveyed about 2000 carefully chosen accessions and summarised the diversity by recognising six isozyme groups which largely corresponded to specific types of varieties from different regions. Thus, groups I and VI correspond to *indica* and *japonica* rices (both tropical and temperate). Group IV, however, is represented by the so-called Rayada varieties, found only in a few villages on the edge of the Madhumati river in Bangladesh (*see* Table 4).

Table 4 Varietal types and isozyme groups in rice

	Isozyme group					
Origin	**I**	**II**	**III**	**IV**	**V**	**VI**
Oka's testers (7)	indica	—	—	—	—	japonica
Iran, Pakistan	—	—	—	—	sadri	—
North-western India	—	—	—	—	basmati	—
Bangladesh	Aman	aus	early deepwater	rayada	—	—
South-eastern Asia	lowland	—	—	—	—	upland
Java, Bali	Tjereh	—	—	—	—	bulu
China	Hsien	—	—	—	—	keng

Source: Glaszmann (1986)

Our current knowledge of rice diversity based on geographic, morphological, agronomic, biochemical and molecular characteristics has resulted in the development of a small core collection of about 270 accessions of *O. sativa* which represents the known genetic diversity of rice (Glaszmann, 1987; Bonman et al., 1990). Using a similar approach, Vaughan (1991a, b) designated a core collection of wild rices to enable researchers to evaluate this germplasm efficiently.

Continued collecting and biodiversity studies require that these core collections are updated. Recently, varieties silent for the allele for the isozyme aminopeptidase-2 have been found in rice varieties from eastern Indonesia (Vaughan and Juliano, 1992). Recent exploration for wild rices has resulted in a new species being added to the collection, and new ecotypes of other species from isolated areas have been found (Vaughan and Sitch, 1991).

The use of these core collections has permitted rapid identification of germplasm for some traits. Thus, all accessions of the small Rayada group of varieties, highlighted in the isozyme study, were found to be resistant to leaf scald, caused by *Microdochium oryzae* (Hashioka and Yokogi) Samuels and Hallett (Bonman et al., 1990). Since there are only 19 Rayada accessions among the 80 000 in the germplasm collection at IRRI, neither random nor stratified sampling would have been likely to include one of these accessions in the designated 270 core accessions.

Broadcasting rice is replacing transplanting rice in many areas of Asia. Consequently, the *O. sativa* core collection was used by Yamauchi et al. (1993) to find varieties suitable for directly sowing under the surface of flooded rice soil. The results showed that deepwater rices and summer (aus) rices from north-eastern India and Bangladesh were most suited to these conditions. This information has led to further in-depth studies comparing these varieties with currently used commercial varieties.

The wild species core collection has been used to find sources of resistance to the tungro virus complex, one of the most serious diseases in Asia. Out of the 208 accessions of 19 species tested, 15 accessions of four species were not infected with one of the two forms of this virus, the rice tungro bacilliform virus (RTBV), and had no or very low infection by rice tungro spherical virus (RTSV). Of these, three *O. rufipogon* accessions were susceptible or moderately susceptible to the green leafhopper vector. Although more than 40 000 rice accessions have so far been screened (Kobayashi et al., 1993a, b), no good sources of resistance to both RTSV and RTBV are available. In addition, two African species, *O. glaberrima* Steud. and *O. barthii* A. Chev., exhibited a previously unreported symptom of tungro infection (Kobayashi et al., 1992). However, use of the wild species core collection has shown that insufficient information is available on intraspecific diversity, and its composition will need revision as more information on this critical issue becomes available.

CONCLUSION

From this review of the relationship between a core collection and the larger collection which it represents, several important points emerge which give an indication of the issues which need to be addressed in the development and use of core collections:

- Germplasm collections are still growing and the studies of biodiversity in germplasm collections are increasing. Both these factors imply that significant new collections and information on the whole collection will require periodic revision of core collections.
- More 'user friendly' databases to guide germplasm users to the germplasm most suited to their needs are now available in many gene banks. Too few users know what is meant by a core collection. The role of the germplasm curator in the development, maintenance and promotion of this part of the whole collection will be necessary if more useful results are to be achieved in a cost-effective way.
- The role of genetic resources centres in conserving products of biotechnology will become increasingly important for some crops. How these will be maintained and whether they become a part of a core collection will need to be addressed. For example, would an alien addition line series of a wild rice in the background of a cultigen be more useful in a core collection than several accessions of the wild species, which is difficult for plant breeders to use directly?
- An economic analysis of the return on investment in rice genetic resources conservation suggests that this return 'far exceeds the cost of managing and collecting' germplasm (Evenson, 1989). However, the same analysis shows that 'the estimated impact of special search landrace materials turns out to be quite large. This has considerable relevance to genetic resource management because these special search materials are found on the fringes of the collection.' Such an analysis is encouraging for those conserving germplasm and underlines the importance of rare genes and appropriate means of finding them.
- The core collection approach to evaluating and understanding conserved genetic resources has been shown to be useful. Sustaining the efforts to evaluate conserved genetic resources for a wide range of traits will increasingly require the use of such an approach as financial and personnel resources for evaluation activities become more scarce.

References

Boerma, H.R. 1979. Comparison of past and recently developed soybean cultivars in maturity groups VI, VII and VIII. *Crop Science* 19: 611-13.

Bonman, J.M., Mackill, A.Q. and Glaszmann, J.C. 1990. Resistance to *Gerlachia oryzae* in rice. *Plant Disease* 74: 306-09.

Brown, A.H.D. 1989. Core collections: A practical approach to genetic resources management. *Genome* 31: 818-24.

Brown, A.H.D., Grace, J.P. and Speer, S.S. 1989. Designation of a 'core' collection of perennial *Glycine*. *Soybean Genetics Newsletter* 14: 59-70.

Chang, T.T. 1985. Principles of genetic conservation. *Iowa State J. Research* 59: 325-48.

Chang, T.T. 1989. The management of rice genetic resources. *Genome* 31: 825-31.

Cox, T.S. 1991. The contribution of introduced germplasm to the development of US wheat cultivars. In Shands, H.L. and Wiesner, L.E. (eds) *Use of Plant Introductions in Cultivar Development (Part 1).* CSSA Special Publication No. 17. Madison, Wisconsin, USA: CSSA.

David, C.C. 1991. The world rice economy: Challenges ahead. In Khush, G.S. and Toenniessen, G.H. (eds) *Rice Biotechnology*. Manila, Philippines: IRRI.

Delannay, X., Rogers, D.M. and Palmer, R.G. 1983. Relative genetic contributions among ancestral lines to North American soybean cultivars. *Crop Science* 23: 944-49.

Dilday, R.H. 1990. Contribution of ancestral lines in the development of new cultivars of rice. *Crop Science* 30: 905-11.

Duvick, D.N. 1984. Genetic contribution to yield gains of US hybrid maize, 1930 to 1980. In Fehr, W.R. (ed) *Genetic Contributions to Yield Gains of Five Major Crop Plants*. CSSA Special Publication No. 7. Madison, Wisconsin, USA: CSSA.

Erskine, W. and Muehlbauer, F.J. 1991. Allozyme and morphological variability, outcrossing rate and core collection formation in lentil germplasm. *Theoretical and Applied Genetics* 83: 119-25.

Evenson, R.E. 1989. Rice Genetic Resources: Economic Evaluation. Mimeograph. Yale University, USA.

Frankel, O.H. 1984. Genetic perspectives of germplasm conservation. In Arber, W., Llimensee, K., Peacock, W.J. and Starlinger, P. (eds) *Genetic Manipulation: Impact on Man and Society*. Cambridge, UK: Cambridge University Press.

Frey, K.J. 1991. Genetic resources of oats. In Shands, H.L. and Wiesner, L.E. (eds) *Use of Plant Introductions in Cultivar Development (Part 1)*. CSSA Special Publication No. 17. Madison, Wisconsin, USA: CSSA.

Glaszmann, J.C. 1986. A varietal classification of Asian cultivated rice (*Oryza sativa* L.) based on isozyme polymorphism. In *Rice Genetics*. Manila, Philippines: IRRI.

Glaszmann, J.C. 1987. Isozymes and classification of Asian rice varieties. *Theoretical and Applied Genetics* 74: 21-30.

Gonçalves Valente, E. 1968. *O Arroz em Moçambique*. Série: Memórias No. 2. (vols. I-IV). Mozambique: Instituto de Investigaçao Agronómica de Moçambique.

Hamon, S. and van Sloten, D. 1989. Characterisation and evaluation of okra. In Brown, A.H.D., Frankel, O.H., Marshall, D.R. and Williams, J.T. (eds) *The Use of Plant Genetic Resources*. Cambridge, UK: Cambridge University Press.

Hildebrand, D.F. and Hymowitz, T. 1981. Two soybean genotypes lacking lipoxygenase-1. *J. American Oil Chemists Society* 58: 583-86.

Hymowitz, T. 1980. Chemical germplasm investigations in soybeans: The flotsam hypothesis. In *Recent Advances in Phytochemistry* 14.

Kato, S., Kosaka, H. and Hara, S. 1928. On the affinity of rice varieties as shown by fertility of hybrid plants. *Bull. Science (Kyushu University, Japan)* 3: 132-47.

Kauffman, H.E., Reddy, A.P.K., Hsieh, S.P.Y. and Merca, S.D. 1973. An improved technique for evaluating resistance of rice varieties to *Xanthomonas oryzae*. *Plant Disease Reporter* 56: 537-41.

Khush, G.S. 1979. Genetics and breeding for resistance to the brown planthopper. In *Brown Planthopper: Threat to Rice Production in Asia*. Manila, Philippines: IRRI.

Khush, G.S. 1991. Redesigning the rice plant. *Shell Agriculture* 10: 23-27.

Kobayashi, N., Ikeda, R., Vaughan, D.A. and Shigenaga, S. 1992. A new symptom of tungro in rice. *International Rice Research Newsletter* 17: 7-8.

Kobayashi, N., Ikeda, R., Domingo, I.T. and Vaughan, D.A. 1993a. Resistance to infection of rice tungro viruses and vector resistance in wild species of rice (*Oryza* spp.). *Japanese J. Breeding* 43: 377-87.

Kobayashi, N., Ikeda, R. and Vaughan, D.A. 1993b. Resistance to rice tungro viruses in wild species of rice (*Oryza* spp.). *Japanese J. Breeding* 43: 247-55.

Lehmann, C.O. 1984. Germplasm evaluation at Gatersleben: The relationship between gene bank and breeder. In Holden, J.H.W. and Williams, J.T. (eds) *Crop Genetic Resources: Conservation and Evaluation*. London, UK: Allen and Unwin.

Meredith, W.R. Jr. and Bridge, R.R. 1984. Genetic contributions to yield changes in upland cotton. In Fehr, W.R. (ed) *Genetic Contributions to Yield Gains of Five Major Crop Plants.* CSSA Special Publication No. 7. Madison, Wisconsin, USA: ASA.

Miller, F.R. and Kebede, Y. 1984. Genetic contribution to yield gains in sorghum, 1950 to 1980. In Fehr, W.R. (ed) *Genetic Contributions to Yield Gains of Five Major Crop Plants.* CSSA Special Publication No. 7. Madison, Wisconsin, USA: ASA.

Moss, J.P., Ramanatha Rao, V. and Gibbons, R.W. 1989. Evaluating the germplasm of groundnut (*Arachis hypogaea*) and wild *Arachis* species at ICRISAT. In Brown, A.H.D., Frankel, O.H., Marshall, D.R. and Williams, J.T. (eds) *The Use of Plant Genetic Resources*. Cambridge, UK: Cambridge University Press.

Oka, H.I. 1958. Intervarietal variation and classification of cultivated rice. *Indian J. Genetics and Plant Breeding* 18: 79-89.

Schmidt, J.W. 1984. Genetic contribution to yield gains in wheat. In Fehr, W.R. (ed) *Genetic Contributions to Yield Gains of Five Major Crop Plants*. CSSA Special Publication No. 7. Madison, Wisconsin, USA: ASA.

Specht, J.E. and Williams, J.H. 1984. Contribution of genetic technology to soybean productivity — retrospect and prospect. In Fehr, W.R. (ed) *Genetic Contributions to Yield Gains of Five Major Crop Plants*. CSSA Special Publication No. 7. Madison, Wisconsin, USA: CSSA.

Vaughan, D.A. 1991a. Choosing rice germplasm for evaluation. *Euphytica* 54: 147-54.

Vaughan, D.A. 1991b. Core collections: Providing access to wild *Oryza* germplasm. In *Rice Genetics II*. Manila, Philippines: IRRI.

Vaughan, D.A. and Sitch, L.A. 1991. Gene flow from the jungle to farmers: Wild-rice genetic resources and their uses. *BioScience* 41: 22-28.

Vaughan, D.A. and Juliano, A. 1992. Silent allele for *Amp-2* found among Sulawesi varieties. *Rice Genetics Newsletter* 9: 107-108.

Yamauchi, M., Aguilar, A.M., Vaughan, D.A. and Seshu, D.V. 1993. Rice (*O. sativa* L.) germplasm suitable for direct sowing under flooded soil surface. *Euphytica* 67: 177-84.

Core Collections of Plant Genetic Resources
Edited by T. Hodgkin, A.H.D. Brown, Th.J.L. van Hintum and E.A.V. Morales

A Wiley-Sayce Publication

5.3

Core collections for gene banks with limited resources

E.A.V. Morales, A.C.C. Valois and I.R.S. Costa

Abstract

Although many collections of genetic resources have been established, the germplasm is still insufficiently used. One way of increasing the use of these collections is to develop a core collection which is about 10-20% of the original collection in size but has 70-80% of its genetic diversity. Such a collection should take into account particular groups representing the genetic diversity of the gene pool and specific and desirable variability required by breeders and research programmes. Gene banks in many developing countries often lack the required financial and human resources to maintain a large collection. This situation increases the need for support from developed countries. In developing a core collection, it is necessary first to check whether an original collection, with an adequate level of desirable genetic variability for the national scientific and technological programmes, is available. If not, the initial work should focus on organising such a collection. The accessions should then be organised into as many groups as necessary to suit users' needs. This requires analysing the representativeness of the collection and determining whether further collection is needed, organising the documentation and information, characterising and evaluating the accessions to establish appropriate differences between them and establishing a core with desirable amounts of different accession groups, making genotypes, genes and alleles available to breeders. Using this methodology, it is possible to establish a useful core collection to preserve the available genetic diversity in a country, even when the amount of evaluation and characterisation data is limited.

The conservation of genetic diversity must be concerned not only with establishing adequate procedures to conserve the natural diversity of species, genes and habitats, but also with identifying and making available potentially useful genetic characteristics. To do this effectively, it is necessary to use modern sciences and new technologies, both to evaluate germplasm and to make material available for use in research and breeding programmes. Such a strategy will reduce the risks associated with genetic uniformity. While such uniformity has led to significant advances in crop production, it may well become an important cause of failure to make further advances (Paterniani, 1989).

To improve the use of plant germplasm, it is important that gene banks, especially those with limited resources, organise their genetic resources into well-structured collections which, based on the gene pool concept of Harlan and de Wet (1971), adequately represent the genetic variation of a crop. A strategy to encourage researchers and breeders to use germplasm collections more frequently could be to establish small collections for a crop or group of crops which, with a minimum of repetiveness, contain as much of the available genetic variability as possible. This type of structure can be achieved by using the approaches involved in developing core collections (Frankel, 1984). In establishing a core collection in the context of limited resources, it is important to take account not only of the amount and viability of the germplasm samples already conserved in national germplasm collections, but also of the importance of the germplasm as a strategic resource, the relationship with research and breeding programmes, the potential genetic diversity available in local or regional biomes, and the strategies required to organise core collections.

APPROACHES IN DEVELOPING GENETIC RESOURCES COLLECTIONS

Genetic resources collections must contain as much as possible of the available genetic diversity required by breeding programmes (Breese, 1989). Thus, an ecological approach must be established for plant collecting and conservation, whereby each ecotype (whether from wild relatives or landraces) is sampled from a distinct population of individuals (Breese, 1989). Samples must also be of sufficient size to conserve the genetic variability of the population. This is more a function of the numbers of individuals that will contribute to the next generation than the numbers of individuals in the population (Vencosky, 1986). The accessions must include landraces or primitive cultivars, genetically related wild and weedy species and obsolete cultivars and genetic stocks (Frankel, 1984).

Germplasm collections have been established using two different approaches:

- A plant breeder's approach: The main objective of this approach is to develop collections of small numbers of accessions with the genetic variability required for a specific research or breeding objective. The accessions are screened, characterised and evaluated for specific traits of interest. Usually, they are obtained from field collections, mutants, lines, elite material produced by the breeder and germplasm obtained from other breeders. The genetic diversity of a particular gene pool is therefore low and loss of accessions and genetic erosion is likely to occur during conservation, regeneration and multiplication.

- A plant germplasm conservation approach: The main objective of this approach is to conserve as much as possible of the available germplasm, primarily to make genetic diversity available for research or breeding programmes. Unfortunately, the development of such collections has tended to have low participation by plant breeders, who often already have their own working collections and easy access to foreign germplasm.

One consequence of the first approach is that although great efforts have been made to collect and to conserve germplasm, the gene pool's full range of genetic diversity in all the main biomes has not always been sampled. The collections are therefore often relatively uniform and do not possess a high level of genetic diversity. Furthermore, while much work is done on morphological characterisation, very little evaluation is carried out. Emphasis is given to preliminary evaluation, without the use of adequate procedures for obtaining data that would be useful in breeding programmes.

Although breeders consider genetic resources collections to be interesting and strategically important, they continue to maintain and use their own working collections as their main source of information in their efforts to obtain rapid solutions to specific problems (Nass et al., 1992). Some surveys have indicated that plant breeders make little use of material from gene banks (Peeters and Williams, 1984; Smith and Duvick, 1989; Nass et al., 1992). It is thus extremely important to improve the links between germplasm curators and germplasm users (usually plant breeders). The core collection concept provides a new approach for achieving this.

It has been reported that only 2-8% of the available germplasm in collections has been used in breeding programmes, and most of this is related to a minimal proportion of the genome (Salhuana, 1985; Tay, 1988; Gill, 1989). This situation, which probably stems from the need for rapid results from breeding programmes, has led some authorities to suggest that germplasm use is too low to warrant the high costs of germplasm collection, screening and maintenance, or that it places a burden on the usually small budgets of national programmes (Lantican, 1988). However, there are enough examples of the use of accessions from gene banks to justify the maintenance of these collections. Exotic germplasm has been the main source of yield improvement for sweet potato cultivars (Kobayashi and Sakamoto, 1988); wild *Arachis* accessions are being used to shorten the cycle of peanuts (Simpson, 1990); resistance to cassava mosaic disease has been incorporated from *Manihot glaziovii* into high-yielding cultivars (Hahn et al., 1980); and resistance to potato late-blight and nematodes has been obtained from *Solanum demissum* and *S. multidessectum* (Watson, 1970). What is needed is a means of encouraging plant breeders to make more use of material in plant gene banks.

A Latin American survey of regional germplasm availability, germplasm use and the genetic diversity of the main cultivated species produced the following results: the level of genetic variability seemed to be high; the maintenance of the autogamous species' genetic variability seemed to be close to the optimum but for allogamous and clonally propagated species it ranged between intermediate to deficient; and the genetic basis seemed to be narrow for autogamous and clonally propagated species but broader for allogamous species (Paterniani, 1985).

Although it has been suggested that available genetic variability is not a serious issue, the low level of demand may be related to the nature of current breeding programmes that emphasise uniformity. This results in low pressure for high levels of genetic diversity. In general, it seems that limited economic resources and the lack of facilities and qualified personnel have had a significant effect on the use of plant germplasm (Williams et al., 1975; Williams and Creech, 1981; Paterniani, 1985; Morales, 1991).

USE OF GERMPLASM

While plant breeders view national germplasm collections as valuable sources of genetic variation, they have commented on the inadequacy of many germplasm collections (Duvick, 1984; Nass et al., 1992). There is an urgent need, therefore, for structured collections which allow breeders easy access to qualitative and quantitative traits for both specific and general situations. The core collection approach appears to offer a structure which meets these needs.

The main barriers to the use of germplasm collections in research and breeding programmes appear to be: inadequate documentation and information on biotic and abiotic factors; lack of germplasm characterisation and evaluation data on adaptative traits relevant to specific agronomic or industrial goals; lack of fast, efficient and cost-effective methods for germplasm screening; lack of adequate measures to increase the use of germplasm in breeding programmes; and difficulty in gaining access to a collection because of distance, expense or quarantine restrictions. These inadequacies often reflect a lack of trained personnel and of the funds required to ensure effective documentation and information management (Frankel, 1982; Singh and Singh, 1982; Salhuana, 1985).

Characterisation and evaluation of genetic diversity (for example, to identify phenotypic variation between and within regions) could be one of the most important ways of improving the use and maintenance of genetic resources (Perry et al., 1991). Once some degree of morphological characterisation and agronomic evaluation is available, allozyme frequencies can be used as a quick and economic way of determining patterns of diversity in germplasm collections (Perry et al., 1991). Complementary characterisation and evaluation can be done using molecular tools such as restriction fragment length polymorphism (RFLP) and random amplied polymorphic DNA (RAPD) (Andersen and Fairbanks, 1990; Dodds and Watanabe, 1990). Multivariate analysis is the main statistical tool for studying the differences between accessions, and discriminant and clustering methods of analysis can be used for selecting parental stocks and revealing evolutionary patterns.

In order to increase the use of germplasm in breeding programmes, it is necessary to pay more attention to plant breeders' requirements. Singh (1982), Harlan (1984), Breese (1989), Gill (1989) and Shands (1990) have suggested that the most important issues in meeting these requirements are:

- maintenance of a structured collection, based on genetic variability, where accessions are classified according to agronomic and other appropriate traits and to response to biotic and abiotic stress
- information on the genetic mechanisms for specific traits, including a description of the genetic architecture of potential parental material
- information on food and agro-industrial uses
- information on current breeding objectives and strategies
- systematic characterisation and evaluation data
- information on the availability of research support, including interdisciplinary research

CONSTRAINTS IMPOSED BY LIMITED RESOURCES

Although gene banks are considered to be important strategic resources, they often suffer from limited technical and financial support. This is related not only to a chronic lack of funds to support their activities, but also to inadequate numbers of trained personnel. In general, these factors apply to gene banks in developing countries. The situation is particularly serious for countries such as Brazil, which have autocthonous germplasm with high levels of genetic diversity but lack the technologies and trained personnel to collect, preserve, characterise and evaluate it and promote its usage.

The strategic role of genetic resources as a developmental tool is well known. In general, however, little support is given to gene bank management and germplasm improvement. This is probably because of the high priority given to critical social needs. As a result, agricultural and development policies may be implemented without an evaluation of their effect on available and potential genetic variation, and thus loss of biological diversity occurs as improved or 'superior' germplasm displaces landraces and original cultivars of autocthonous crops (Breese, 1989; Clark and Juma, 1991).

The international programmes for agricultural development, conducted by international research institutions, may be another contributing factor to this situation. These programmes have supported large amounts of work on genetic resources without reciprocal actions by national programmes, with the result that some countries do not consider it a priority to support, establish or improve national actions on genetic resources. One of the most important consequences of this is an increasing dependence of national programmes upon international ones. An improved strategy would be to continue to support such work but to develop complementary mechanisms which encourage developing countries to recognise and solve their own problems and needs.

Despite these general considerations, two important factors remain as major constraints for gene banks with limited resources: the lack of trained personnel to improve the germplasm quality; and the lack of funds for gene bank management. The first point affects policies and procedures; the second limits the efficiency of strategies already adopted. As a result, knowledge of the variation in large collections is poor, thus reducing the usefulness of such collections. Here again, the core collection approach appears to offer an effective strategy for improving this situation.

THE CORE COLLECTION AS A RATIONAL GENE BANK STRATEGY

A core collection which is 10-20% of the size of the original collection and contains at least 70% of the genetic variation must be established in a stratified manner (Brown, 1989a, b). Once a core is organised, it is possible to define and establish priorities for management, maintenance, research and use and to allocate human resources to ensure that these goals are achieved. A well-structured core permits easy retrieval of information and of the accessions needed for research or breeding programmes, not only of specific genes and alleles but also of desirable gene blocks and genotypes.

The main factors to take into account in organising a core collection are the level and type of genetic variability to be conserved. Although a core collection could be organised with the genetic variability available in the original collection, the main task must be to sample not only the available national or regional genetic variation for autocthonous germplasm, but also as much as possible of the gene pools' total genetic diversity. For many economically important species, it seems advisable to incorporate when possible some foreign elite germplasm and genetic stocks developed outside of the centres of diversity.

The development of a core collection requires a strong collaborative effort involving not only germplasm curators, plant breeders and other researchers, but also such specialists as biologists, molecular biologists and statisticians. It also requires careful costing and the use of clearly defined procedures. Five major steps need to be carried out, in the following order:

1 Determine the genus, taxonomic position and evolution patterns to be considered.

2 Evaluate the genetic diversity, as well as the geographical distribution, of the gene pools to be used.

3 Organise the documentation and information in an accessible way.

4 Determine whether a desirable level of genetic variability, adequate for national scientific and technological programmes, is present in the original collection to be used as a major source of the core. If such a collection does not exist, it will be necessary to develop one or to locate a collection that can be used.

5 Organise the accessions in a stratified manner with as many groups as necessary. This requires organising the documentation and information data; analysing the available genetic variation to check the representativeness of the final core and establish needs for further germplasm collection or documentation; defining the number of desirable strata that will permit genotypes, genes and alleles to be available; and characterising and evaluating the accessions in a manner which allows them to be grouped.

It may be important to consider as sources of germplasm not only the original germplasm collection but also any available working collections and collections of genetic stocks. Since the stratification and the sampling procedures will strongly influence the final core, it is essential to assess the genetic basis and structure of the original germplasm sources. Here it is necessary to determine the level of representativeness of the gene pools' genetic diversity, the availability of lines, mutants and genetic stocks, the level of demand from current research programmes and the size of the original collections from which the core is drawn.

It is unnecessary to establish a core in the following situations: where there is not a significant amount of the genetic diversity in the gene pools; where the germplasm does not have the specific traits required for research and breeding; where the size of the original collection is small (less than 1000 accessions); or where the original collection is well characterised and evaluated and easily accessible to users in its current form.

With regard to the original germplasm collection to be used, it is most appropriate to utilise the base collection. This will ensure that accessions have a low level of genetic drift or genetic erosion. The organisation of the core collection should be based on at least four main strata:

1 Adaptative traits from landraces, wild relatives, and collections obtained on ecogeographical basis. For these, the allocation should be proportional to the number in the original collection, using random sampling.

2 Specific traits from haploids, polyploids, mutants and genetic stocks. For these, the allocation of the accessions should be systematic, selected according to the research requirements of plant breeders.

3 Elite germplasm such as superior landraces and cultivars. For these, the allocation should be proportional to the number in the original collection, using random sampling.

4 Improved germplasm or breeding material, such as preliminary or advanced lines, pure lines or desirable cultivars. For these, the allocation of accessions should be systematic, selected according to the research requirements of plant breeders.

The main criteria and procedures for organising a core collection have been established (Brown, 1989a, b). However, it is necessary to ensure that the final size is 10-20% of the original size and that the methods of allocating accessions follow those described by Brown (1989a, b). If large differences in numbers of accessions exist between different groups, it is advisable to use an allocation in proportion to the logarithm of the number of accessions in each group.

If some reliable passport data are available, it is possible to organise a core by choosing the main groups on an ecogeographic basis and by sampling the accessions using proportional sampling. In a subsequent step, selected accessions are characterised and evaluated in order to obtain better data and to define and eliminate any duplication or genetic redundancies. Using this procedure, it is necessary to note that different genetic forms will remain in the original collection and that work with the original collection to preserve it for future action should continue.

For countries with considerable genetic diversity in a given species or crop, it seems advisable to group landraces and obsolete cultivars on an ecogeographical basis, which will provide groups that reflect adaptative traits. However, exotic germplasm could be organised using geopolitical strata, particularly where there is evidence that different countries have different approaches to the collection, maintenance and use of germplasm.

Once the core is established in preliminary form, several studies are needed before establishing the definitive core collection. To identify duplications and genetic redundancy, it seems important to complement the morphological characterisation with specialised evaluation (for example, agronomic or pharmaceutical) and to use specialised techniques such as isozyme analysis. It will be necessary to use multivariate and discriminant analysis to confirm the proposed strata (Peeters and Martinelli, 1989).

Important matters to consider in efforts to improve the scientific basis of core collections include: methodologies for the rapid measurement and determination of genetic diversity distribution over populations and geographic areas; sampling techniques to capture a gene pool with a minimum of repetitiveness; documentation and information on agroecological aspects of the genetic distribution; and potential use for breeding programmes (CGIAR, 1991). To ensure maximum use by breeders, specific points that need attention include: the source of material and its potential for establishing heterotic combinations; the general combining ability of the accessions; the detection of adaptative environmental traits, including the degree of stress tolerance; and specific characteristics of high value, such as resistance to diseases and insect pests (Salhuana, 1985). It is also important to initiate economic and social studies which will provide the justification for establising and maintaining a core collection.

A suitable strategy for adoption by developing countries with impressive amounts of genetic diversity would be to use the core collection approach in association with *in situ* conservation initiatives. Here, populations conserved *in situ* can be characterised and evaluated, and selected material can then be removed to form an *ex situ* core collection. This strategy will allow the *in situ* collection to serve as the main collection, and the *ex situ* collection to represent 70-80% of the available genetic variation but with only 10-20% of the population samples.

An important component of organising of a core collection is the development of complementary studies linked with the release of a core collection. Among the studies suggested by Cordeiro et al. (*Chapter 3.5, this volume*) are: evaluation of crosses between accessions selected from different strata, in order to evaluate the use of hierarchical cluster analysis; use of genetic markers to detect duplication and redundancy; evaluation of the level of genetic variability in each stratum, in order to define possible future germplasm collection needs; and evaluation of genotype x environment interaction, mainly for clonal crops.

CONCLUSION

A core collection is an appropriate way of establishing a useful germplasm collection for gene banks which have limited resources. It should reflect most of the genetic diversity in the crop gene pool and contain the specific traits required by research and breeding programmes. Because of its small size, the funds required to manage a core collection will be easier to acquire than those needed to manage a large collection.

The main aims of a core collection are not only to preserve germplasm, but also to use it. Thus, collaborative participation by plant breeders and germplasm curators is essential in establishing such a collection and promoting its usage. As the size of a core is significantly smaller than the original collection, it is feasible for researchers and plant breeders to be involved in its continuous characterisation and evaluation. However, establishing and maintaining a core collection is not only the task of curators and plant breeders; it also requires inputs by a multidisciplinary team of specialists.

The establishment of a core collection requires a clear assessment of the institutional integration and collaboration required and the need for technical, administrative and political support. In order to know the full value of the germplasm through characterisation, evaluation and screening, and to implement sound conservation procedures, it is necessary to have adequate funding, facilities and trained personnel. This may be beyond the capacity of some national and regional programmes. The most difficult task for developing countries is to find ways to organise, establish, evaluate, conserve and use a core collection in a collaborative and integrative programme, both within the country and through partnership between countries in the same region.

Although a core will never substitute for a base collection, it may form an effective active collection. In critical circumstances, the structured core provides the last genetic refuge against significant genetic erosion and provides a basis for future scientific and technological programmes. Nevertheless, it is important to point out that once the original collection is lost, a significant amount of the variability in it will also be irrevocably lost.

References

Andersen, W.R. and Fairbanks, O.J. 1990. Molecular markers: Important tools for plant genetic resource characterisation. *Diversity* 6 (3&4): 51-53.

Breese, E.L. 1989. *Regeneration and Multiplication of Germplasm Resources in Seed Gene Banks: The Scientific Background.* Rome, Italy: IBPGR.

Brown, A.H.D. 1989a. The case for core collections. In Brown, A.H.D., Frankel, O.H., Marshall, D.R. and Williams, J.T. (eds) *The Use of Plant Genetic Resources.* Cambridge, UK: Cambridge University Press.

Brown, A.H.D. 1989b. Core collections: A practical approach to genetic resources management. *Genome* 31: 818-24.

CGIAR. 1991. *Report of the 3rd External Review of the International Board for Plant Genetic Resources (IBPGR).* Rome, Italy: CGIAR-TAC/FAO.

Clark, N. and Juma, C. 1991. *Biotechnology for Sustainable Development. Policy Options for Developing Countries.* Nairobi, Kenya: African Centre for Technology Studies.

Dodds, J.H. and Watanabe, K. 1990. Biotechnological tools for plant genetic resources management. *Diversity* 6 (3&4): 26-28.

Duvick, D.N. 1984. Genetic diversity in major crops on the farm and in reserve. *Economic Botany* 38: 161-78.

Frankel, O.H. 1982. Genetics resources and the plant breeder: Introduction. In Singh, R.B. and Chomchalow, N. (eds) *Genetics Resources and the Plant Breeder*. Bangkok, Thailand: IBPGR.

Frankel, O.H. 1984. Genetic perspectives of germplasm conservation. In Arber, W.K., Llimensee, K., Peacock, W.J. and Starlinger, P. *Genetic Manipulation: Impact on Man and Society.* Cambridge, UK: Cambridge University Press.

Gill, K.S. 1989. Role of plant genetic resource collections in research and breeding. In Brown, A.H.D., Frankel, O.H., Marshall, D.R. and Williams, J.T. (eds) *The Use of Plant Genetic Resources.* Cambridge, UK: Cambridge University Press.

Hahn, S.K. Terry, E.R. and Leuschner, K. 1980. Breeding cassava for resistance to cassava mosaic disease. *Euphytica* 29: 673-83

Harlan, J.R. 1984. Evaluation of wild relatives of crop plants. In Holden, J.H.W. and Williams, J.T. (eds) *Crop Genetic Resources: Conservation and Evaluation.* Rome, Italy: IBPGR.

Harlan, J.R. and de Wet, J.M.J. de. 1971. Toward a rational classification of cultivated plants. *Taxon* 20: 509-17.

Kobayashi, M. and Sakomoto, S. 1988. Utilisation of exotic germplasm in sweet potato breeding. In Suzuki, S. (ed) *Crop Genetic Resources of East Asia*. Rome, Italy: IBPGR.

Lantican, R.M. 1988. Recent developments in plant genetic resources conservation work in Southeast Asia. In Suzuki, S. (ed) *Crop Genetic Resources of East Asia*. Rome, Italy: IBPGR.
Morales, E.A.V. 1991. *Propuesta de Plan de Trabajo para una Acción Integrada en el Manejo y Conservación de los Recursos Genéticos del Trópico Suramericano*. Brasília, Brazil: IICA.
Nass, L.L., Pellicano, I.J. and Valois, A.C.C. 1992. *Utilizaçao de Recursos Genéticos no Melhoramento de Milho e Soja no Brasil*. Brasília, Brazil: EMBRAPA-CNPR.
Paterniani, E. 1985. An evaluation of the genetic diversity in the varieties currently utilised. In *Report of the Latin American Plant Breeding Research Forum — Latin America's Plant Resources: Abundant Food Supplies for the Future*. Caracas, Venezuela: Pioneer Hi-Bred International.
Paterniani, E. 1989. Diversidade genética em plantas cultivadas. In *Encontro sobre Recursos Genéticos*. Anais, Jaboticabal, Brazil: UNESP/FCAV-EMBRAPA/CENARGEN.
Peeters, J.P. and Williams, J.T. 1984. Towards better use of gene banks with special reference to information. *Plant Genetic Resources Newletter* 60: 22-32.
Peeters, J.P. and Martinelli, J.A. 1989. Hierarchical cluster analysis as a tool to manage variation in germplasm collections. *Theoretical and Applied Genetics* 78: 42-48.
Perry, M.C., McIntosh, M.S. and Stoner, A.K. 1991. Geographical patterns of variation in the USDA soybean germplasm collection. II. Allozyme frequencies. *Crop Science* 31: 1356-60.
Salhuana, W. 1985. Strategies for increasing the use of germplasm. In *Report of the Latin American Plant Breeding Research Forum — Latin America's Plant Resources: Abundant Food Supplies for the Future*. Caracas, Venezuela: Pioneer Hi-Bred International.
Shands, H. L. 1990. Plant genetic resources conservation: The role of the gene bank in delivering useful genetic materials to the research scientist. *J. Heredity* 81: 7-10.
Simpson, C.E. 1990. Introgression of early maturity into *Arachis hypogae* L. In *American Peanut Research and Education Society Proceedings*. Stone Mountains, USA: APRES.
Singh, R.B. 1982. Crop genetic resources and their utilisation in Southeast Asia: An overview. In Singh, R.B. and Chomchalow, N. (eds) *Genetic Resources and the Plant Breeder*. Bangkok, Thailand: IBPGR.
Singh, R.K. and Singh, M. 1982. The use of grain legume germplasm in India. In Singh, R.B. and Chomchalow, N. (eds) *Genetic Resources and the Plant Breeder*. Bangkok, Thailand: IBPGR.
Smith, J.S.C. and Duvick, D.N. 1989. Germplasm collections and the private plant breeder. In Brown, A.H.D., Frankel, O.H., Marshall, D.R. and Williams, J.T. (eds) *The Use of Plant Genetic Resources*. Cambridge, UK: Cambridge University Press.
Tay, C. 1988. Present status, management and utilisation of tropical vegetable genetic resources at AVRDC. In Suzuki, S. (ed) *Crop Genetic Resources of East Asia*. Rome, Italy: IBPGR.
Vencosky, R. 1986. *Tamanho Efetivo Populacional na Coleta e Preservação de Germoplasma de Espécies Alógamas*. Brasília, Brazil: EMBRAPA-CENARGEN.
Watson, I. 1970. The utilisation of wild species in the breeding of cultivated crops reistant to plant pathogens. In Frankel, O.H. and Bennet, E. (eds) *Genetic Resources in Plants — Their Exploration and Conservation*. Oxford, UK: Blackwell.
Williams, J.T. and Creech J.L. 1981. *Crop Genetic Resources of the Far East and the Pacific*. Bangkok, Thailand: IBPGR.
Williams, J.T., Lamourex, C.H. and Wulijarni-Soetjipto, N. (eds). 1975. *South East Asian Plant Genetic Resources*. Bogor, Indonesia: IBPGR-SEAMEO/BIOTROP-LIPI.

Part 6

CONCLUSION

Core Collections of Plant Genetic Resources
Edited by T. Hodgkin, A.H.D. Brown, Th.J.L. van Hintum and E.A.V. Morales

A Wiley-Sayce Publication

6.1

Future directions

T. HODGKIN, A.H.D. BROWN, Th.J.L. VAN HINTUM and E.A.V. MORALES

After a fairly lengthy gestation period (8 years passed between the publication of the original papers on core collections and the workshop on which this book is based), core collection work has moved firmly into the practical arena. A number of core collections have been established and more are being developed. The work involved in some of these initiatives has been described in the preceding chapters.

During the workshop, a group of participants met informally to identify the major issues that had been raised during each day's presentations. Their conclusions were presented at the workshop as part of a final discussion on future work needed on core collections. This discussion led to the development of a series of recommendations (*see* Appendix, *page 261*). Four working groups were also convened to discuss topics which were considered to be particularly important to the development and use of core collections. The topics were:

- the role of core collections in gene bank management
- core collections and molecular genetics
- core collections in developing countries
- core collections and the plant breeder

In this chapter we draw on the discussions of the working groups and the final workshop session and discuss issues that we believe will be of most significance for future core collection work.

SOME GENERAL CONSIDERATIONS

A considerable diversity of approach and procedure is apparent from even the limited number of core collections described in this book. The Barley Core Collection (Knüpffer and Hintum, *Chapter 4.1, this volume*) depends upon international collaboration, others are based on collections in single international or national gene banks. The methodology used varies significantly, as does the emphasis given to different kinds of information or the relative size of the resulting cores. This is not surprising and is, in fact, almost certainly desirable. The need is to develop a body of practical experience based on the establishment of core collections for a variety of crop species and their wild relatives. Gene bank managers and their collaborators involved in developing core collections face a range of problems which vary depending upon the character of their collections, the nature of the material with which they are working and the resources they have at their disposal. The chapters in this book capture some, but not all, of the possibilities and, we hope, will allow those interested in developing cores to consider various options and possibilities and develop an approach which will best fit their needs.

There is now a substantial measure of agreement concerning the underlying objectives and processes of core collection development. The definition by Brown (*Chapter 1.1*) provides an appropriate and practical approach to core collections. It places the development of a core collection in the context of the management of the whole collection as a way of improving overall organisation and use of that collection. The importance of recognising that the core is a component of the whole collection with which it is intimately connected has been emphasised by many authors. It is worth re-emphasising that the core collection is not an entity on its own; it is a guide and entry point to the whole collection. Core collections can therefore be expected to increase the value of the whole collection rather than reduce the value of the non-core element.

In the core collections described to date, the processes followed have been generally similar. The development of groups using a process of hierarchical stratification has been followed by sampling of the accessions within each group to give the core. Variations around this basic unity of approach will undoubtedly occur, reflecting the availability of different kinds of data (passport, pedigree, characterisation, evaluation, biochemical or molecular), the nature of the accessions in the source collection (such as breeders' lines, landraces and wild species), the biological character of the species (whether self- or cross-pollinated, vegetatively reproducing or apomictic) and the interests of the users (be they breeders or other research scientists).

An area where much work is needed is the development of core collections for clonally propagated crops. At present, experience is limited to cassava (Cordeiro et al., *Chapter 3.6*). Genetic data are usually more limited for such crops, and the way in which accessions in the core represent and act as a point of entry to the accessions in the source collections is much less clear. Collections of clonally propagated crops are usually much smaller and the percentage of accessions that are included in the core may be higher than the 10% sometimes suggested as the norm. However, the practical consequences of developing a core for a clonally propagated crop may be much higher than for a seed-propagated species. It can be more immediately beneficial for gene bank managers in identifying accessions which have priority for maintenance *in vitro* or in field gene banks in different countries. This development would be particularly beneficial where exchange of clonally propagated crops is beset with problems of potential disease transmission and the facilities available for disease indexing have to be limited to a few accessions.

Another general issue faced by those forming core collections for use in specific countries or areas of the world is the extent to which accessions that are clearly not adapted to their specific environment are included. Does the core collection for a seed crop such as sorghum or sesame include material which, by reason of photoperiod response for flowering, will never be productive in the area for which the core is intended? Users of the material may regard such accessions with disfavour and avoid them if possible. However, they may well possess characters, such as disease resistance, not present in other accessions and their omission might result in excluding a significant fraction of the total variation. Practical considerations will be important in making decisions on such material. For example, gene banks that have the facilities to induce flowering in photoperiod-sensitive material may well include them in the core, while others may choose to exclude them in the first instance.

These different starting points and approaches in developing a core collection indicate that there will not be a perfect core collection for a specific crop. Size and content of the core will depend, apart from the availability of information and material, upon the specific demands and possibilities of the scientist compiling it.

It is worth noting, in passing, that it may not always be either appropriate or necessary to develop a core collection. Where collections are small and can be handled without difficulty by a gene bank, and where sufficient infrastructure exists to ensure that all accessions can be made available to users on demand, there is clearly no compelling reason to develop a core collection. Work can proceed at the desired level on the whole collection, and should do so. However, it is also worth noting that the analytical procedures for identifying and studying the structure of variation in the collection and for sampling from a collection may well be useful for such collections.

HIERARCHY AND THE DEVELOPMENT OF GENETICALLY MEANINGFUL GROUPS

A central theme which runs through many of the chapters in this book concerns the development of an effective way of grouping the accessions in a collection or, as in the case of the Barley Core Collection, several different collections. The approaches described and used to date are essentially hierarchical. They follow the suggestions made by Brown (*Chapter 1.1*), successively dividing the collection on the basis of taxonomic, geographic and ecological information concerning the accessions. The presumption is that this will lead to groups of accessions which share common characteristics such that variation between groups will be increased and variation within groups decreased. As noted by Hintum (*Chapter 2.1*), there is considerable evidence from the literature on the distribution of genetic diversity in crop species and their wild relatives to support the general approach. As information on the distribution of variation in different gene pools becomes more complete, the ways in which grouping may be achieved become more sophisticated. An example of this is the work described by Tohme et al. (*Chapter 3.1*) on the development of the CIAT *Phaseolus* core. This case combined information on known patterns of diversity with agroecological data and genetic information to identify the groups and hence accessions that entered the core collection.

Further work in this area is likely to be particularly rewarding and relevant, not only for the development of core collections but also for our understanding of the extent to which genetic diversity is structured. In turn, we can learn how users should most effectively employ different indicators of the distribution of variation in a collection to identify material for their own programmes.

For core collections, passport data are likely to provide much of the information needed to identify groups, and immediate questions concern the amount of such data that is needed and the way in which it should be used. Is geographic distribution sufficient for effective grouping or does the addition of data on characteristics such as altitude, soil type and annual rainfall result in the development of more genetically meaningful groups? Although passport data are often incomplete, they may be relatively complete with respect to site of origin of the accessions in a gene bank. Where sites of origin are known, it is often possible to identify a variety of climatic and geological characteristics for the sites which could be used to modify the groups identified, but it remains to be determined whether this improves the grouping for core collection development.

Another area of interest concerns the way in which genetic data may best be combined with passport data. For the *Phaseolus* core (Tohme et al., *Chapter 3.1*), genetic information was used initially as a weighting device so that material from known areas of high diversity was given preference. After agroecological classification, steps were taken to ensure that different character states were included for certain key characters. This is not the only approach possible and there are a number of interesting possibilities for exploring the effect of different procedures on the construction and balance of the groups. Should geographic distribution take precedence over major morphological divisions within a crop in forming groups or should it be the other way round? What morphological or physiological criteria are meaningful in this respect? Sometimes specific criteria will represent a natural grouping of material for the users (for example, some breeders might prefer to work within spring or winter types of the crop).

It would be valuable to explore these and related issues for different species with relatively well-defined collections so that the robustness of the hierarchical grouping procedure can be established. The result may vary depending upon the crop selected, its history and current distribution and the extent to which material in different areas has been reproductively isolated and selected for different purposes. So far, there has been almost no work on exploring the relative success of different grouping strategies to partition variation within the context of developing core collections. There is a need to explore this both for 'useful' genetic variation concerning characters of morphological and agronomic importance and for biochemical and molecular markers. It is important that the different classes of characters are included in such studies in order to integrate information of importance to users with the more basic genetic analysis of the consequences of the different strategies that are used to develop groups for core collections.

SAMPLING WITHIN GROUPS

The evidence presented by Schoen and Brown (*Chapter 2.3*) suggests that using available information on the extent and distribution of variation within groups can increase the diversity of the core. In many cases such information may not be available and alternative solutions may be required. If no data are available, the logarithmic or proportional strategies may be used and would appear likely to give satisfactory results in many cases. Alternatively, the technique described by Hamon et al. (*Chapter 3.3*) may be used for identifying the entries to be selected from within groups rather than using random sampling. This may be particularly useful for collections for which evaluation data exist but where there has been little formal genetic analysis of the collection.

Once groups have been established, it may be possible to survey a limited number of accessions from each group for variation at isozyme loci or other biochemical markers so as to obtain preliminary estimates of the distribution of variation within the different groups. This information would then allow judgements to be made on the numbers of samples to be included in the core for the different groups. In practice, the relative importance of certain groups to the users will also often determine the number of accessions from these groups to be included in the core. Although cultivars represent a relatively small part of the diversity of the total *Hordeum* gene pool, they comprise a large part of the Barley Core Collection, especially when compared to the wild species (Knüpffer and Hintum, *Chapter 4.1*).

INTEGRATING DIFFERENT TYPES OF DATA

One of the most significant features of the work on core collections has been the impetus that it is giving to exploring the ways in which genetic resources data can be collated, analysed and used more efficiently. Thus, there is considerable interest in combining data from evaluation work with that obtained from biochemical or molecular studies and in carrying out multivariate analyses of large data sets involving many hundreds of accessions. As experience with such data analysis develops, it may be possible to identify more or less useful characters for studies to develop core collections, or more or less useful methods of data analysis where both quantitative and qualitative data may be combined.

There continues to be much interest in the potential for molecular markers in the development of core collections and this is reflected by Gepts (*Chapter 3.4*). Techniques continue to evolve rapidly and to become increasingly cost effective and easy to use. The numbers and heterogeneity of accessions involved in most genetic resources operations still present problems for the use of molecular techniques and it is probable that these techniques will be used on small samples of the total collection to test specific procedures or hypotheses. In particular, studies on the relationship between diversity as detected by molecular markers and that detected using other methods are needed.

One of the criticisms of core collections that is often cited is that the development of a core requires an impossible amount of data on the accessions in the total collection, and the emphasis of the above paragraphs on data collection, analysis and use might be seen to support such an argument. In reality, the process is rather different. Experience now shows that the development of a core collection increases the use of, and interest in, data already existing and places in sharp focus certain data management issues for gene bank managers.

Broadly speaking, genetic data on accessions may be of two kinds: qualitative (concerned with characters controlled by known genes with identified alleles and simple Mendelian inheritance, such as marker genes, isozymes and molecular markers) or quantitative (concerned mainly with agromorphological characters). There is some evidence from the data given in this book (for example, Hamon et al., *Chapter 3.3*, and Cordeiro et al., *Chapter 3.6*) that those working with quantitative characters find that the percentage of the number of accessions that should be included in a core collection is higher than the 10% level advocated on the basis of qualitative characters. In this area, there is a particularly urgent need for studies to compare the information from both types of data obtained on the same sets of accessions. With respect to quantitative characters, a residual question remains as to whether it is desirable to include a range of states for a character such as plant height or simply to include in the core those accessions that, when intercrossed, will generate all the desired variation of a character.

MANAGING AND USING CORE COLLECTIONS

To date, we have not acquired much experience on how core collections are best managed in gene banks or on their use to meet the various needs of plant breeders, research scientists and other users. Once a core has been established the first question that will be asked concerns the extent to which it achieves its aims. How much of the diversity of a collection or gene pool is contained in the core? Galwey (*Chapter 4.3*) gives some general guidelines on how this question may be answered but, in general, it is likely that surveys of the core and whole collection using biochemical or molecular markers will be necessary to establish whether there are significant pockets of variation that have escaped selection. Although the suggestion has been made that a core should also reflect the *distribution* of variation in the whole collection, the primary concern should be to represent the *range* of variation. Frequencies of different genotypes in the core collection can differ considerably from those in the main collection, provided that the desired types are present.

Conservation is a dynamic process. Knowledge of a gene pool is always incomplete and is continually improving. This raises the question of the way in which changes to the constitution of the core collection should be managed. Accessions shown to posses significant new variants absent from the core should undoubtedly be added to it, but on what basis and in what way? Should some existing entries be discarded in order to keep the total number of accessions at the same level? Should this occur on a continuing basis or should there be periods in which the core is held constant? These questions are best answered by gene bank managers on a case-by-case basis and are likely to reflect the size of the core and the management system chosen. Changing the constitution of the international Barley Core Collection is likely to occur fairly infrequently given the management complications of such a change, whereas a small core developed by a specific gene bank for its own user community could be changed more often.

All gene bank managers strive to maintain high standards for the maintenance and regeneration of their accessions even when resources limit their ability to use optimum procedures. Core collections will undoubtedly require the best standards of maintenance available to the gene bank. Thus, where possible, it may be desirable to use larger populations for regeneration than for other accessions and to give priority to the core with respect to safety duplication and other appropriate management practices. This is not to suggest that the rest of the collection has to be neglected. It is essential that this does not occur if the core is to realise its full value.

If core collections are to be effective as part of a process of improving the conservation and use of plant germplasm, they should be properly documented. In some senses core collections can act as communication devices and, for this to be effective, their data need to be widely accessible and widely used. This requires database management procedures which allow optimal and full exchange and use of the information. Another aspect of the use of the documentation of the core collection is the selection of accessions from the main collection. How can the information on the core be extrapolated to the main collection? How can searches in the main collection be optimised using information on the core?

CORE COLLECTIONS FOR GENE BANKS WITH LIMITED RESOURCES

The development, management and use of core collections make demands on resources which may be difficult for gene banks with limited resources, particularly those in developing countries, to mobilise. The issues involved are dealt with in detail by Morales et al. (*Chapter 5.3*) and it is not necessary to repeat them here. Nevertheless, it is worth emphasising that core collections are likely to result in the more effective and more economically efficient use of the total resources allocated to plant genetic resources conservation in a number of ways:

- The development of a core collection relieves gene banks from the need to hold large or unstructured collections of many crops. A core collection from another gene bank may be sufficient for their purposes.

- As has recently been suggested by an international workshop on sesame, a core collection may be an effective way of rapidly identifying that range of variation which is likely to be of most value in meeting plant breeders' urgent demands for variation in a crop that receives little international research support from organisations such as the international research centres.

- As emphasised by Morales et al. (*Chapter 5.3*), a basic component of the development of core collections is collaborative work between gene bank managers, plant breeders and other agricultural research scientists. This collaboration will increase qualitatively the effectiveness of the conservation effort as well as the effectiveness of the breeding and research work necessary for crop improvement. It will also draw on resources that may not otherwise be available to those concerned with conservation.

FUTURE POSSIBILITIES

The basic concepts that have been developed to identify accessions for a core collection are likely to find other applications. Most obviously, they can be used when new collections are established to ensure that the variation in such collections is maximised and genetic redundancy minimised. The procedures that allow groups to be identified and that ensure effective sampling within groups should also ensure that new collections are not burdened with large numbers of accessions that add little to the overall objective of conserving the maximum possible variation present in a gene pool.

Could the same argument be taken further? In recent years there has been an increasing interest in *in situ* conservation, particularly of those wild relatives of crops that constitute the primary, secondary and tertiary gene pool. It is tempting to suggest that a strategy based on using the techniques developed for identifying core collections could also be used to identify populations which should have highest priority for *in situ* conservation programmes. Herbarium and ecogeographic surveys could be used to identify populations of the target taxa which could then be grouped according to taxonomic, geographic and ecological criteria. Priority for *in situ* conservation would go to populations from different groups whenever practical considerations (such as management, social and political issues) allowed, thus attempting to maximise the conserved diversity.

Undoubtedly, many questions concerning the most effective development, management and use of core collections remain. However, we hope that the chapters in this book confirm the value of core collections, identify the major principles involved in their development and provide a secure foundation for workers wishing to develop their own collections and researchers interested in investigating the many unresolved issues.

Appendix

CORE COLLECTIONS WORKSHOP: RECOMMENDATIONS FOR A GLOBAL RESEARCH AGENDA

At the final session of the workshop, the participants discussed the research work needed on core collections. It was agreed that the most important research areas for core collections were:

1 Model studies to assess the specific problems of developing core collections of particular types of crops (such as cross-pollinated annuals and clonally propagated perennials).

2 Tests of the quantity and quality of information needed to develop core collections, with particular reference to the use of passport data and the advantages gained from additional agroecological, characterisation and genetic data in terms of core representativeness and value.

3 Investigation of the apparent contrast in results from using quantitative and qualitative approaches to develop core collections in order to determine the basis for the observation that more accessions are needed to include a sufficient amount of quantitative variation than are needed to include observed qualitative variation.

4 Studies on the concepts of hierarchical structuring of genetic diversity and of stratified sampling which are central to developing core collections. The effect of using different sampling and structuring strategies on the diversity retained in the core needs to be thoroughly researched.

5 Investigation of the consequences of using different statistical methods in developing core collections, particularly with respect to different approaches to data analysis and sampling methodology.

6 The development of compatible and optimal information systems which allow different gene banks to compare information on accessions; these are particularly important in developing core collections through networks.

7 The most effective and practical ways in which isozyme or molecular data can be used in developing or validating core collections.

8 The use of the core collection concept in assembling a new germplasm collection, with particular reference to the use of stratified sampling techniques and hierarchical structuring of variation in designating key areas for collection or germplasm for conservation.

9 Studies on the regeneration and dynamics of a core collection which have as yet received little or no attention and where practical results are required.

10 Applications of the core concept to *in situ* conservation through the use of agroecological or ecogeographic approaches to designated areas of dissimilarity which should be included in any integrated *in situ* programme.

Acronyms

ARS	Agricultural Research Service (USDA)
ASIC	Association Scientifique Internationale du Café (France)
AWCC	Australian Winter Cereals Collection
BCC	Barley Core Collection
BIOTROP-LIPI	Regional Center for Tropical Biology (Indonesia)
CATIE	Centro Agronómico Tropical de Investigación y Enseãnza (Costa Rica)
CENARGEN	Centro Nacional de Pesquisa em Recursos Genéticos e Biotecnologia (Brazil)
CGIAR	Consultative Group on International Agricultural Research
CIAT	Centro Internacional de Agricultura Tropical
CIMMYT	Centro International de Mejoramiento de Maíz y Trigo
CIP	Centro Internacional de la Papa
CNPMF	Centro Nacional de Pesquisa de Mandioca e Fruticultura (Brazil)
CPATSA	Centro de Pesquisa Agropecuária do Trópico Semi-Arido (Brazil)
CPATU	Centro de Pesquisa Agropecuária do Trópico Umido (Brazil)
CSSA	Crop Science Society of America
CSIRO	Commonwealth Scientific and Industrial Research Organisation (Australia)
ECP/GR	European Cooperative Programme for the Conservation and Exchange of Crop Genetic Resources
EEA	Estación Experimental de Agricultura
EMBRAPA	Empresa Brasiliera de Pesquisa Agropecuaria (Brazil)
EPAGRI	Empresa de Pesquisa Agropecuária e Difusao de Tecnologia (Brazil)
FAO	Food and Agricultural Organisation of the United Nations
IAC	Instituto Agronômico de Campinas (Brazil)
IAPAR	Instituto Fundação Instituto Agronômico do Paraná (Brazil)
IBPGR	International Board for Plant Genetic Resources
ICARDA	International Center for Agricultural Research in the Dry Areas
ICRISAT	International Crops Research Institute for the Semi-Arid Tropics
IFCC	Institut Français du Café, du Cacao et d'Autres Plantes Stimulants
IICA	Interamerican Institute for Cooperation in Agriculture
INIAP	Instituto Nacional de Investigaciones Agropecuarias (Ecuador)
INTA	Instituto Nacional de Tecnologia Agropecuario (Argentina)
IPA	Instituto de Pesquisas Agronómicas (Brazil)
IPGRI	International Plant Genetic Resources Institute
IRCC	Institut de Recherche sur le Café et le Cacao (France)
IRRI	International Rice Research Institute
JUST	Jordan University of Science and Technology
NAS	National Academy of Sciences (USA)
NOAA	National Oceanic and Atmospheric Administration (USA)
ORSTOM	Institut Français de Recherche Scientifique pour le Développement en Coopération (formerly Office de la Recherche Scientifique et Technique Outre Mer) (France)
PGRU	Plant Genetic Resources Unit (Cornell University, USA)
PUDOC	Centre for Agricultural Publishing and Documentation (Netherlands)
SEAMEO	Regional Center for Tropical Biology (Indonesia)
TAC	Technical Advisory Committee (CGIAR)
UNEP	United Nations Environmental Programme
USAID	United States Agency for International Development
USDA	United States Department of Agriculture

Index